Overtraining Syndrome in Athletes

Flavio Cadegiani

Overtraining Syndrome in Athletes

A Comprehensive Review and Novel
Perspectives

 Springer

Flavio Cadegiani
Federal University of São Paulo
São Paulo
São Paulo
Brazil

ISBN 978-3-030-52630-6   ISBN 978-3-030-52628-3   (eBook)
https://doi.org/10.1007/978-3-030-52628-3

This Springer imprint is published by the registered company Springer Nature Switzerland AG
The registered company address is: Gewerbestrasse 11, 6330 Cham, Switzerland

To my mom, Ana Dorinda, who always supported me.

To my dad, Flavio, for his unique way of loving.

To my stepdad, Weber; stepmom, Mara; sister, Pietra; my maternal grandparents, Maria del Carmen and Mariano; and paternal grandparents, Alice and Pietro, in memoriam.

To the athletes that participated in the EROS study, who opened the doors to a new era in understanding overtraining syndrome.

To the non-athletes that participated in the EROS study, who opened the doors to the cognizance of endocrinology and metabolism in the healthy athlete.

To Prof. Dr. Claudio Kater, my PhD guide and tutor forever, who supported me wherever I wanted to go, providing me the most expert opinions, and who has become the reference of my professional and scientific life.

To my colleagues from the PhD program in Clinical Endocrinology at the Adrenal and Hypertension Unit, Department of Medicine, Federal University of São Paulo, Brazil.

To all editors who gave me the chance to publish my work, opening the doors to the world.

To all reviewers who gave me true lessons and guidance to never stop improving the quality of what I publish.

To the team at my private clinic, who helped me make last-minute changes and supported me with all my non-professional tasks when I was too busy researching and writing this book.

To all my patients, who showed me how to constantly improve my medical practice, and who opened my eyes to the gaps in the understanding of impaired athletes and overtraining syndrome.

To all those who had no idea of what I was writing—and may still not understand—but encouraged me in this endeavor.

To my giant cats, Ken and Barbie, who jumped on my hands while I was writing the book and gave me their unconditional love throughout this journey.

Preface

Overtraining syndrome is among the diseases with the largest prevalence in elite athletes, but not limited to this group.

Its occurrence was first hypothesized when athletes began to present unexplained decrease in sports performance in the 1940s and 1950s. By that time, the concept "the more the better" in terms of training was prevailing, and competitiveness was on the rise due to recommencement of the Olympics after the World War II (WWII) and the Cold War, when performance in sports had unprecedented political weight. Some athletes started to show a progressive loss of performance that responded paradoxically to increase of training load, that is, the more they trained, the worse they performed, and were unresponsive to recovery interventions.

This new paradox observed in some athletes was unsuccessfully investigated for an extensive range of causes could also justify the underperformance.

A syndrome of unexplained underperformance in athletes then emerged, which was strictly related to excessive training. Hence, the term overtraining syndrome was intuitively coined, although only decades after the first descriptions of overtraining.

Once excessive training was assumed to be the sole trigger of overtraining syndrome, changes in training patterns towards periodization, avoiding excessive training, was expected to mitigate this syndrome. However, periodization of training sessions as the only intervention unexpectedly failed to reduce the incidence overtraining syndrome. Indeed, looking backward, the identification of excessive training as the intrinsic factor that leads to loss of performance, as well as the underlying mechanisms to justify reduction of physical capacity in response to over-exercising, were not elucidated by that time, and the hypothesis that overtraining syndrome was actually caused by overtraining was merely empirical.

Since changes in the paradigm of training programs were not enough to resolve overtraining syndrome, other triggers were likely present, although these have not been extensively investigated. End result, a gap in the knowledge of what really triggered overtraining syndrome remained unsolved.

While the majority of elite athletes experience at least one episode of overtraining syndrome, the growing number of non-professional athletes that practiced sports at extremely intense levels, resembling professionals, with concurrent rigid

practices and living patterns, led to an increase of overtraining syndrome incidence—and recognition.

The use of the term overtraining has spread among athletes, sports coaches, and
more recently in specific and general media. However, almost always this expression was used to describe conditions other than overtraining syndrome, while
unsubstantiated characteristics and manifestations alleged to be part of overtraining
syndrome that been universally accepted led to an almost general misinterpretation
and misunderstanding of overtraining syndrome, with consequent misdiagnosis in
the majority of the cases. The two hallmarks of overtraining syndrome, the loss of
performance and the lack of an apparent explanation for impaired performance,
have become secondary in the context of the broad use of the term overtraining.

With general misuse of the term overtraining, the loss of significance of overtraining syndrome, and the reduction in scientific publishing on the field in the
1990s, those suffering from this condition remained underappreciated, which ended
the career of several professional athletes.

Fortunately, interest in this syndrome recovered recently, with the publication of
the first and only guidelines on overtraining syndrome, in 2013.

The guidelines on overtraining syndrome helped educate sports-related health
providers of its actual characteristics, and the unexpected decrease of sports performance has become central again.

However, unanswered questions regarding overtraining syndrome pathophysiology, triggers, mechanisms, and effects, which have only been partially addressed by
classical theories, still prevailed.

In this context, several novel studies have recently had the opportunity to revisit
overtraining syndrome from a more comprehensive perspective, which comprised
wide, different aspects to be potentially correlated with the development of the syndrome, and did not limit the occurrence of overtraining syndrome to training
patterns.

Consequently, multiple novel findings have been unveiled, which provided
embodiment for novel insights and novel concepts that have resulted in a novel
understanding of overtraining syndrome. The aim of the book is to present the novel
insights, hypotheses, and proposed functioning of overtraining syndrome arisen
from the novel uncovered mechanisms.

In short, the recent discoveries showed that overtraining syndrome resulted from
sum, respective intensity, and interactions between a combination of chronic deprivations and excessive efforts, including relatively insufficient caloric and protein
intake, non-repairing sleep, lack of compensatory reduction of training load, and
concurrent intense physical and cognitive demands. These chronic deprivations
build a hostile tissue environment chronically depleted from energy and repairing
mechanisms, which forces the occurrence of multiple adaptations aiming to keep
surviving and functioning. The outgrowth adaptations are overwhelmingly dysfunctional ("*maladaptations*"), resulting in multiple hormonal, metabolic, immunologic,
inflammatory, and muscular abnormalities, which eventually lead to the main manifestations of overtraining syndrome, including mental and physical exhaustion,
pathological muscle soreness, loss of multiple abilities in physical performance,

increased predisposition to overall infections, and burnout-like signs and symptoms. Hence, a chronically misadjusted routine, rather than excessive training alone, is the key trigger of overtraining syndrome.

The more comprehensive understanding of overtraining syndrome as being a condition that encompasses a broader number of factors, mechanisms, and characteristics is not limited to its intrinsic features. In a bigger picture, overtraining syndrome is highly representative of an ample and diverse range of aspects.

Overtraining syndrome is the representative dysfunction resulting from a society that increasingly worships high performance in major aspects of life simultaneously—physical, intellectual, social financial, professional, sexual, and psychologically, towards an unachievable perfection and artificial robotization of human being.

Since high performance depends on overcoming specific deprivations that occur at great extent, overtraining syndrome is also the representative condition of a cluster of diseases derived from multiple deprivations—deprivation of food, sleep, rest, and indulgences—that generate a state of low energy availability (LEA) and consequent chronic energy deprivation, which eventually induces a decrease of sports performance and shares multiple similarities with burnout syndrome. These conditions comprise the female athlete triad and its expanded concept of relative energy availability of the sport (RED-S), the burnout syndrome of the athlete, and *pseudoovertraining* and overtraining-like states.

Collectively, the multiple similarities allow to hypothesize that all these dysfunctions may be different poles of a same, broader condition, which has been termed as the "impaired athlete" although extensive research is needed to confirm this hypothesis. Indeed, if athletes diagnosed with the female athlete triad or relative energy availability of the sport were assessed for overtraining syndrome, a good part would also have diagnostic criteria for this syndrome.

Overtraining syndrome is the representative consequence of the current pathological anthropological context that considers strict eating as a quality to be worshiped, with an implicit sense of superiority of those that can fully control their dietary patterns. From a societal perspective, the more implicit and camouflaged version of eating behavior disorders is notorious, and can be easily recognized by the increasing veneration of excessively lean bodies allied to the growing fat-phobia and obesophobia.

The presence of apparent body fat, even without overweight, has become an indirect and almost unconscious marker of weakness from excessive indulgences and allowances, as "denounced" by the presence of larger (but not large) skin folds.

Correspondingly, unlike before, when body fat was secondary if one aimed to improve performance, the simultaneous seek for both performance and decreased body fat due to cultural reasons propitiates stricter regimens that naturally lead to deprivations.

Overtraining syndrome is the representative dysfunction of decreased performance as the final result of dysfunctional adaptive processes, which may also occur under states that are not necessarily labeled as a specific diagnosis.

Despite the description of burnout syndrome of an athlete as a distinct condition, overtraining syndrome better represents the model of burnout syndrome in athletes, in particular regarding the lack of simple, frank, and ubiquitous markers or mechanisms that explain the challenges to perform a correct diagnosis and the high individuality of the clinical, biochemical combination of manifestations. From a more comprehensive analysis, athletes with OTS present features equivalent to those in the recently recognized burnout syndrome.

From a comprehensive perspective, overtraining syndrome represents the complexity of when adaptive models in athletes get disrupted.

Overtraining syndrome is also the representative example that one-size-fits-all is by far the most inappropriate approach to athletes. The inappropriateness of the excessively protocoled assessment as the only tool to diagnose more complex syndromes can be extended to the general population. While attempts to fit overtraining syndrome into few patterns of presentation failed due to its complexity, a more accurate comprehension of overtraining syndrome may have been precluded.

Overtraining syndrome is the demonstrative representation that the lack of overt biochemical abnormalities does not necessarily mean lack of dysfunctions; on the contrary, dysfunctions tend to be harder to be detected because of the complexity of non-obvious alterations. And indeed, when we go deeper in the analysis, and only after a multiple-comparative analysis, we realize that dysfunctions do exist, but they are hidden by the previous optimizations these parameters underwent in this population of athletes. Perhaps, its complexity associated with several types of "unexplained complaints," including those in joints and muscles, has spread its common yet inappropriate employment of diagnosis of overtraining syndrome in clinical practice.

The early stages of overtraining syndrome, termed as functional (FOR) and non-functional overreaching (NFOR), represent the continuum that this syndrome undergoes before its end-stage and demonstrates that overtraining syndrome results from persistent ignoring of the warning signs that occur through repeated episodes of overreaching, for the sake of an objective, masked by massive motivation that eventually becomes toxic. The point of no return, after which athletes develop overtraining syndrome, is reached when conditioning processes and adaptations to sport become dysfunctional, triggering multiple novel pathological pathways that start a process of self-destruction which is extremely burdensome and effortful to overcome. At least one episode of overreaching has occurred in virtually all patients, and does not necessarily represent an alert sign when it occurs singly and is fully addressed.

The recent lessons from the novel insights in overtraining syndrome allow to reframe overtraining syndrome from a simplistic understanding of the triggers as being excessive training to a complex sum and web of interactions of dysfunctional changes that occur in response to a sum, respective intensities, and synergistic interactions between deprivations. Conversely, from historical lessons to future perspectives, the comprehension that oversimplified attempts to explain the pathophysiology of overtraining syndrome will always be unsuccessful is key, since overtraining syndrome is a result of the exact opposite: complex interactions that consequently lead to complex dysfunctions with a pitch of enigma.

The swell version of overtraining syndrome is a representation of the necessary improvements that need to be performed on the methodology and design of the research on endocrinology of physical activity and sport, since inconsistent findings on hormonal responses in healthy athletes as well as the lack of determination of sports-, sex-, age-, and conditioning level-specific hormonal and metabolic findings prevented the disclosure of abnormalities in these respective systems in overtraining syndrome once the basis for comparative analyses was undetermined.

The book will introduce you to a completely novel, more comprehensive view on overtraining syndrome that moves from a training-centered to an athlete-centered perspective. It is a practical approach that will help you understand the topic step by step, simplifying this inherently complicated disease.

Overtraining Syndrome in Athletes: A Comprehensive Review and Novel Perspectives brings the scientific background to demonstrate that overtraining syndrome should not be assessed as a distinct disorder, and also proposes the concept of Paradoxical Deconditioning Syndrome (PDS) as an expression that better reflects the central characteristic of overtraining syndrome.

In conclusion, this book goes beyond: beyond the classical view on overtraining syndrome, which naturally encompasses a broader range of aspects, beyond the classical and somewhat biased characterization of athletes suspected of overtraining syndrome, beyond overtraining syndrome per se, and beyond the classical proposed mechanisms for overtraining syndrome, since the growing understanding that overtraining is the end of a pathological process that comprises several other related conditions is making these deprivation-derived, LEA-generated, and burnout-related dysfunctions progressively indistinguishable between them, while the specific location of the line that divides between physiological and pathological is becoming blurrier.

We learned to respect the timing of training. Now we need to learn to respect our limits. And listen to them.

São Paulo, São Paulo, Brazil   Flavio A. Cadegiani

About the Book

Overtraining syndrome (OTS) is a dysfunction initially alleged to result from a combination of excessive training and lack of recovery, leading to a paradoxical decrease of sports performance associated with fatigue unresponsive to a period of resting, unable to be explained by alterations, or dysfunctions that could also lead to impairment of training performance. OTS is a complex and multifactorial disorder, in which the understanding of its pathophysiology and biomarkers lacked, despite the high incidence among elite athletes.

With the new insights discovered by the Endocrine and Metabolic Responses on Overtraining Syndrome (EROS) study, OTS was shown to be even more complex, and not necessarily triggered by excessive training, but by a combination between inappropriately low protein, carbohydrate, and/or overall caloric intake; excessive concurrent intellectual activities; insufficient physical and/or mental resting; and unrefreshing and/or bad sleep quality, which takes to multiple negatively synergistic interactions between chronically energy deprivation and shortage of mechanisms of repair, leading to dysfunctional maladaptive processes, in which the sum and intensity of these processes can trigger OTS.

An improved and more comprehensive understanding of OTS has arised, which includes multiple losses of previous adaptive conditioning processes that occur in athletes ('*deconditioning* processes'), with consequent decrease in pace, intensity and volume of training, in association with an overall state of decreased metabolic rate and expenditure ("*hypometabolism*"), physical and mental burnout, loss of libido, paradoxical body fat gain and muscle loss, and blunted hormonal responses to stimulations, in order to save energy, as a sort of a diseased derived from chronic energy deprivation.

Hence, the name "Overtraining syndrome" was found to be inaccurate, as it describes a trigger that is now less consistent than other risk factors and does not describe the key aspects of this syndrome. Instead, we suggest the name "Paradoxical Deconditioning Syndrome" (PDS), which better describes the key pathophysiology that leads to decreased performance, the hallmark of this syndrome.

The book depicts all of the novel findings, concepts, hypotheses, mechanisms, risk factors, pathogenesis, and clinical, biochemical, and metabolic manifestations that lead to a new understanding of OTS in an illustrative, step-by-step, easy-to-understand, and logical manner that will allow readers to learn about OTS, naturally infer insights from a logical perspective, and guide researches for efficient models of research protocols. In addition, the highly practical clinical applications provided in the book will certainly help athletes and sport coaches to improve the assessment and detection of athletes at high risk and earlier stages of OTS, when recovery improves dramatically, saving many athletes careers.

The key objectives of the book include:

1. Build a novel understanding of overtraining syndrome, including the proposal of a novel and more intuitive name (that has been highly accepted by the scientific community)
2. Provide readers an understandable way to read overtraining syndrome, since its syndrome is highly complex, and the information from previous texts and studies were disconnected between them, which precluded from a logical understanding of what overtraining syndrome is
3. Describe the novel concepts, biomarkers, diagnostic tools, and characteristics of overtraining that have been recently uncovered
4. Help understand the reasons why so many athletes are still developing overtraining syndrome, despite the adequate training volume
5. Unprecedentedly provide an effective preventive approach, which is highly desirable, since overtraining syndrome can rarely fully remit.

This book should be a game changer in the field of Sports Science and Medicine. It should also redirect further researches in the field. And will likely help reduce the incidence of OTS, if properly spread among health-related professionals.

There are unique and unprecedented characteristics of the book that are remarkable for the advances in clinical practice and research not only in OTS, but in overall endocrinology of physical activity and sport. This is the first book to explain Overtraining Syndrome from a truly scientific perspective, with thorough and comprehensive description of its key characteristics. This is at present a logical, intuitive, and highly illustrative book. The book brings an unprecedented view on Overtraining Syndrome as not being caused by excessive training, but by a combination of different factors instead. This is actually the first scientifically accepted understanding of overtraining. We are the first to demonstrate that Overtraining Syndrome is not caused by overtraining and why Overtraining Syndrome is in fact the Burnout Syndrome when manifested in athletes. This is the first time when a book proposes effective preventive approaches, before overtraining syndrome becomes challenging to heal and unrecoverable, and provides highly practical diagnostic scores for Overtraining Syndrome that can be widely used in clinical practice. Finally, this book will fill several knowledge gaps and answer questions researchers and sports health providers have.

Main Disciplines and Fields Assessed in the Present Book

Sports Science
Sports Medicine
Endocrinology
Nutrition
Metabolism
Psychology
General/Internal Medicine
Sports Coaching
Sports Endocrinology
Endocrinology of Physical Activity and Sport
Clinical Endocrinology

Contents

About the Author

Flávio A. Cadegiani, MD, MSc, PhD   obtained his medical degree at the University of Brasília. He then completed his medical residency in Internal Medicine, immediately followed by a fellowship in Endocrinology and Metabolism. After the fellowship, which is also termed as a specialization medical residency in some countries, he obtained the Board Certification in Endocrinology and Metabolism and began two concurrent careers: as an MSc and then PhD student along with clinical practice at his own clinic.

During his research on fatigue-related conditions, which resulted in the publication of the systematic review "Adrenal Fatigue Does Not Exist: A Systematic Review," published at the BMC Endocrine Disorders and has become a world reference to debunk the existence of this fake disease, Dr. Cadegiani realized that those states that were markedly affected by fatigue had methodological issues in the assessment of hormones and metabolic parameters. Among these states, he has called special attention for overtraining syndrome, for which endocrinology aspects were claimed to be central for the occurrence and as a major manifestation in this disease.

Dr. Cadegiani then performed a systematic review on hormonal aspects of overtraining syndrome, from which he not only concluded that this syndrome lacked massive gaps in the pathophysiological understanding and biomarkers, but also that the research on the endocrinology of physical activity and sport required substantial improvements in the methodological design and assessment methods.

By addressing these methodological issues, and aiming to answer the major questions on overtraining syndrome, as well as on healthy athletes, he conducted the Endocrine and Metabolic Responses on Overtraining Syndrome (EROS) study, which evaluated more than 110 parameters in more than 50 participants, including two control groups, and revealed multiple novel biomarkers and mechanisms not only in overtraining but also in healthy athletes, and its longitudinal follow-up was still ongoing when the present topic was written. This study has become the largest in its field, with multiple branches, and has been unfolded in more than 10 published papers.

Dr. Cadegiani currently has the largest number of published research papers in the field of Overtraining Syndrome and Sports Endocrinology and has been invited to the editorial board of a large number of high-impact PubMed indexed journals and has been a reviewer of approximately 90% of the manuscripts in the field of Sports Endocrinology submitted to important journals.

Finally, as the first board-certified endocrinologist to obtain a PhD degree in the Endocrinology of Physical Activity and Sport in a doctoral program of Clinical Endocrinology, Dr. Cadegiani has become an expert in both Endocrinology and Sports Medicine fields, providing a unique endocrinological perspective to Overtraining Syndrome. With his studies and now with this book, Dr. Cadegiani helped to unveil multiple characteristics of this syndrome and to understand the exact pathological mechanisms that affected athletes undergo.

Chapter 1
Introduction, Historical Perspective, and Basic Concepts on Overtraining Syndrome

Outline

Introduction, in which overtraining syndrome (OTS) is briefly described, but in a manner that many readers will finally understand of the syndrome really is.

Historical perspective, which brings a deep historical perspective on OTS, which has not been performed before, by rescuing decades-old preliminary researches and theories and putting all them into a chronological and logical perspective. This aims to better understand the progressive understanding of OTS until its current format and gives readers a deeper contextualization that will further help them to truly understand the rationale behind OTS.

Basic concepts on overtraining syndrome, where the concepts on this syndrome are depicted in an almost illustrative manner, since the understanding at this part is key for a good comprehension of the book and the syndrome.

1.1 Introduction

Overtraining syndrome (OTS) is one of the most frequent sport-related dysfunctions in athletes, leading to a paradoxical decrease of sports performance, associated with physical and mental fatigue unresponsive to a period of resting, and not explained by alterations or dysfunctions that could also impair training performance [1–3]. OTS is classically alleged to be resulted from a combination of excessive training and lack of recovery, and its pathophysiology is complex, multifactorial, and still unclear, despite the high incidence among elite athletes.

OTS prevalence varies between 15% and 60% of elite athletes during their career, or an incidence of 50,000 to 60,000 athletes a year, which will depend on the type, frequency, and intensity of the sport(s) practiced, as well as external factors [1–3].

© The Editor(s) (if applicable) and The Author(s), under exclusive license
to Springer Nature Switzerland AG 2020
F. Cadegiani, *Overtraining Syndrome in Athletes*,
https://doi.org/10.1007/978-3-030-52628-3_1

This figure may be growing as a consequence of the increasing number of elite, semi-professional, or amateur athletes, particularly those who become highly physically active and participate in competitions, despite sports not being their main activity.

1.2   Historical Perspective

The understanding of stress-derived syndromes triggered by long-term exposure to inhospitable biological environments was first proposed in 1936, from the "stress theory" from Selye [4], in which the author proposed a sequence of events that results in stress-driven syndromes: a "stress syndrome," which could be either "functional," with further improvement, or "dysfunctional," without further recovery, leading to an irreversible progression to a chronic non-recoverable energy-deprived and mechanisms of repair-depleted environment, eventually causing stress-derived syndromes.

However, the correlation between this theory and the dysfunctions found to occur in elite athletes under chronic exposure to extremist and degradable physical conditions only arose later in the twentieth century, after an exponentially growing number of athletes and sports practitioners undergoing increasingly extreme physically challenging conditions (high mountains climbing, deep diving, ultramarathons, triathlon).

From a sociological perspective, strong sports competitiveness can be interpreted as a healthy channeling process of once deleterious process of human violence or compulsive behavior, which is an inherent and primitive human mechanism and behavior [5–7]. The beginning of the modern Olympic games, in Athens, Greece, in 1896, allied with the exponential growth of all sorts of competitions, and the emergence of multiple types of sports enhanced the sense of competitiveness in humans [5, 7].

The growing importance of sports competitions since the early twentieth century stimulated a parallel run for a continuous improvement in sports performance. Since then multiple training methods have been attempts. In common, these methods included a continuous increase of training intensity, volume, and pace, since continuous changes in any of the training patterns would hypothetically lead to highly functional and beneficial stress-triggered adaptive processes, with further improvement of performance improvement.

However, once it was assumed that good sleep quality would be sufficient for proper recovery, progressively longer and more intense training sessions were employed, regardless of non-sleeping resting periods, relaxation, and appropriate eating patterns. Non-interrupting training without resting days with continuous progression of at least one of the training patterns was believed to be the most efficient method to outstrip previous record marks and performance. The concept of "the more, the better" was strongly printed in sports coaches, without the realization that a proper recovery with a minimal interval between training sessions would lead to compensatory further improvement and which the duration of an appropriate resting between training sessions depended on the level of volume, pace, intensity, and strength of the last trainings.

Consequently, these athletes unfailingly reached a "plateau," which failed to fade with additional progressive training load. Despite the inability to overcome the "plateau" with progressive trainings, increase of training intensity and volume was persistently attempted, instead of considering a prolonged resting. With the overwhelmingly physical efforts, some of these athletes developed a paradoxical decrease in performance, which were not explained by any infection, inflammation, metabolic, hormonal, cardiovascular, autoimmune, or pulmonary dysfunction. Despite the paradoxical response to training, new intensifications of the training patterns were attempts that even with resting periods, no longer responded adequately, but with a further paradoxical exacerbation of the loss of performance instead. These were the first cases of a new issue that was emerging: an unexplained loss of performance, without any apparent etiology, that responded paradoxically to new attempts of increasing training load.

Meanwhile, the term "overtraining" had begun to be used in the beginning of the 1960s to nominate a scientific type of experimental interventional approach, aimed to fiercely induce cognitive abilities in animals and in children (as a sort of "mental intensive training" to increase cognitive abilities), without any correlation with its current use [8–11].

The first scientific use of the term "overtraining" to designate its current meaning dates from 1978 [12] and started to be increasingly used in the 1980s, while its initial meaning as challenging psychological tasks has faded. A theoretical model of psychological and physical men strength and power ("pyramid model of health manpower") proposed by McTernan in 1982 [13] was adopted to connect the intensified psychological training with the intensified physical training, since both excessive physiological cognitive effort in animals and kids and the excessive physical effort led to unexplained paradoxical effects. Since then, several studies started to adopt the term "overtraining" semantically as "excessive physical training" [14, 15]. This term was then correlated with other "syndromes," including the "muscular overuse syndrome" [16]. The first definition for OTS as we know nowadays was proposed by Prof. Kreider and then revisited by other authors [17].

The first description of the presence of hormonal dysfunctions in OTS was made in 1985, by Barron et al., and published at the important *Journal of Clinical Endocrinology and Metabolism* (JCEM) [18]. In this study, athletes diagnosed with OTS underwent an insulin tolerance test (ITT), which showed blunted hormonal responses, in comparison to healthy athletes, which means that hormonal responses to stimulations may be compromised under OTS, possible leading to loss of optimized physical responses. However, besides the absence of a second control group of healthy sedentary controls, this study was not further reproduced, as further studies did not employ standardized endocrinological functional tests and was not further explored by endocrinology research groups. Instead, further tests on OTS were mostly performed as exercise-dependent tests, which is heavily dependent on athletes' level of performance.

Hundreds of following publications in the decades of 1980, 1990, 2000, and 2010 attempted to find useful and accurate biomarkers of OTS [1–3, 20], alongside with proposal of different hypotheses as potential underlying mechanisms of OTS. Once the lack of an explanation for the underperformance is inherently part of this syndrome, attempts to unveil the exact pathophysiology of OTS were naturally challenging.

1.3 Basic Concepts on Overtraining Syndrome

The classic and still the most accepted understanding of OTS is that the paradoxical reduction of performance, the hallmark of OTS, is resulted from a combination between excessive training and lack of proper recovery. The imbalance between excessive activities and inappropriate repairing leads to chronic deprivation of energy and shortage of mechanisms of repair, which forces the occurrence of dysfunctional adaptations in multiple systems, in order to keep survival and minimum functionality under an inhospitable environment [1, 20]. These forced adaptations, although were among the only feasible ways to maintain the organism surviving and functioning under a hostile state, are highly dysfunctional and are also termed as *maladaptations*. The multiple "maladaptive" processes lead to multiple immunological, hormonal, neurological, muscular, metabolic, and psychological alterations, which eventually causes the key features of OTS.

Additional factors other than excessive training could presumably trigger OTS, including bad sleep quality and lack of sleep hygiene, personal and social factors, and inappropriate food intake, as all these circumstances may also induce or enhance the state of chronic stress with depleted energy and mechanisms for self-protection [1–3].

OTS is the term designated for one of three different states altogether termed as "overtraining," which also includes the functional *overreaching* (FOR) and non-functional *overreaching* (NFOR) states [1–3]. FOR and NFOR are earlier states of OTS, directly related to training overload, which differ from OTS; these states lead to temporary and reversible reduction of sports performance, with complete recovery after a period of resting. In common, both states lead to acute physical signs of exhaustion and inability to perform expectedly, while psychological symptoms tend to be milder or barely present. FOR has a very short duration (of few days) with overcompensatory improvement of performance, achieving better results when compared to the performance prior to the episode of FOR. FOR seems to be a natural cycle between periods of important training overload and appropriate moments of resting, when acute signs for the need of recovery arise. This functional twist logic leads to repeated and continuous processes of "overcompensation" with progressive improvement in pace, strength, and other abilities required for the performance. Ultimately, FOR seems to be the healthy and functional expression of overtraining load and compensatory repairing, including appropriate sleeping, eating, resting, and relaxing, with long-term benefits. Conversely, after an episode of

NFOR, although performance is fully recovered, it does not improve, while the period of underperformance can last for up 2 to 3 weeks. NFOR seems to be a non-functional state of overreaching, i.e., there is not such an ulterior improvement or functionality in response to the episode of NFOR, but is at least recoverable. FOR and NFOR can be differentiated through two different aspects: (1) duration of the episode, whenever shorter than 1 week, it is likely FOR, and if longer than 1 week, it should be an episode of NFOR, and (2) how the athlete performed after the recovery, if better than before, it should be FOR, while a lack of further improvement is likely a diagnose of NFOR. However, there is not a clear boundary between these two states.

Whenever the underperformance persists after weeks, and the athlete is unable to fully recover this previous sports performance, this turns out to be a true and overt OTS. Oppositely to overreaching states, in OTS there the duration of the unexplained underperformance may last until months to years, unresponsive to long periods of recovery, commonly associated with more intense and severe mental exhaustion, and associated with significant psychological issues [1–3]. Unlike NOR and NFOR, athletes affected by OTS may never fully recover and develop a sort of post-traumatic stress syndrome, or a burnout syndrome of the athlete, which will be further discussed, which in turn compromises the athlete career.

While overcompensatory responses leading to improvement of performance are prevailing in FOR and NFOR, in true OTS overcompensations become dysfunctional (termed as 'maladaptations'). and presumably causes a combination between systemic subclinical inflammation (herein called as "subclinical," as classic inflammatory markers, C-reactive protein (CRP) and erythrocyte sedimentation rate (ESR), are not typically altered in OTS), increased levels of reactive oxygen species (ROS) due to chronic shortage of tissue mechanisms of repair, and chronic severe energy deprivation generates multiple harmful effects on the nervous central system (NCS), including worse humor and energy levels, mental fatigue, and neurohormonal dysfunctions [1, 20]. These effects lead to multiple longterm consequences, particularly in a broad number of psychological aspects that become resistant to any sort of treatment. Therefore, athletes with severe mental manifestations of OTS may never fully recover, which force them to interrupt their career.

In summary, overreaching states is the result of a physiological response to a training accumulation, leading to a short-term and reversible decrement in performance, whereas overtraining leads to a long-term and is unlike to fully reversible loss in performance.

Despite the classification into FOR, NFOR, and OTS, many consider overreaching and overtraining as a *continuum* process of a same condition, which means that whenever overreaching state becomes more severe and prolonged, does not respond to resting and proper nourishment, and is associated with additional stressful factors, *overtraining* is more likely to occur [2, 19]. In addition, a clear differentiation between NFOR and OTS is sometimes challenging and not always feasible, and OTS can only be properly diagnosed when an athlete does not fully recover after a resting period [1, 16]. However, a true OTS could be sometimes recognized without the need to follow the underperformed athlete for months, as OTS many times tends to present more severe fatigue, psychological disturbances, and more impact on sports performance than FOR and NFOR since its beginning [1, 3, 20].

With multiple attempts to define, differentiate, and characterize the states of FOR, NFOR, and OTS, these proposed concepts are not fully consistent and have conflicting statements, due to the complex characteristics of these states [1–3]. Ultimately, although attempts to classify *overtraining* in one of the three states may be important, once each athlete exhibit a unique combination of characteristic, symptoms, and alterations and that many athletes may ignore signs of overtraining due to their high motivation to keep training, it is unlikely that athletes can be precisely diagnosed into any specific state.

Despite the growing understanding of OTS, the clarification of its pathophysiology, triggers, biomarkers for its early identification, and tools for its management lack extensively. For example, the learning of the need for training periodization, initially predicted to be an effective solution to prevent OTS, since OTS was alleged to be majorly caused by excessive training without compensatory resting, unexpectedly failed to lead to a reduction in OTS incidence. This attempt to shrink OTS has been frustrating, as one have realized that other factors than excessive training may also have important roles and induce OTS.

The complexity of the features of OTS, and at times unfeasible differentiation between OTS, FOR, and NFOR, and the highly individual clinical and biochemical presentation of OTS in each affected athlete make the identification of OTS to remain challenging, despite all advances in the research in this field, and its diagnosis is still dependent on a multiple-step exclusion criteria flowchart, highly time-consuming, and most of the times unable to be performed in most sports centers [1, 4].

Owing to the high and growing incidence of OTS, and its long-term and potentially irreversible psychological, professional, and social consequences, a scientific investigation for its prevention, early identification, and effective management is of extreme relevance. Furthermore, all learnings on OTS should be widespread for all sports-related health practitioners and even for those professionals who are not directly related to athletes, since there is a fast-growing number of extreme physically active subjects that practice as an additional activity or hobby, who are not regularly followed by any sports coach or other health provider, and who seek for general practitioners, instead of assistance from specific sports-related health professionals.

1.4   Conclusion

Overtraining syndrome (OTS) emerged in the beginning of the twentieth century with the increasing popularity of sports, leading to progressive increase of training load, until a point where athletes reached a "plateau," followed by a paradoxical decrease in performance, without any apparent etiology.

OTS is currently one of the most frequent sport-related dysfunctions in athletes resulting from a combination of excessive training and lack of recovery that leads to paradoxical decrease of sports performance, unresponsive to long periods of resting, and unexplainable by dysfunctions that could also lead to impaired performance.

Its pathophysiology is complex, multifactorial, and unclear, although it is likely that the imbalance between excessive activities and inappropriate resting leads to chronic deprivation of energy and mechanisms of repair, forcing adaptations to maintain survival and functionality, which attend to be highly dysfunctional (*maladaptations*), leading to multiple abnormalities in the immunological, hormonal, neurological, muscular, metabolic, and psychological systems and eventually OTS.

References

1. Meeusen R, Duclos M, Foster C, et al. European College of Sport Science; American College of Sports Medicine. Prevention, diagnosis, and treatment of the overtraining syndrome: joint consensus statement of the European College of Sport Science and the American College of Sports Medicine. Med Sci Sports Exerc. 2013;45(1):186–205.
2. Lehmann M, Foster C, Keul J. Overtraining in endurance athletes: a brief review. Med Sci Sports Exerc. 1993;25(7):854–62.
3. Rietjens GJ, Kuipers H, Adam JJ, et al. Physiological, biochemical and psychological markers of strenuous training-induced fatigue. Int J Sports Med. 2005;26(1):16–26.
4. Selye H. A syndrome produced by diverse nocuous agents. Nature. 1936;138:32.
5. Dolan P, Connolly J. Emotions, violence and social belonging. An Eliasian analysis of sport spectatorship. Sociology. 2014;48(2):284–99.
6. Thing LF. Quest for excitement: sport and leisure in the civilising process. Ann Leis Res. 2016;19(3):368–73.
7. Dunning E, Rojek C. Sport and leisure in the civilizing process. Critique and counter-critique. Toronto: University of Toronto Press; 1992.
8. Youniss J, Furth HG. Reversal learning in children as a function of overtraining and delayed transfer. J Comp Physiol Psychol. 1964;57:155–7.
9. Marsh G. Effect of overtraining on reversal and nonreversal shifts in nursery school children. Child Dev. 1964;35:1367–72.
10. Mackintosh NJ. Overtraining, reversal, and extinction in rats and chicks. J Comp Physiol Psychol. 1965;59:31–6.
11. Ohlrich ES, Ross LE. Reversal and nonreversal shift learning in retardates as a function of overtraining. J Exp Psychol. 1966;72(4):622–4.
12. Smit PJ. Sports medicine and 'overtraining'. S Afr Med J. 1978;54(1):4.
13. McTernan EJ, Leiken AM. A pyramid model of health manpower in the 1980s. J Health Polit Policy Law 1982 Winter. 6(4):739–51.
14. Stamford B. Avoiding and Recovering From Overtraining. Phys Sportsmed. 1983;11(10):180.
15. (No authors listed). Overtraining of Athletes. Phys Sportsmed. 1983;11(6):92–202.
16. Dressendorfer RH, Wade CE. The Muscular Overuse Syndrome in Long-Distance Runners. Phys Sportsmed. 1983;11(11):116–30.
17. Kreider R, Fry AC, O'Toole M. Overtraining in sport: terms, definitions, and prevalence. In: Kreider R, Fry AC, O'Toole M, editors. Overtraining in sport. Champaign (IL): Human Kinetics; 1998. p. VII–IX.
18. Barron JL, Noakes TD, Levy W, Smith C, Millar RP. Hypothalamic dysfunction in overtrained athletes. J Clin Endocrinol Metab. 1985;60(4):803–6.
19. Kreher JB, Schwartz JB. Overtrining syndrome: a practical guide. Sports Health. 2012;4(2):128–38.
20. Fry RW, Morton AR, Keast D. Overtraining in athletes. An update. Sports Med. 1991;12(1):32–65.

Chapter 2
Classical Understanding of Overtraining Syndrome

Outline

Initial theories: in which the author describes the major theories proposed by the most relevant researchers of the field.

   Preliminary biomarkers: in which the author enumerates the previous attempts to identify specific biomarkers of OTS and why most attempts have failed. This part includes both biochemical and also clinical proposed markers.

   Current diagnostic criteria: the book brings a logical description of the current diagnostic criteria, why each criterion has been proposed, as well as its limitations, particularly in the clinical practice.

2.1 Initial Theories

As OTS is by definition an unexplained underperformance syndrome, there is inherently not an apparent identified factor or trigger other than an alleged imbalance between training and resting. None of the proposed theories could fully encompass the pathophysiology of OTS, but at the same time some of these theories had high plausibility to explain at least part of the underlying mechanisms of OTS [1–6]. In addition, none of the evaluated potential markers or mechanisms was able to provide a definite background for OTS. As none of the proposed theories, evaluated markers, or mechanisms could fully explain this syndrome, but at the same time some of them could have a partial role in its pathophysiology, these theories and mechanisms do not preclude the other proposed mechanisms, but they may synergistically complement each other instead [1, 2, 4, 7].

   As a background for the comprehension of the meaning of the classical theories on OTS, a more comprehensive understanding of OTS is important. OTS syndrome is a consequence of a chronic exposure to a hostile environment, fully depleted of energy

© The Editor(s) (if applicable) and The Author(s), under exclusive license to Springer Nature Switzerland AG 2020
F. Cadegiani, *Overtraining Syndrome in Athletes*,
https://doi.org/10.1007/978-3-030-52628-3_2

9

and mechanisms of repair, with massive levels of oxidative stress reactive species. This combination forces the organism to undergo multiple adaptive changes in a broad variety of tissues and systems that are strictly necessary in order to maintain survival and functionality, even under highly hostile environment. These adaptations occur and are successful in terms of viability to main both living and functioning, but are highly dysfunctional and troublesome, as they are progressively harmful in the long run and extremely hard to be fully overcome, i.e., the reestablishment of the normal previous reactions become unlikely. This means that athletes with OTS, which are exposed to these aberrant adaptive reactions, are truly difficult to have these *maladaptations* reversed, which may explain the almost unfeasible complete reversibility of a true OTS process. These maladaptive processes cause several aberrant reactions in all systems, leading to multiple dysfunctions in muscular, hormonal, immunologic, neurological, psychiatric, autonomic, neuromuscular, cardiovascular systems, among which many of these dysfunctions are actually losses of the benefits acquired with exercising, and eventually leading to the key features of OTS. This sequence of processes, from the perspective of the current understanding on the syndrome, is illustrated in Fig. 2.1.

In addition to the classic theories, in the guideline on OTS [1], chronic stress and the subsequent chronic increase of cortisol secretion could theoretically play an important role in the loss of sensitivity of the central response to stressors, since it has been shown that in acute and chronic stress, the responsiveness of CRH-producing neurons decreases quickly after a short period of persistent stress.

The understanding of OTS as a syndrome related to chronic biological stress can be more comprehensively understood through the perspective of Selye's general adaptation syndrome [3], which states extreme situations can force the occurrence of functional and positive adaptations, in order to get better prepared for further similar extreme situations. In athletes, extreme situations correspond to repeated training overloads, leading to adaptations that eventually enhance athletes' performance. The improvement of the performance is the athletes' response to get better prepared for further training overloads, which also finds exact correspondence in Selye's general adaptation syndrome. These adaptations include multiple conditioning processes in the cardiovascular, musculoskeletal, neuromuscular, and autonomic systems; increased pulmonary, ventilation, and respiration capacity; optimization of the overall metabolism; and energy utilization, among other mechanisms that eventually lead to improved physical capacity. Conversely, oppositely to physiological adaptations, when extreme demands are followed insufficient repairing capacity, shortage of mechanisms of repair, and depleted energy availability, forced adaptations still occur to maintain survival and functionality despite the hostile environment. However, these adaptive processes become highly dysfunctional, and for this reason the term "maladaptations" can be coined for this process. These *maladaptations* are the critical triggers of the cascade of dysfunctions in multiple systems. Analogically, in athletes that undergo extreme physical demands without adequate resting and energy intake, maladaptations occur in order to tolerate further physical efforts even in the absence of repairing mechanisms and energy availability, and cause multiple secondary dysfunctions in a vareity of systems, eventually leading to OTS.

Some of the main theories regarding the pathophysiology are listed in Table 2.1 [7–32]. As previously mentioned, none of the theories may fully explain OTS, and all have some flaws. Improvement of these theories and more scientific data to support the proposed hypotheses are recommended and will be further discussed.

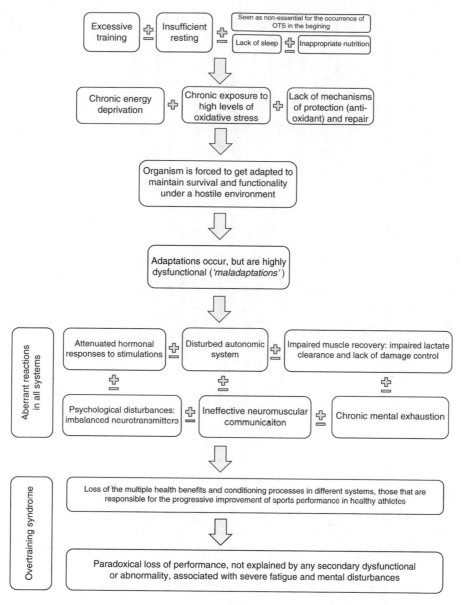

Fig. 2.1 Sequence of processes that lead to OTS, from the current understanding on the syndrome

2.2   Biomarkers in Overtraining Syndrome

Since the first description of OTS, multiple clinical and biochemical markers have been evaluated for their applicability for the identification and early diagnosis of OTS, including metabolic, hormonal, immunologic, muscular, and inflammatory parameters, and early psychological or performance signs [1, 4].

Table 2.1 Classical theories on OTS

Theory/hypothesis	Background	Pros	Cons	Suggestions for the improvement of the theory
Chronic glycogen depletion theory (BCAA hypothesis) [10–14]	Depleted glycogen would lead to fatigue and reduced performance, particularly in the muscle tissue. Insufficient energy for muscle needs would increase oxidation and consequently decrease BCAA levels. Alterations in BCAA levels would lead to alterations in neurotransmitters that are related to central fatigue	Low glycogen storage is associated with fatigue	Studies failed to demonstrate correlations between glycogen levels and performance. OTS-affected athletes tend to disclose normal glycogen levels. Decreased glycogen may lead to fatigue, but not necessarily to decreased performance. This theory does not explain all symptoms	*Global chronic energy-deficit theory* Glycogen levels are not appropriately evaluated when performed in a transversal manner. Only studies that evaluate the area under the curve (AUC) of glycogen levels would be able to conclude whether glycogen is depleted or not in OTS, since glycogen levels vary widely along the day. Glycogen levels could be indirectly assessed using diet record and indirect calorimetry
Chronic glutamine depletion theory [5, 13–20]	Glutamine is essential for immune function, DNA and RNA syntheses, nitrogen transportation, and glyconeogenesis. Depleted glutamine may lead to immunological dysfunctions and increased risk of infections	Glutamine levels are decreased after prolonged trainings, which also increases the risk of upper respiratory tract infections (URTIs). In vitro, level of activity of immune cells gets compromised in the absence of glutamine	Glutamine levels are not correlated with its level of activity and availability for immune cells. Glutamine intake did not lead to reduction in the risk of any infection. Increased risk of infections is not a specific characteristic of OTS. Very few researches focused on this hypothesis	*Chronic glutamine decreased tissue activity and availability and relative immunodeficiency* This revisited theory describes that the level of glutamine activity and its availability within immunologic cells may be better predictor of immunological activity than its serum levels. From a more comprehensive perspective, OTS can be understood as a multiple immunologic disorder, which may combine reduced availability and/or altered activity

Central fatigue theory [13–19, 21–23]	Increased non-conjugated tryptophan availability in the central nervous system (CNS), which induces increase of serotonin, would induce the perception of mental fatigue	There are direct correlations between tryptophan, 5-HT, and fatigue. Serotonin reuptake inhibitors (SSRIs) may decrease performance. Use of BCAAs in particular cases, which could directly compete with non-conjugated tryptophan, may increase energy and clearness	There is no evidence that alterations in 5-HT are specific to OTS. This theory does not justify symptoms present in the majority of athletes with OTS. Decrease of 5-HT levels after resting does not lead to subsequent improvement in performance. While healthy athletes would have decreased sensitivity to 5-HT, leading to fewer symptoms, OTS athletes would have a paradoxically increased sensitivity, although this is yet to be demonstrated	*Revisited and amplified central fatigue theory* It is unlikely that a single neurotransmitter status could fully predict OTS, but a balance between neurotransmitters instead. Also, this balance would not predict alone, but always associated with other factors, such as lack of glycogen or glucose availability, misalignment between those places that demand energy and those that have storage of energy. Availability of BCAA and glycogen should preferably be evaluated together with 5-HT levels
Autonomic nervous system (ANS) theory [5, 13–16, 21–26]	An imbalance between sympathetic and parasympathetic activities would lead to signs of disautonomy, mixed fatigue, and aberrant responses to adrenergic stimulations	Heart rate variability (HRV) responds to physical activity and could be used as a parameter for the analysis of the balance of the ANS. HRV is alleged to be reduced in OTS	The majority of studies on OTS failed to find alterations in the HRV. Nocturnal urinary catecholamines (NUC) were found to be inconsistent along different studies. Sensibility of adrenergic receptors has not been addressed in this theory, neither they have been evaluated	ANS was shown to have a role in the attempt to maintain functionality in athletes under OTS, despite the chronic energy depletion. Changes in the level of sensitivity of alpha- and beta-adrenergic receptors, limiting or enhancing their activities, may have important role in the autonomic dysfunction alleged to be present in OTS
Oxidative stress theory [21, 22, 27, 28]	Excessive oxidative stress without sufficient mechanisms of repair could lead to unrepaired inflammations and damages, muscular dysfunctions, inability to recover and to perform as before, and fatigue	Increase of oxidative stress during exercise was exacerbated in OTS, compared to healthy athletes with similar training levels, for a same training load. In animal models, citrate synthase, a largely used marker inversely correlated to oxidative stress, was found to be reduced in animals with corresponding alterations of OTS	Reduction of oxidative capacity is found to occur under any model of excessive training, independently from the occurrence of OTS, when studied in humans, and is also reduced in healthy and asymptomatic athletes. Whether increased oxidative stress in OTS is a consequence or its cause is unknown. Studies in humans lack	Exacerbated oxidative stress is likely an important part, but not the full pathophysiology of OTS, which may act synergistically with other factors to cause OTS

(continued)

Table 2.1 (continued)

Theory/hypothesis	Background	Pros	Cons	Suggestions for the improvement of the theory
Cytokines theory [5, 13–19, 20–26]	Increased release of cytokine and activity of non-classic inflammation pathways would lead to most of the key effects of OTS	This is a "unified" theory that could explain both what causes and the consequences of OTS. Cytokines have important actions in the hypothalamus to control and regulate multiple metabolic and hormonal pathways, including the feeling of fatigue and level of humor. Cytokines decrease muscular glucose uptake by a downregulation of the GLUT-4 glucose transporter, leading to inability to recover properly, muscular fatigue, and lower physical capacity. Some cytokines have clinical implications, including depression and sleep disorders, even in previously healthy subjects, and also to anorexia, which enhances the glycogen deprivation, in the case of OTS. The non-classic inflammatory pathways also enhance the hypothalamic-pituitary-adrenal (HPA) axis responses, while blunt the hypothalamic-pituitary-gonadal (HPG) axis activity. When chronically stimulated, certain cytokines may increase the risk of infections due to reduction of both cell-mediated and humoral immunity. Increased IL-1beta, IL-6, IL-10 and TNF-alpha levels have been reported in OTS	Evidence on increase of cytokines in OTS is still weak and inconsistent	Alterations in cytokines could be the underlying mechanism of some aspects of the pathophysiology of OTS. Alterations in the production and activity of myokines, the cytokines produced by muscle tissue, may also have important roles in OTS. Intra-tissue exposure to certain cytokines, independently of their levels in the human fluids, may be more important than the levels per se. Evaluation of the level of activity of each cytokine in each tissue may also bring more answers. There is the example of the IL-6 produced in the muscle tissue, which has substantial differences in its activity, compared to the IL-6 produced in the adipocytes. This should be further explored

| Hypothalamic theory [1, 5, 29–32] | Imbalance between hypothalamic-pituitary (HP) axes could lead to many of the characteristics of OTS | Hypothalamic-pituitary (HP) adrenal (HPA), lactotrophic (prolactin), and somatotrophic (GH) axes responses to repeated exercises are blunted in OTS, while exacerbated in overreaching states, and normal in healthy athletes | Previous studies have never compared the results with healthy sedentary controls. All tests were dependent on exercise performance, which is by definition compromised in OTS. Hence, differences in the performance, leading to sub-optimized stimulation of the HP axes, and not their level of responsiveness, could justify the blunted hormonal responses found in OTS | "Loss of hormonal conditioning" would better justify the loss of performance in OTS. While athletes have been shown to have optimized responses of the HP axes to stimulations, independently of exercise, when compared to healthy sex-, age-, and BMI-matched sedentary, which could be directly correlated with the progressive improvement of sports performance. The optimization of the hormonal responses found to occur in athletes seems to be lost under OTS, which could at least partially justify the loss of physical performance |

Candidate biomarkers for OTS should accurately differentiate OTS from confounding disorders and OTS-affected from healthy athletes, should be present in earlier stages of the disease, i.e., be detectable prior to the paradoxical decrease in sports performance, since the identification of OTS before the development of clinical and severe manifestations is critical for successful and complete recovery, and should be easily assessed and widely available, as not all sports centers have prompt access to medical services, and should preferably disclose a well-established cutoff.

Overall, all markers previously evaluated failed to be suitable for the diagnosis or prevention of OTS [1]. Despite the lack of accurate biomarkers, some offer clues for the directions of further studies on OTS. It is noteworthy that only few studies on OTS have been conducted since 2008–2010 [1], particularly those on biochemical parameters.

2.2.1 Biochemical Markers

In regard with biochemical markers, glycogen storage and serum glucose were shown to be normal [16, 33–35], which weakens the theory of chronic depletion of glycogen. However, conclusions regarding these two parameters are limited, once studies measured them isolated, instead of employing a continuous monitorization or a systematic and regular assessment, for the evaluation of their area under the curve (AUC). In particular, a single glycogen measure is inaccurate due to its high and fast variability in both hepatic and muscular tissues, particularly in athletes which limits the conclusions.

Despite normal glucose levels, its utilization may be relatively compromised since decreased insulin sensitivity has been reported [33], reinforcing the theory that proposes that muscular GLUT-4 insulin receptor is downregulated in OTS.

Maximum lactate levels have been shown to be reduced, while its sub-maximum levels were unaltered. However, lactate basal levels measured at least 24 hours after last training session, which would indirectly estimate the speed of lactate clearance, had not been reported [1].

Creatine kinase (CK) levels were found to be prolongedly high in OTS, when in response to a certain amount of eccentric effort, pathologically until up to 2 weeks [50], which is a typical abnormal response in the absence of glycogen for the recovery process [33, 35].

Although plasmatic glutamine levels were suggested to be reduced by preliminary studies [1, 20, 36, 37], they were not further confirmed [1, 34–38]. Despite the inconsistent findings on glutamine levels, it has not been demonstrated to have a role in the likely relative immunosuppression in OTS. The glutamine-to-glutamate ratio could be a potential marker, but confirmatory studies are lacking [1].

Among hormonal markers, none of the basal hormonal levels, including total testosterone, insulin-like growth factor 1 (IGF-1), dehydroepiandrosterone sulfate (DHEA-S), thyroid-stimulating hormone (TSH), adrenocorticotropic hormone (ACTH), cortisol, and prolactin, were altered in OTS [1, 16, 32, 35, 39–43].

The testosterone-to-cortisol ratio, alleged to be an accurate marker of the balance between anabolism and catabolism, and extensively used as a marker of excessive training, has been improperly used as a diagnostic or inclusion criteria for OTS, since this ratio does not present any actual scientific support in the literature. Besides, studies that evaluated this ratio failed to demonstrate alterations in OTS, when compared to healthy athletes [32, 42].

In relation to catecholamine excretion, as part of the autonomic nervous system (ANS), both plasmatic and urinary results were conflicting [32, 40, 41, 43].

Oppositely to basal levels, hormonal responses to stimulations were found to be blunted in OTS, while unaltered or exacerbated in earlier states of overreaching, in particular the ACTH, growth hormone (GH), cortisol, and prolactin, in response to maximum exercise or two-bout exercise protocols (with 4 hours difference between the first and second exercise sessions) [29–32, 41, 42]. These studies added considerable information in terms of the potential roles of hormones on OTS. However, important limitations have been identified, including inherent differences in performance between healthy and affected athletes, which could be the underlying reason for sub-optimized hormonal responses, instead of blunted hormonal responsiveness; the lack of a second control group of non-physically active subjects, in order to obtain the baseline hormonal responses (without the influence of the adaptive changes from exercising); and a low number of participants in the studies. Further studies addressing these methodological issues are highly recommended. The summary of the hormonal findings on OTS identified in a systematic review is shown in Fig. 2.2.

Other markers, including leptin, interleukin-6 (IL-6), adiponectin, tumor necrosis factor alpha (TNF-alpha), and insulin-like growth factor binding protein 1 (IGFBP-1), were shown to be reduced in pathological catabolic states, like OTS.

Hormone	Quantity of studies	Normal response	Blunted response	Exacerbated response
Cortisol	26 (65.0%)	18 (69.2%)	8 (30.8%)	-
Total T	18 (45.0%)	14 (77.8%)	3 (16.7%)	1 (5.9%)
GH	13 (32.5%)	7 (53.8%)	5 (38.4%)	1 (7.8%)
T/C	10 (25.0%)	5 (50.0%)	4 (40.0%)	1 (10.0%)
ACTH	10 (25.0%)	6 (60.0%)	4 (40.0%)	-
Plasma catecholamines	9 (22.5%)	4 (44.4%)	4 (44.4%)	1 (11.1%)
Insulin	8 (20.0%)	6 (75.0%)	2 (25.0%)	-
LH	8 (20.0%)	7 (87.5%)	1 (12.5%)	-
Prolactin	6 (15.0%)	4 (66.7%)	2 (33.3%)	-
FSH	6 (15.0%)	5 (83.3%)	1 (16.7%)	-
IGF-1	5 (12.5%)	3 (60.0%)	1 (20.0%)	1 (20.0%)
Aldosterone	5 (12.5%)	1 (20.0%)	2 (40.0%)	2 (40.0%)
Free T	3 (7.5%)	1 (33.3%)	1 (33.3%)	1 (33.3%)
T4	2 (5.0%)	2 (100%)	-	-
T3	2 (5.0%)	2 (100%)	-	-
NUC	2 (5.0%)	-	2 (100%)	-
Salivary C	2 (5.0%)	-	2 (100%)	-
Salivary T	2 (5.0%)	1 (50.0%)	1 (50.0%)	-
NUC	2 (5.0%)	1 (50.0%)	1 (50.0%)	-
IGFBP-3	2 (5.0%)	1 (50.0%)	1 (50.0%)	-
IGF-1/IGFBP-3	2 (5.0%)	2 (100%)	-	-
SHBG	1	1	-	-
DHEA-S	1	1	-	-
Peptide F	1	1	-	-
Fertility	1	-	1	-
Beta-2 muscle receptor	1	-	1	-

NUC Nocturnal urinary catecholamines, T/C testosterone/cortisol ratio, T/SHBG testosterone/sexhormonebindingglobulin, T3 thyronine, T4 thyroxine, salivary C salivary cortisol, salivary T salivary testosterone, IGF-1/IGFBP-3 IGF-1/IGFBP3 ratio, fertility evaluation by semen analysis

Fig. 2.2 Summary of the findings in the systematic review on the hormonal aspects of overtraining syndrome

These reductions are possible consequences of chronic energetic deficit and have important effects on hormonal and energetic regulations [1, 16, 35, 36, 39–41, 44].

A transient immunosuppression is physiologically expected after a training overload, independently from OTS, what may justify the increased incidence of upper respiratory tract infections (URTIs) [45–48]. The increased occurrence of infections, particularly URTIs, probably results from a reduced response of neutrophil degranulation to bacterial stimulation found to occur in elite athletes during intensified training; reduced neutrophil, monocyte, and natural killer (NK) cells activity; abnormal lymphocyte proliferation; imbalanced proportion between CD4+ and CD8+ cells; decreased natural killer (NK) cells activity; and reduced salivary IgA [46–48]. However, despite all the immunologic alterations reported in athletes exposed to prolonged and intensified trainings, specific data in OTS is scarce and inconsistent, even though URTIs in OTS-affected athletes tend to be more severe and prolonged, compared to those in healthy athletes. This is possibly due to an exacerbation of the already expected decrease of both cell-mediated and humoral immune responses [49].

Overall, the most remarkable biochemical findings on OTS include blunted GH, prolactin, and cortisol responses to exercise-dependent stimulation tests and exacerbated and prolonged increase of CK levels. However, none of these markers were able to distinct OTS from healthy athletes accurately, to provide practical reasonable cutoffs, and were helpless to prevent or detect early stages of OTS.

2.2.2 Clinical Markers

From a perspective of changes in performance, besides the sine qua non presence of at least 10% reduction in pace, intensity, strength, and/or volume of training for the diagnosis of OTS, a reduction of more than 20% in the time-to-fatigue, i.e., for a same training load, athletes get fatigued at least 20% faster than previously. This has been shown to better identify OTS compared to the 10% reduction in performance, as in the continuous intensity increment test, affected athletes have weaker responses and shorter time until fatigue [16, 26].

Among the types of methods for psychological assessment, the Profile of Mood States (POMS) questionnaire was shown to be more consistently altered for fatigue, depression, anger, confusion, tension, and vigor levels in OTS, compared to healthy athletes. POMS was also slightly altered in athletes under important training overload, irrespective of the presence of OTS [50–53], in a linear correlation with training intensity, which started to alter as early as 2 days of training overload [51], however which quickly improves after a short period of resting. Besides being less altered, athletes under intensified training yet unaffected by OTS have normal anger, confusion, tension, and depression levels, whereas in OTS all subscales tend to be affected, particularly in the depression subscale [1, 50–55].

As attempts to improve the sensitivity for an early identification of OTS, more specific questionnaires have been proposed, including the POMS-based OT, the

Training Distress Scale (TDS), and the RESTQ-Sport [1]. However, once OTS may be influenced by factors other than training patterns, a non-specific questionnaire like the POMS could be more appropriate, in addition to the fact that POMS has been extensively validated. Hence, for researches in OTS, POMS should always be employed, which does not exclude the concurrent use of other questionnaires (which could be valuable to perform correlations between their results) [1, 50–55].

The quality of the application of the psychological assessment methods is key to determine reliable results. Unless performed by a third party, unaware of who belongs to each group, the person who performs the questionnaire needs to be fully focused on avoiding differences in terms of non-verbal communication and body behavior, including voice tone, body language, and slight changes in the formulation of the questions, which could induce changes in responses, leading to inaccurate results. The "faking good" bias for healthy athletes and "faking bad" bias for affected athletes, i.e., questions are performed in a way that may induce normal responses from healthy athletes, while negatively abnormal responses from athletes affected by OTS are likely the most important biases, which can be avoided by a proper understating, familiarity, and training for the questions to be performed, in order to achieve high levels of standardization for the questions [1, 56].

Conversely, psychomotor speed psychological tests are promising, once they are simple to be performed; not human-dependent, which decreases the human behavior-related biases; of low cost; and noninvasive and cannot be consciously manipulated by athletes. Fatigue-related syndrome, like chronic fatigue syndrome (CFS), burnout syndrome, and OTS, lead to temporary reductions in some of the cognitive functions, including memory, attention, concentration, and time-to-reaction (reflection for responses) [1, 2, 5, 53–55].

The heart rate variability (HRV), an indirect effect of the autonomic balance and variability control over the heart, and a potential marker of a balance between the parasympathetic (mainly through vagus nerve) and sympathetic autonomic nervous systems (ANS), showed inconsistent results in OTS [1, 4, 23–26, 56].

2.3   Current Diagnostic Criteria

The diagnosis of OTS is challenging due to the lack of specific marker, the complex and highly heterogeneous clinical and biochemical manifestations, and the multiple possible confounding disorders and alterations, which are not always overtly present.

For this reason, the diagnosis of OTS should still be performed using the latest guideline on OTS [1], which is based on the clinical assessment of the athlete suspected for OTS, followed by the exclusion of alterations and dysfunctions that could explain the decreased performance. These are two key steps for the diagnosis of OTS since the sine qua non criteria for OTS includes a decrease in sports performance, which cannot be explained by any clinical or biochemical apparent dysfunction, i.e., OTS is by definition an unexplained underperformance syndrome.

However, the latest guideline on OTS has some important limitations. First, it does not provide a formal list to be considered as possible confounding disorders, since an accurate diagnosis of OTS requires the exclusion of a wide range of dysfunctions that may be present uncharacteristic and unspecific manifestations. Good illustrating examples of disorders unrelated to physical activity that may lead to reduced performance are subacute or post-acute infections, which are many times unnoticed and asymptomatic at resting, but may arise under physical effort by easier, exacerbated, and prolonged fatigue and muscle soreness. The list of infections that require investigation in non-specific clinical states include *Epstein-Barr*, myocarditis, hepatitis, and *Lyme* disease (*Borrelia*). Non-infectious diseases that must be excluded prior to the diagnosis of OTS include vitamin B12 and iron deficiency (as well anemia due to any of these deficiencies), metabolic dysfunctions like new onset or uncontrolled types 1 and 2 diabetes mellitus, overt hormonal alterations, including hypothyroidism and hypogonadism (except when hormonal dysfunctions are consequences of OTS, which the differentiation will be further discussed), autoimmune disorders including fatigue-related autoimmune disorders and those types of arthritis that compromise movements, cardiovascular diseases (congestive heart failure, clinically significant valvar disorders), pulmonary conditions (chronic obstructive pulmonary disease (COPD), poorly controlled asthma, rhinitis), and renal or hepatic failure.

In addition to the lack of a formal list of confounding diseases to be excluded, the guideline did not provide a list of specific exams in which alterations could explain the loss of performance. Furthermore, several other conditions have been more recently correlated with decreased performance and also should also be excluded for the diagnosis of OTS.

Additional signs may also be helpful for the identification of OTS, including fatigue that does not improve after a resting period, psychological disturbances, easier infections (particularly upper respiratory tract infections – URTIs), flu-like symptoms typically present in more severe fatigue states, decreased libido, muscle soreness, and increased sense of effort to perform a same training load, besides a broad spectrum of other symptoms, since OTS is complex and multi-systemic, in which the combination of manifestations is unique to each affected athlete.

2.4   Conclusion

Although there are not theories or concepts fully able to explain its pathophysiology, OTS, which is an unexplained underperformance syndrome, can be comprehensively understood as a consequence of chronic exposure to a hostile energy-deprived environment without mechanisms of repair, which forces adaptations to maintain survival and functionality under critical circumstances. In order to be successful, these adaptations that occur tend to be highly dysfunctional and aberrant, which are truly hard to be reversed, and may explain the almost irreversibility of true OTS.

Its biochemical markers include decreased muscular GLUT-4 insulin sensitivity; reduced maximum lactate, leptin, interleukin-6, adiponectin, and tumor necrosis factor alpha (TNF-alpha); prolonged increase of creatine kinase (CK); and decreased growth hormone (GH), cortisol, and prolactin responses to exercise, while clinical manifestations include increased fatigue, depression, anger, confusion, and tension and reduced energy levels and abnormalities in the heart rate characteristics, including heart rate variability (HRV).

OTS should only be diagnosed after appropriate detection of reduced performance and exclusion of confounding disorders or factors that could also lead to reduced performance.

References

1. Meeusen R, Duclos M, Foster C, European College of Sport Science, American College of Sports Medicine, et al. Prevention, Diagnosis, and Treatment of the Overtraining Syndrome: Joint Consensus Statement of the European College of Sport Science and the American College of Sports Medicine. Med Sci Sports Exerc. 2013;45(1):186–205.
2. Rietjens GJ, Kuipers H, Adam JJ, et al. Physiological, biochemical and psychological markers of strenuous training-induced fatigue. Int J Sports Med. 2005;26(1):16–26.
3. Selye H. A syndrome produced by diverse nocuous agents. Nature. 1936;138:32.
4. Stamford B. Avoiding and recovering from overtraining. Phys Sportsmed. 1983;11(10):180.
5. Overtraining of athletes. Phys Sportsmed. 1983;11(6):92–202.
6. Kreher JB, Schwartz JB. Overtraining syndrome: a practical guide. Sports Health. 2012;4(2):128–38.
7. Nederhof E, Lemmink KA, Visscher C, Meeusen R, Mulder T. Psychomotor speed: possibly a new marker for overtraining syndrome. Sports Med. 2006;36(10):817–28.
8. Kentta G, Hassmen P, Raglin J. Overtraining and staleness in Swedish age-group athletes: association with training behavior and psychosocial stressors. Int J Sports Med. 2001;2001:460–5.
9. Mergan W, O'Connor P, Ellickson K, Bradley P. Psychological characterization of the elite female distance runners. Int J Sports Med. 1987;8(Suppl 2):124–31.
10. Raglin J, Sawamura S, Alexious S, Hassmen P, Kentta G. Training practices and staleness in 13-18 year old swimmers: a cross-cultural study. Pediatr Sports Med. 2000;12:61–70.
11. Snyder AC. Overtraining and glycogen depletion hypothesis. Med Sci Sports Exerc. 1998;30(7):1146–50.
12. Gastmann UA, Lehmann MJ. Overtraining and the BCAA hypothesis. Med Sci Sports Exerc. 1998;30(7):1173–8.
13. Budgett R. Fatigue and underperformance in athletes: the overtraining syndrome. Br J Sports Med. 1998;32:107–10.
14. Budgett R, Hiscock N, Arida R, et al. The effects of the 5-HT2C agonist mchlorphenylpiperazine on elite athletes with unexplained underperformance syndrome (overtraining). Br J Sports Med. 2010;44:280–3.
15. Smith LL. Cytokine hypothesis of overtraining: a physiological adaptation to excessive stress? Med Sci Sports Exerc. 2000;32:317–31.
16. Castell LM, Poortmans JR, Leclercq R, et al. Some aspects of the acute phase response after a marathon race, and the effects of glutamine supplementation. Eur J Appl Physiol. 1997;75:47–53.
17. Davis JM. Carbohydrates, branched-chain amino acids, and endurance: the central fatigue hypothesis. Int J Sport Nutr. 1995;5(Suppl):S29–38.
18. Meeusen R, Watson P, Hasegawa H, Roelands B, Piacentini MF. Brain neurotransmitters in fatigue and overtraining. Appl Physiol Nutr Metab. 2007;32(5):857–64.

19. Walsh NP, Blannin AK, Robson PJ, Gleeson M. Glutamine, exercise and immune function. Links and possible mechanisms. Sports Med. 1998;26(3):177–91.
20. Halson SL, Jeukendrup AE. Does overtraining exist? An analysis of overreaching and overtraining research. Sports Med. 2004;34(14):967–81.
21. Petibois C, Cazorla G, Poortmans JR, Déléris G. Biochemical aspects of overtraining in endurance sports: a review. Sports Med. 2002;32(13):867–78.
22. Hohl R, Ferraresso RL, DeOliveira RB, et al. Development and characterization of na overtraining animal model. Med Sci Sports Exerc. 2009;41(5):1155–63.
23. Kajaia T, Maskhulia L, Chelidze K, Akhalkatsi V, Kakhabrishvili Z. The effects of nonfunctional overreaching and overtraining on autonomic nervous system function in highly trained athletes. Georgian Med News. 2017;264:97–103.
24. Kiviniemi AM, Tulppo MP, Hautala AJ, Vanninen E, Uusitalo AL. Altered relationship between R-R interval and R-R interval variability in endurance athletes with overtraining syndrome. Scand J Med Sci Sports. 2014;24(2):e77–85.
25. Uusitalo AL, Uusitalo AJ, Rusko HK. Heart rate and blood pressure variability during heavy training and overtraining in the female athlete. Int J Sports Med. 2000;21(1):45–53.
26. Margonis K, Fatouros IG, Jamurtas AZ, et al. Oxidative stress biomarkers responses to physical overtraining: implications for diagnosis. Free Radic Biol Med. 2007;43(6):901–10.
27. Tanskanen M, Atalay M, Uusitalo A. Altered oxidative stress in overtrained athletes. J Sport Sci. 2010;28(3):309–17.
28. Meeusen R, Nederhof E, Buyse L, Roelands B, De Schutter G, Piacentini MF. Diagnosing overtraining in athletes using the two-bout exercise protocol. Br J Sports Med. 2010;44(9):642–8.
29. Meeusen R, Piacentini MF, Busschaert B, Buyse L, De Schutter G, Stray-Gundersen J. Hormonal responses in athletes: the use of a two bout exercise protocol to detect subtle differences in (over)training status. Eur J Appl Physiol. 2004;91(2–3):140–6.
30. Urhausen A, Gabriel HH, Kindermann W. Impaired pituitary hormonal response to exhaustive exercise in overtrained endurance athletes. Med Sci Sports Exerc. 1998;30(3):407–14.
31. Cadegiani FA, Kater CE. Hormonal aspects of overtraining syndrome: a systematic review. BMC Sports Sci Med Rehabil. 2017;9:14.
32. Lemon P, Mullin J. Effect of initial muscle glycogen on protein catabolism during exercise. J Appl Physiol. 1980;48:624–9.
33. Xiao W, Chen P, Dong J. Effects of overtraining on skeletal muscle growth and gene expression. Int J Sports Med. 2012;33(10):846–53.
34. Angeli A, Minetto M, Dovio A, Paccotti P. The overtraining syndrome in athletes: a stress-related disorder. J Endocrinol Investig. 2004;27(6):603–12.
35. Rowbottom D, Keast D, Goodman C, Morton A. The haematological, biochemical and immunological profile of athletes suffering from the overtraining syndrome. Eur J Appl Physiol. 1995;70:502–9.
36. Robson P, Blannin A, Walsh N. The effect of an acute period of intense interval training on human neutrophil function and plasma glutamine in endurance-trained male runners. J Physiol. 1999;515:84–5.
37. Lancaster G, Halson S, Khan Q, et al. Effect of acute exhaustive exercise and a 6-day period of intensified training on immune function in cyclists. J Physiol. 2003;548P:O96.
38. Halson S, Lancaster G, Jeukendrup A, Gleeson M. Immunological responses to overreaching in cyclists. Med Sci Sports Exerc. 2003;35(5):854–61.
39. Joro R, Uusitalo A, DeRuisseau KC, Atalay M. Changes in cytokines, leptin, and IGF-1 levels in overtrained athletes during a prolonged recovery phase: a case-control study. J Sports Sci. 2017;35(23):2342–9.
40. Steinacker J, Lormes W, Reissnecker S, Liu Y. New aspects of the hormone and cytokine response to training. Eur J Appl Physiol. 2004;91:382–93.
41. Urhausen A, Gabriel H, Kindermann W. Blood hormones as markers of training stress and overtraining. Sports Med. 1995;20:251–76.

42. Wittert G, Livesey J, Espiner E, Donald R. Adaptation of the hypothalamo–pituitary adrenal axis to chronic exercise stress in humans. Med Sci Sports Exerc. 1996;28(8):1015–9.
43. Verde T, Thomas S, Shephard RJ. Potential markers of heavy training in highly trained endurance runners. Br J Sports Med. 1992;26:167–75.
44. Bishop NC, Gleeson M. Acute and chronic effects of exercise on markers of mucosal immunity. Front Biosci. 2009;14:4444–56.
45. Gleeson M. Mucosal immune responses and risk of respiratory illness in elite athletes. Exerc Immunol Rev. 2000;6:5–42.
46. Gleeson M, Bishop N, Oliveira M, McCauley T, Tauler P, Muhamad A. Respiratory infection risk in athletes: association with antigen-stimulated IL-10 production and salivary IgA secretion. Scand J Med Sci Sports. 2012;22(3):410–7.
47. Neville V, Gleeson M, Folland J. Salivary IgA as a risk factor for upper respiratory infections in elite professional athletes. Med Sci Sports Exerc. 2008;40(7):1228–36.
48. Reid V, Gleeson M, Williams N, Clancy R. Clinical investigation of athletes with persistent fatigue and/or recurrent infections. Br J Sports Med. 2004;38:42–5.
49. Morgan W, Brown D, Raglin J, O'Connor P, Ellickson K. Psychological monitoring of overtraining and staleness. Br J Sports Med. 1987;21:107–14.
50. Morgan W, O'Connor P, Sparling P, Pate R. Psychological characterization of the elite female distance runner. Int J Sports Med. 1987;8(Suppl 2):124–31.
51. Raglin J, Morgan W, Luchsinger A. Mood state and selfmotivation in successful and unsuccessful women rowers. Med Sci Sports Exerc. 1990;22:849–53.
52. Raglin J, Morgan W, O'Connor P. Changes in mood state during training in female and male college swimmers. Int J Sports Med. 1991;12:585–9.
53. Raglin J, Wilson G. Overtraining and staleness in athletes. In: Hanin YL, editor. Emotions in sports. Champaign: Human Kinetics; 2000. p. 191–207.
54. O'Connor P. Overtraining and staleness. In: Morgan WP, editor. Physical activity & mental health. Washington, D.C.: Taylor & Francis; 1997. p. 145–60.
55. Rice S, Olive L, Gouttebarge V, et al. Mental health screening: severity and cut-off point sensitivity of the athlete psychological strain questionnaire in male and female elite athletes. BMJ Open Sport Exerc Med. 2020;6(1):e000712.
56. Hynynen E, Uusitalo A, Konttinen N, Rusko H. Cardiac autonomic responses to standing up and cognitive task in overtrained athletes. Int J Sports Med. 2008;29(7):552–8.

Chapter 3
Methodological Challenges and Limitations of the Research on Overtraining Syndrome

Outline

Methodological challenges and limitations of the research of overtraining syndrome: OTS is a complex and multi-factorial disorder, and does not show any marker universally present. Due to its high complexity and lack of an accurate understanding of the pathophysiology, the author explains, point-by-point, the challenges in the research on OTS, as well as why most of the past researches had some methodological flaws that limited the findings and proposals to overcome these challenges.

3.1 Overtraining Syndrome as a Challenging Disorder

Overtraining syndrome (OTS) is characterized by being highly complex and multi-factorial, without a clear pathophysiology, grueling to be appropriately diagnosed, and with multiple possibilities of complex interactions between risk factors, wide different dysfunctional pathways, and abnormal reactions, which in turn leads to a wide variety of combinations of clinical and biochemical alterations. OTS does not exhibit any ubiquitously present dysfunction, does not have a single marker that accurately distincts OTS from unaffected athletes, and whose combination of metabolic, biochemical, and clinic manifestations is truly individual and therefore unlikely present in two different affected athletes [1–3].

With these characteristics, research on OTS turns out to be greatly challenging, deeply subjective, and ultra-sensitive to minor flaws and roughens the assessment for its selection, diagnosis, and evaluation. In fact, studies on OTS showed inconsistent results and were unsuccessful to elucidate in its markers and pathophysiology, essentially due to some unintentional flaws and impreciseness [4].

© The Editor(s) (if applicable) and The Author(s), under exclusive license to Springer Nature Switzerland AG 2020
F. Cadegiani, *Overtraining Syndrome in Athletes*,
https://doi.org/10.1007/978-3-030-52628-3_3

3.2 Remarkable Challenges and Limitations of the Research on Overtraining Syndrome

Some of the most remarkable challenges and limitations researchers face when studying OTS are:

1. Several studies evaluated OTS in athletes that underwent a training load program, in which to them this would inherently lead to a state of OTS. However, OTS is not an automatic or expected consequence of excessive increase in training intensity or volume, but resulted from a complex interaction of factors, in which excessive training may not always be present. An intensification of the training sessions does not lead to OTS, as training overload does not generate equal effects in all athletes. Besides, excessive training tends to lead to an overreaching state, not OTS, which does not always happen and can only be diagnosed when participants start an acute paradoxical decrease in performance. In these cases, OTS is not an appropriate diagnosis, since it can only be considered as possible OTS when symptoms get prolonged, worse, with more severe psychological disturbances and unresponsive to a period of recovery. Hence, intensification of training programs should be employed for the evaluation of the direct effects of excessive training and should not be extrapolated to any of the OTS/FOR/NFOR states. In these cases, overreaching should only be considered in case of documented decrease of performance.

   Indeed, it has been observed that studies that evaluated OTS as a natural consequence of a process of progressive incrementing of training intensity and/or volume, regardless of decreased performance or signs for the diagnosis of any of the underperformance state, presented distinct results from those that evaluated athletes with natural-occurring OTS [4]. Studies on actual OTS had very few similar results than those that evaluated the effects of an intensified training load, which may help justify the inconsistency in the findings between different studies, and precluded from more strength conclusions regarding biochemical and hormonal markers of OTS. It has also been observed that majority of the studies alleged to evaluate OTS actually assessed the effects of intensified training, not OTS properly diagnosed [4]. Due to the unique and complex characteristics of OTS, which are distinct from the natural responses to intensified training, it is strongly recommended that OTS should be diagnosed from real-life cases and following all the criteria proposed for its diagnosis – an unexplained underperformance unresponsive to recovery [5].

2. Similarly to the attempts to artificially induce a state of OTS, the use of forced physical trainings in animal models aiming to understand the pathophysiology of OTS does not consider the facts mentioned above. To simulate a more precise overreaching or OTS state, an animal model should simultaneously increase training load; decrease the carbohydrate, protein, and overall calorie intake; induce a disturbed sleeping; and induce concurrent cognitive efforts.

3. The precise definitions and criteria for the diagnosis of OTS, as well as the differentiation between FOR/NFOR and OTS, were extremely heterogeneous [6, 7]. For an adequate diagnosis of OTS for research purposes, decreased performance or increased effort for a same training load, or reduced time-to-fatigue, should be documented. To differentiate between overreaching and overtraining, a persistence of any of these characteristics after a certain period of resting would lead to the diagnosis of OTS in animal models. However, many of these suggestions are based on the human behavior of the disease, and other adaptations are likely needed to be employed.

4. The appropriate inclusion and exclusion criteria for OTS are hard, time-consuming, and onerous. It has been noticed a very limited number of studies on OTS designed their selection criteria based on the OTS diagnostic flowchart and excluded confounding disorders prior to the diagnosis of OTS [4]. In addition, many of the studies based the diagnosis of OTS on inaccurate criteria, including the use of CK levels, testosterone-to-cortisol ratio, or the presence of fatigue as the only clinical sign of OTS, without a proper verification and quantification of the level of loss of performance by a skilled sports coach. Besides the nonnatural occurrence of OTS, the inappropriate diagnosis of OTS is a second important reason that may justify why many of results were highly conflicting and at times misleading. The inherent complexity of the diagnosis of OTS [1] precludes studies from employing precise selection criteria, and, when studies do so, they unlikely have a large sample size.

5. For the evaluation of potential biomarkers for OTS, several studies employed tests in which the responses were highly dependent on exercise, aiming to identify suboptimal or abnormal reactions to physical activity in OTS. However, since the performance of OTS affected athletes are by definition compromised, differences between the performances during an exercise stimulation test will mislead results. In this case, differences in the responses between OTS-affected and healthy athletes may result from a relative sub-utilization of the musculoskeletal and cardiovascular capacity in affected athletes, compared to unaffected ones, rather than responsiveness or impaired responses to stimulation.

   Moreover, adjustment of the results for the level of performance seems unfeasible, once responses are rarely linear with the training intensity and peak.

   Hence, differences in results are not able to elucidate exact etiology or the primary source of possible alterations to be identified in OTS. Attempts to identify the exact site of the dysfunction that may have caused possible alterations are only feasible when tests are performed independently from exercise, avoiding the expected inherent differences in performance already discussed.

6. A general observation of the findings on OTS is that alterations are not overt, i.e., results in OTS-affected athletes are unaltered when compared to the normal ranges of the general population, but are not absent either, since results are different from those in healthy athletes. It means that alterations in OTS are relative, as they cannot be identified using the regular ranges, but are altered than the expected levels for athletes, or when compared to healthy athletes with similar characteristics.

However, the absence of the standardization of adaptations of ranges adapted for physically active subjects, and the lack of a general understanding of the fact that athletes may yield distinct results from sedentary and be yet normal, precludes a proper and precise analysis of the results to be found in athletes from the perspective of what to be expected for these athletes.

In addition, healthy athletes may not always represent the actual appropriate levels to be used as the only control group, since many of the still unrecognized adaptive alterations in a wide variety of systems likely occur in response to training.

Hence, researchers struggle to find biomarkers and to understand the underlying mechanisms of OTS. And comparisons with a single control group of healthy athletes, which would provide the perspective of the expected findings for athletes, would only allow conclusions regarding the direct comparisons to what to be expected in athletes, not whether dysfunctions are frankly present. A solution for this challenge is the regular inclusion of a second control group, of healthy non-physically active subjects, with similar baseline characteristics (sex, age, and BMI), for a double perspective of the analysis of the findings on OTS: the understanding of this syndrome as an overt condition and/or as a disease resulted from alterations of the adaptations expected to be present in athletes. This double standard would allow more precise conclusions, since the comparison with the two control groups allow to identify whether and which dysfunctions found in OTS are relative, that is, altered when compared to the expected to athletes but similar to general population, or overtly altered, i.e., abnormal when compared to both healthy athletes and sedentary.

7. OTS is a multi-factorial and multi-systemic disorder, resulting from a combination of dysfunctions of a broad spectrum of systems and pathways, with multiple interactions that play synergistically to induce OTS. Hence, the exclusive evaluation of only one cluster of markers, such as hormones, cytokines, immune markers, or muscle markers, will unlike to be able to unveil the mechanisms on OTS. Instead, the simultaneous evaluation of parameters of different aspects and their correlations is highly desirable for the elucidation of the paths that lead to OTS.

8. The quality of the methodology used for an appropriate sample and control for variables is crucial for proper findings on OTS. The samples to be tested, which are many times collected on the site of the trainings, should be under a strict pre-analytic protocol including transport, storage, and centrifugation, as some hormones and other biochemical markers may easily decay or be highly sensitive to environmental conditions and movements. Also, acute responses to exercise may last as long as 24 hours, and for this reason resting levels should be collected at least 24 hours (preferably a minimum of 48 hours) since the last training session.

9. Besides training patterns, other variables may largely influence development or severity of OTS, including dietary and sleeping patterns, and period (whether off, pre-, or during season). Thus, all these characteristics must be assessed, controlled, and specifically described in every study on OTS, once results are

largely dependent on each of these factors. Also, whenever suitable, parameters that are known to be influenced by any of these characteristics should be adjusted aiming to reduce biases of external factors.

10. Many hormones have specific releasing patterns, including pulsatile, seasonal, and circadian variations. Additionally, hormonal alterations do not exclusively occur due to changes in its release, but also due to abnormalities in its actions, including alterations in its receptor and coupling, including down- or upregulation of the receptors, or reduced affinity, or impaired secondary signaling after the coupling between the hormone and its receptor. The inherent variation in the hormonal levels and the technical difficulties in the evaluation of hormonal receptors and actions challenges the evaluation of the endocrine aspects of researches in athletes. Indeed, owing to the severe alterations in the tissue environment due to chronic energy deficiency and lack of mechanisms of repair, it is likely that athletes affected by OTS present disturbances not only in hormonal responses but also in hormonal receptors and specific intracellular signaling pathways – which is not as easily assessed and elucidated.

11. In the presence of hormonal alterations, the differentiation between whether these are secondary to OTS and whether the hormonal dysfunctions are the primary cause of the loss of performance is many times unfeasible. While actual hormonal dysfunctions may lead to decrease of performance, which excludes the diagnosis of OTS, since OTS is by definition a primary condition, OTS may itself lead to hormonal alterations. As both situations lead to a sort of a vicious cycle effect, when both are present, it is hard to define which is the actual trigger, and under- or overdiagnosis of OTS may occur [1, 4].

In general, when the organism is exposed to metabolic dysfunctions, including obesity, diabetes, a state of hypogonadotrophic hypogonadism originated in the hypothalamus (also termed as tertiary hypogonadism) is induced, and is resulted from subclinical inflammatory process in the non-classical pathways. In a similar manner, circumstances that lead to highly stressful, hostile, and chronically depleted energy environments, like anorexia nervosa, bulimia, cachexia, and female athlete triad, also tend to induce tertiary hypogonadism, as a maladaptive response to energy-deprived states, for energy preservation. Since OTS can be considered as a sort of athlete-specific burnout state in which chronic lack of energy is key, central hypogonadism is expected to occur earlier or later during the natural disease course.

In the case of metabolically induced hypogonadism, deficiency in males tends to be mild (in males, testosterone above 180–220 ng/dL), which is commonly associated with low-normal sex hormones binding globulin (SHBG). Under mild low testosterone levels, it is more probable that these slight abnormal levels are consequences to OTS, which confirms the diagnosis of OTS, when the other criteria are met. Indeed, mild hypogonadism alone is unlikely able to induce and justify the reduction of sports performance.

Conversely, under moderate-to-severe hypogonadism (testosterone levels below 180 ng/dL), which may also be hypergonadotrophic/primary, in which the testicles are the primary site and that may be accompanied by normal-high or high SHBG levels, decreased performance may be a consequence of the hypogonadism, and therefore OTS is excluded. In the case of hypogonadism as the primary cause of underperformance, and particularly when it is severe or new-onset, further investigation is recommended and should include research for organic, anatomical, genetic, or post-traumatic dysfunctions.

However, in many times a pre-existent but asymptomatic and unnoticed state of hypogonadism only becomes clinically relevant when training is intensified, and slight differences in hormonal levels becomes important enough to preclude from a progressive performance improvement. Moreover, as athletes tend to be highly motivated, mild symptoms are usually ignored, avoiding the detection of milder or earlier stages of hypogonadism.

As mentioned, SHBG tends to be lower in hypogonadotrophic states, i.e., due to lack of central stimulation, while higher when the dysfunction is located in the testicles. In primary hypogonadism, luteinizing (LH) and follicle-stimulating (FSH) hormones are increased, due to the lack of feedback responses of testosterone and inhibin A from the testicles, but SHBG levels are higher, since LH and FSH concurrently stimulate the production of SHBG in the liver, and for this reason, increased LH and FSH tend to lead to increased SHBG. SHBG is the major protein that binds to testosterone, its primary function is to stabilize testosterone in serum, but at the same time, it inactivates its action while they are linked.

Hence, exceptionally, in mild hypogonadism states with high SHBG levels, there is a more prominent reduction of free testosterone, which could eventually lead to more severe symptoms including reduced performance. In this specific, the diagnosis of primary hypogonadism as the etiology of decreased performance, not OTS, is more likely.

In summary, unveiling the causality relationship between hormonal dysfunctions, OTS, and decreased performance may be highly challenging.

12. The level of standardization of the tests to be performed is essential for proper conclusions. As reported [4], only one study on OTS, conducted in 1985 [8], employed hormonal functional tests that had been validated and recommended by endocrinology societies. Since 1985, none of the studies on OTS based their hormonal functional tests on standardized tests or guidelines from an endocrinology perspective. Endocrine functional tests that evaluated specific hormonal responses in each level (primary, secondary or pituitary, and tertiary or hypothalamic), and which have been extensively validated and reproduced by endocrinology research groups, can be employed in athletes without the need of major adaptations, whenever the objective is to identify changes in hormonal secretions and responsiveness. Interpretation and conclusions from the results of functional tests tend to be more reliable scientifically when performed according to scientific standards, particularly when one searches for the actual etiology of hormonal alterations. Besides the functional tests, resting basal hormonal and other biochemical parameters should also be performed according to a minimum level of standardization and previous validation, not randomly or according to an alleged plausible mechanistic explanation. The great heteroge-

neity between the hormonal assessments of different studies also explains the inconsistent results found in OTS [6].

13. The characteristics of the population to be studied, particularly in terms of type(s) of sport(s) practiced, should be defined prior to the beginning of the selection of participants and be thoroughly described during the presentation of the results, due to the particularities in the timing of collection of the samples and the chronology of the procedures. Furthermore, acute and chronic responses may be largely dependent on the sort of sports practiced, including endurance (triathlon, cycling, long-distance running), strength (body building, weight lifting), explosive (or "anaerobic" or "stop-and-go") (ball games, including basketball, rugby, soccer, American football, squash, tennis), or mixed (high-intensity functional training (HIFT), CrossFit, army training), which also influences the interpretation of the findings, which reinforces the specification of the type of sport(s) practiced. Differences of the responses and the mechanisms that lead to OTS according to the sort of physical effort are seemingly substantial, at an extent that two types of OTS were proposed in the past: the parasympathetic, induced by endurance training, and the sympathetic, induced strength and explosive exercises [1, 4, 9]. Nonetheless, the parasympathetic and parasympathetic types of OTS have not been further confirmed, due to the low number of studies on strength-related OTS.

Owing to the potential important differences in the characteristics of OTS according to the type of sport, findings on OTS should be interpreted as sports-specific alterations and should not be extrapolated to other types of sports. Whether the alterations of OTS in a certain sport are also present in the others, this should be studied and reproduced for each type of exercise.

3.3 Design Characteristics of Studies on Overtraining Syndrome

Taking into account all the abovementioned challenged and limitations, to be considered minimally adequate, a study on OTS must have at least the following design characteristics:

(a) Recruitment of natural-occurring OTS candidates.
(b) Performance of the full diagnostic flowchart for those suspected of OTS.
(c) Delimitation of the type(s) of sport(s) to be evaluated, and divide into groups accordingly.
(d) Inclusion of a control group(s) of athletes with similar training, performance, and baseline characteristics. For each type of sport(s) to be evaluated, there should be two groups: the group affected by OTS and the group of healthy athletes.
(e) Inclusion of an additional second control group of sex-, age-, and BMI-matched non-physically active subjects.

(f) Preferably, whether there are three (in case of one type of sport to be evaluated), five (in case of two types of sport to be evaluated), seven (in case of three types of sport to be evaluated), or nine (in case of four types of sport to be evaluated) groups, all should have similar baseline characteristics and undergo the same tests.

(g) Employment of standardized and validated tests, with at least some tests independent from exercise. This is an effective way in order to equalize OTS-affected and healthy groups of athletes, from different types of sports, and the group of sedentary controls.

(h) Evaluation of the broadest number of different aspects, including those from immunologic, classical, and non-classical inflammatory pathways, glucose and lipid metabolism, muscular and fat tissue parameters, hormonal, hematological, body composition, and metabolism, among others, aiming to correlate the multiple sorts of dysfunctions that occur concurrently in OTS.

(i) After a transversal analysis, a sequence with a longitudinal follow-up to identify the markers of early, late, and lack of recovery in athletes with OTS.

3.4   Conclusion

Research on overtraining syndrome is highly challenging due to its extreme complexity and potential broad variety of combinations of clinical and biochemical dysfunctions and manifestations. Besides, because of its subjective nature, the design of research on OTS is hard to be driven, while the assessment for its selection, diagnosis, and evaluation is hard and minor changes may lead to great differences in results.

Challenges on the studies on OTS include its unpredictable occurrence, unique manifestations, lack of a single marker, a multiple-step diagnostic flowchart, relatively normal levels when compared to general population, its multi-factorial and multi-systemic characteristics, the need for controlling for different sorts of variables, and the causality and consequence differentiation between hormonal dysfunctions and OTS.

Researches on actual OTS must recruit natural-occurring OTS; diagnose OTS adequately; delineate the population of athletes; include control groups of healthy athletes with similar training patterns and sedentary matched for sex, age, and BMI; employ validated tests; and evaluate different sorts of clinical, biochemical, and metabolic parameters within the same athletes.

In conclusion, an appropriate study on OTS is highly toilsome, challenging, and costly, which may justify the shortage of studies on the field, despite the high and likely growing incidence of OTS and its popularity.

References

1. Meeusen R, Duclos M, Foster C, et al. European College of Sport Science; American College of Sports Medicine. Prevention, diagnosis, and treatment of the overtraining syndrome: joint consensus statement of the European College of Sport Science and the American College of Sports Medicine. Med Sci Sports Exerc. 2013;45(1):186–205.
2. Lehmann M, Foster C, Keul J. Overtraining in endurance athletes: a brief review. Med Sci Sports Exerc. 1993;25(7):854–62.
3. Rietjens GJ, Kuipers H, Adam JJ, et al. Physiological, biochemical and psychological markers of strenuous training-induced fatigue. Int J Sports Med. 2005;26(1):16–26.
4. Cadegiani FA, Kater CE. Hormonal aspects of overtraining syndrome: a systematic review. BMC Sports Sci Med Rehabil Aug. 2017;2(9):14.
5. Cadegiani FA, Kater CE. Novel causes and consequences of overtraining syndrome: the EROS-DISRUPTORS study. BMC Sports Sci Med Rehabil. 2019;11:21.
6. Vrijkotte S, Roelands B, Pattyn N, Meeusen R. The overtraining syndrome in soldiers: insights from the sports domain. Mil Med. 2019;184(5–6):e192–200.
7. Grandou C, Wallace L, Impellizzeri FM, Allen NG, Coutts AJ. Overtraining in resistance exercise: an exploratory systematic review and methodological appraisal of the literature. Sports Med. 2020;50(4):815–28.
8. McTernan EJ, Leiken AM. A pyramid model of health manpower in the 1980s. J Health Polit Policy Law. 1982;6(4):739–51.
9. Stamford B. Avoiding and recovering from overtraining. Phys Sportsmed. 1983;11(10):180.

Chapter 4
New Findings on Overtraining Syndrome

Outline

Advances in the methodology: the author explains how himself addressed the major challenges in the study he conducted (the EROS study), as well as in the most recent studies performed by other research groups.

Lessons from the Endocrine and Metabolic Responses on Overtraining Syndrome (EROS) study, which is subdivided in:

Summary of the results: in this part the author describes the main findings of the EROS study that uncovered more than 50 new biomarkers and serendipitously unveiled novel hormonal processes that healthy athletes undergo.

Analysis of the results: mechanisms and aspects unveiled – the mechanisms and aspects unveiled by the EROS study, in which the author links the findings in a logical and big perspective and learns several novel mechanisms that were found to occur in overtraining syndrome, in particular as part of its pathogenesis.

Unexpected remarkable findings: the EROS study identified some interesting findings that had never been considered or hypothesized as being present in OTS and unexpectedly helped improve the level of the comprehension of this syndrome.

New markers of overtraining syndrome: the author describes all the new biomarkers identified in OTS, from the results of the studies published in the recent years, in addition to the previous findings.

© The Editor(s) (if applicable) and The Author(s), under exclusive license
to Springer Nature Switzerland AG 2020
F. Cadegiani, *Overtraining Syndrome in Athletes*,
https://doi.org/10.1007/978-3-030-52628-3_4

4.1   Advances in the Methodology

The Endocrine and Metabolic Responses on Overtraining Syndrome (EROS) study was designed to address some of the challenges and limitations of the assessment methods of the studies on OTS, to identify the correlations between biochemical parameters and OTS, and to unveil the complex pathophysiology of OTS, through the elucidation of the roles of each dysfunctional pathway and system in the pathogenesis of OTS, as well as the multiple interactions between these dysfunctions [1–10], as described in Fig. 4.1.

Aiming these objectives, the EROS study evaluated a total of 117 parameters, including basal and stimulated hormonal responses; muscular, immunologic, classic inflammatory, lipid, and hematologic parameters; body composition and metabolic rates; and psychological, sleeping, and detailed eating patterns. The attempts to overcome some of the methodological limitations, among which some were mentioned above, included:

(a) Employment of two control groups, of healthy athletes and also of healthy sedentary, which allowed the analysis of the results from a more comprehensive perspective, since the simultaneous evaluation of the influence of the physical activity under healthy state and how this influence is altered under OTS is possible due to the concurrent comparisons with sedentary controls. In addition, findings on healthy athletes, when compared to non-active participants, were also relevant.

(b) Recruitment of athletes suspected of OTS in real life, aiming to evaluate actual and natural-occurring OTS, and a strict differentiation between actual OTS and overreaching states.

(c) Utilization of the full diagnostic criteria proposed by the latest guideline on OTS [1], including the exclusion of confounding diagnoses and the sine qua non presence of the key criteria of a minimum of 10% reduction in sports-specific performance, which should be verified by a specialized sports coach.

(d) Exclusive employment of extensively validated and standardized tests and endorsed by specialized societies, in order to have reliable results and conclusive meanings.

(e) Performance of exercise-independent stimulation tests to avoid suboptimized responses due to compromised performance inherently present in OTS and to allow comparisons with non-physically active controls.

(f) Tests that evaluated direct hormonal responses, without the need of intermediates, to avoid differences in the level of signaling from external systems, including those from cardiovascular, musculoskeletal, autonomic, and neuromuscular systems.

(g) Concurrent evaluation of multiple and broad different aspects for the identification of which sorts of dysfunctions are present in OTS, how these dysfunctions correlate and interact for the development of OTS, and the detection of independent predictors of OTS and possible causality relationships between these parameters evaluated.

Characteristic	Prevailing studies in the past	Current study	Implications of the differences for the strengthen of the hypothesis of the present study
Dependence on physical stimulation and the level of intensity during the test for hormonal responses	Essentially dependent, and responses are highly dependent on the performance during the test	Independent from physical activity. Physical performance is not necessary, and therefore does not influence the interpretation of the results	Athletes and sedentary have clear differences in physical conditioning. The employment of exercise-independent stimulation avoided the differences in hormonal responses due to differences in performances, which could bias the avoiding the interpretation of the results
Influence of the level of physical conditioning	Large influenced by the level of physical conditioning	Responses are independent from the level of physical conditioning	Differences in physical conditioning or muscle tissue did not influence the results, which avoided the bias of superior physical conditioning in athletes
External interferences from other systems (e.g., cardiovascular, musculoskeletal) in hormonal response during stimulation tests	Interferences from the level of stimulation from muscle, cardiovascular tissue and other metabolic, and autonomic systems	No interferences from other systems, except for minor percentages from the test and muscle tissues that are constitutively produced	Responses from the endocrine system occur entirely independent from the characteristics of any other system. Conditioning processes of some of these systems would affect the responses, and not allow appropriate conclusion
Direct (i.e., does not require any intermediate response), or indirect stimulation and evaluation of the hypothalamic-pituitary axes	Hypothalamic-pituitary responses depend a sequence of indirect stimulations, i.e., are secondary to those tissues stimulated by exercise	The stimulation (hypoglycemia) directly stimulates the hypothalamic-pituitary axes	The avoidance of the influence of any intermediate reaction or response precluded from known and unreported differences in any of these between sedentary and athletes
Variables concurrently evaluated and adjusted	Sex, type of exercise (within athletes group), and sometimes age	Age, sex, body mass index (BMI), eating, sleeping, social patterns, and type of exercise (within athletes group)	Despite the lack of evidence for influences on hormonal responsiveness according to age or BMI, the study avoided any possible hypothetical factor that could alter the responses
Selection criteria of the participants	Current orthopedic lesions and at times specific conditions. Lack of systematic clinical or biochemical exclusion.	Exclusion criteria for more than 30 diseases and 20 biochemical abnormalities as possible	Several clinical conditions and biochemical dysfunctions are known to alter the hormonal responsiveness, even those that do not seem to be directly related. Hence, strict inclusion and exclusion criteria prevented less direct biases
Aspects and parameters evaluated others than those present in the studies (for possible adjustments)	Usually, only the factors to be analyzed are evaluated	A total of 117 parameters of body metabolism and composition, psychological, basal muscular, inflammation, of macronutrient intake, sleeping, working/studying, among other aspects were concurrently evaluated	Psychological, eating, metabolic, and biochemical characteristics may also influence on the responses. Adjusting them is key for appropriate conclusion.
Level of standardization of the hormonal tests performed	Tests performed in previous small studies, but not systematically validated	Largely standardized and endorsed by specific guidelines of the respective fully satisfy criteria for endocrine functional tests	Standardization of endocrine functional tests through extensive testing is essential to avoid previewed and unconsidered biases. The non-validation of a test of hormonal responses may lead to misleading findings.
Presence and characteristics of groups of comparison	Control groups are not commonly used	Healthy sedentary with similar characteristics as control group	Conclusions regarding hormonal adaptations to exercise can only be inferred when athletes are compared to sedentary with similar baseline characteristics, or longitudinally, within the same subjects, from a sedentary to an active state
Average number of participants	90% of previous studies that evaluate hormonal responses are performed with four to 15 participants in total	51 participants, including 39 in the healthy groups	A relatively large number of participants, particularly when taking into account the high level of similarity of two completely different groups in terms of fitness level, increases the value and the validity of the findings
Determine the exact level of the origin of the differences in responses	No, could be the hypothalamic, pituitary, peripheral, or not endocrine	No, could be either hypothalamic or pituitary	The lack of differences in direct adrenal stimulation allowed the conclusion of the central location of the hormonal adaptation, but comparisons between direct pituitary and hypothalamic stimulations are needed to understand which level adaptations occur
Characteristics of the papers that demonstrated the existence of novel mechanisms in sports	Smaller number of participants and less control for variables and confounding factors were necessary for the demonstration. Proposed conclusions.	Relatively larger number of participants and stricter control for biases. Provided evidence for the hypothesis, did conclusion.	There is sufficient data to confirm the hypothesis of the presence of a hormonal conditioning process in athletes. For an appropriate comparison between athletes and sedentary, all the characteristics of the present study should be present

Fig. 4.1 Differences of the characteristics between the current and the past studies and how these differences implied in the evidence of the existence of an inherent, independent, and diffuse hypothalamic-pituitary hormonal conditioning process in athletes

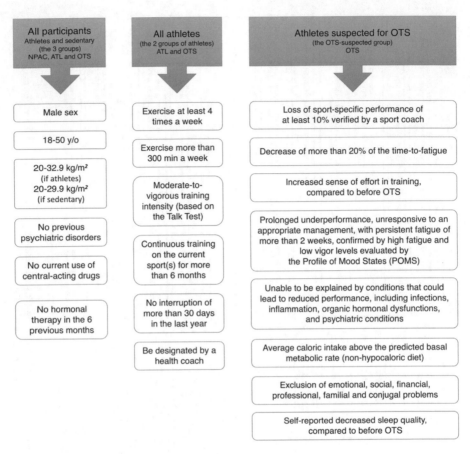

Fig. 4.2 Inclusion and exclusion criteria for the EROS study

The strict criteria used for the selection of participants lead to an exclusion rate of approximately 90% among those suspected for OTS, due to the presence of multiple confounding diagnoses, and misconceptions and erroneous perceptions on the definitions of OTS among those who work with athletes. Figure 4.2 summarizes the inclusion and exclusion criteria for all three groups, whereas Fig. 4.3 depicts the selection process.

Only male athletes were included, to avoid the influences of menstrual cycle or the use of oral contraceptives on hormonal responsiveness.

Additional factors that were not present in the flowchart for the diagnosis of [11], but which could influence any of the parameters to be tested, were also excluded, including current or recent use of centrally acting drugs or hormones and the diagnosis of conditions that were not listed as possible confounding disorders, including metabolic, endocrinological, inflammatory, infectious, or psychiatric dysfunctions.

Finally, baseline characteristics were designed to be similar between the three groups, including age and BMI, as well as training patterns between the two groups

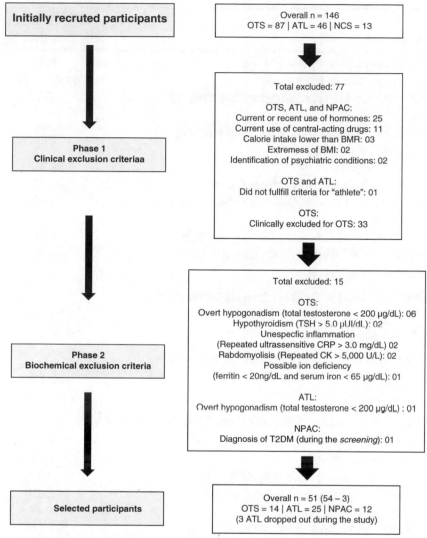

Fig. 4.3 Selection process of the EROS study

of athletes. All athletes included in the OTS group had actual and naturally occurring OTS, as shown in Fig. 4.4.

The EROS study attempts to follow characteristics of an ideal marker proposed by the latest guideline in OTS [1]. An ideal marker of OTS should be present before the establishment of the OTS, distinguishable from chronic changes, easy to measure, available, not invasive, and inexpensive. However, the same guideline requires that markers of OTS should be derived from exercise-stimulation tests, which has been shown to be no longer mandatory [1–10].

Parameter	OTS-Affected Athletes (n = 14)
Fatigue lasting more than 2 wk	100
Mean duration of fatigue, d	44.3 ± 23
Performance fully recovered by the time of the study, %	0
Increased intensity and volume of training in the last 3 months, %	100
Training monotony, %	14.30
Increased frequency of infections, %	21.40
Sleep disturbance started or worsened with OTS, %	42.90
Increased effort for same training load, %	100
Increased sensitivity to heat or cold, %	42.90
Profile of Mood Scales	
Fatigue subscale (0–28; the higher, the worse)	
>14, %	85.70
Mean score	19.9 ± 6.1
Vigor subscale (0–28; the lower, the worse)	
<14, %	85.70
Mean score	9.9 ± 5.6
Specific tests performed (coach verified), % of athletes (No.)	
Pace (compared with previous pace)	100 (14)
Volume of training	92.8 (13)
Highest speed or intensity achieved	21.4 (3)
Maximum strength at a 1-repetition maximum strength test	14.3 (2)
Maximum number of repetitions for the same weight lifted	7.1 (1)

Fig. 4.4 Characteristics of the OTS group

4.2 Lessons from the Endocrine and Metabolic Responses on Overtraining Syndrome (EROS) Study

4.2.1 Summary of the Results

The changes in the methodology of the EROS study allowed the identification of novel findings and the clarification of previously inconsistent results. Below is the summary of the most remarkable, among which those that are unprecedent are marked as "N" (from novel) and those which were unexpected are marked as "U" [1–10]. The summary of the findings is also described in Fig. 4.5.

1. Through a 7-day thorough and precise diet record, athletes affected by OTS had a prior diet of approximately two times less carbohydrates, two times less protein, and two times less overall caloric intake shown as g/kg/day, g/kg/day, and

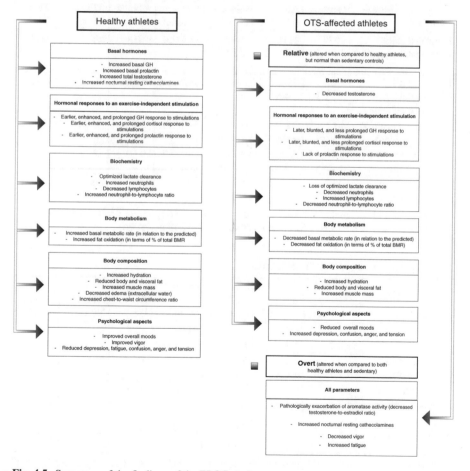

Fig. 4.5 Summary of the findings of the EROS study

kcal/kg/day, respectively), when compared to healthy athletes, and three times less carbohydrate than sedentary controls (N). Details of the eating patterns are described in Table 4.1.

2. OTS-affected athletes had worse sleep quality (but not shorter duration) and had longer working or studying duration (h/day) than healthy ones and at similar extent than non-physically active participants (N), as shown in Table 4.2.

3. At an insulin tolerance test (ITT), performed to evaluate hormonal responses to a stressful stimulation (hypoglycemia), affected athletes disclosed blunted and delayed responses of the hypothalamic-pituitary-adrenal (HPA) (cortisol, Table 4.3, and ACTH, Table 4.4), somatrotrophic (GH) (Table 4.5), and lactotrophic (prolactin) (Table 4.6) axes, but similar to the responses of sedentary

Table 4.1 Eating patterns

Mean ± SD	OTS (N = 14)	ATL (N = 25)	NPAC (N = 12)	P-value overall	ATL vs OTS	ATL vs NPAC	OTS vs NPAC
Estimated daily calorie intake (kcal/day)	2673 ± 1677	4114 ± 719	4473 ± 2265	<0.001	<0.001	n/s	0.002
Estimated daily calorie intake per weight (kcal/kg/day)	29.7 ± 12.4	53.2 ± 4.8	54.7 ± 3.9	<0.001	<0.001	n/s	0.001
Daily protein calorie intake (g/kg/day)	1.48 ± 0.59	2.87 ± 0.92	0.92 ± 0.49	<0.001	<0.001	<0.001	0.013
Daily CHO calorie intake (g/kg/day)	3.76 ± 2.96	7.26 ± 3.07	9.85 ± 4.49	<0.001	0.003	n/s	<0.001
Daily fat calorie intake (g/kg/day)	0.97 ± 0.47	1.41 ± 1.21	1.29 ± 0.91	n/s	n/s	n/s	n/s

OTS group of athletes affected by overtraining syndrome, *ATL* group of healthy athletes, *NPAC* non-physically active controls, *CHO* carbohydrate; *SD* standard deviation

Table 4.2 Sleeping and social characteristics

Mean ± SD	OTS (N = 14)	ATL (N = 25)	NPAC (N = 12)	P-value overall	ATL vs OTS	ATL vs NPAC	OTS vs NPAC
Sleep quality (0–10)	6.2 ± 2.1	8.3 ± 1.7	7.0 ± 0.9	0.005	0.004	n/s	n/s
Duration of night sleep (hours)	6.2 ± 1.1	6.5 ± 0.8	6.7 ± 0.9	n/s	n/s	n/s	n/s
Daily number of hours of work or study	8.8 ± 2.6	6.8 ± 1.5	9.3 ± 2.9	<0.001	<0.001	<0.001	n/s
Libido (0–10)	6.1 ± 3.0	7.9 ± 1.6	8.8 ± 1.0	0.003	0.024	n/s	0.004

OTS group of athletes affected by overtraining syndrome, *ATL* group of healthy athletes, *NPAC* non-physically active controls, *SD* standard deviation

Table 4.3 Mean (± standard deviation) cortisol response to insulin tolerance test (ITT)

Cortisol (μg/dL) (Mean ± SD)	OTS athletes (OTS) ($n = 14$)	Healthy athletes (ATL) ($n = 25$)	Sedentary subjects (NPAC) ($n = 12$)
Basal	11.6 ± 2.5	12.5 ± 3.1	10.9 ± 2.8
During hypoglycemia	12.4* ± 3.3	15.9& ± 5.3	11.8 ± 3.1
30′ after hypoglycemia	17.9*** ± 2.9	21.7&&&& ± 3.1	16.9 ± 4.1
Absolute increase from basal to 30′ after hypoglycemia	6.3** ± 2.3	9.2& ± 3.7	5.9 ± 3.9

Differences between **OTS** and **ATL**: *$p < 0.05$; **$p < 0.01$; ***$p < 0.005$; ****$p < 0.001$
Differences between **ATL** and **NPAC**: &$p < 0.05$; &&$p < 0.01$; &&&$p < 0.005$; &&&&$p < 0.001$
OTS group of athletes affected by overtraining syndrome, *ATL* group of healthy athletes, *NPAC* non-physically active controls, *SD* standard deviation

Table 4.4 Median (95% confidence interval) adrenocorticotropic hormone (ACTH) response to insulin tolerance test (ITT)

ACTH (pg/mL) (Median – 95% CI)	OTS – affected athletes ($N = 14$)	ATL – healthy athletes ($N = 25$)	NPAC – sedentary subjects ($N = 12$)	Statistical signif. (P)
Basal	19.6 (11.4–32.9)	18.7 (6.5–37.8)	21.4 (8.7–37.8)	n/s
During hypoglycemia	28.2 (8.4–238.9)	57.8 (7.3–229.5)	29.5 (14.8–191.7)	n/s
30′ after hypoglycemia	30.3**** (9.8–93.7)	59.9 (22.1–195.7)	51.4 (22.7–137.5)	0.006
Absolute increase	9.7**** (−14.4–64.4)	45.1 (22.1–195.7)	38.0 (0.5–108.8)	0.004

Differences between **OTS** and **ATL**: *$p < 0.05$; **$p < 0.01$; ***$p < 0.005$; ****$p < 0.001$
Differences between **ATL** and **NPAC**: &$p < 0.05$; &&$p < 0.01$; &&&$p < 0.005$; &&&&$p < 0.001$
ACTH adrenocorticotropic hormone, *OTS* group of athletes affected by overtraining syndrome, *ATL* group of healthy athletes, *NPAC* non-physically active controls, *CI* confidence interval

Table 4.5 Median (95% confidence interval) growth hormone (GH) response to insulin tolerance test (ITT)

GH (μg/L) (median – 95% CI)	OTS – affected athletes ($N = 14$)	ATL – healthy athletes ($N = 25$)	NPAC – sedentary subjects ($N = 12$)	Statistical signif. (P)
Basal	0.10** (0.05–0.87)	0.26&&& (0.1–1.26)	0.06 (0.03–0.47)	0.003
During hypoglycemia	0.40* (0.05–4.68)	2.50&& (0.08–40.94)	0.16 (0.05–8.13)	0.006
30′ after hypoglycemia	3.28**** (0.03–13.95)	12.73& (1.1–38.1)	4.80 (0.33–27.36)	0.001

Differences between OTS and ATL: *$p < 0.05$; **$p < 0.01$; ***$p < 0.005$; ****$p < 0.001$
Differences between ATL and NPAC: &$p < 0.05$; &&$p < 0.01$; &&&$p < 0.005$; &&&&$p < 0.001$
GH growth hormone, *OTS* group of athletes affected by overtraining syndrome, *ATL* group of healthy athletes, *NPAC* non-physically active controls, *CI* 95% confidence interval

Table 4.6 Median (95% confidence interval) prolactin response to insulin tolerance test (ITT)

Prolactin (ng/mL) (median – 95% CI)	OTS – affected athletes ($N = 14$)	ATL – healthy athletes ($N = 25$)	NPAC – sedentary subjects ($N = 12$)	Statistical signif. (P)
Basal	9.20*(5.27–19.46)	12.10 (7.19–23.0)	10.65 (7.91–15.69)	0.048
During hypoglycemia	8.95****(4.72–47.22)	17.85&&(9.99–63.39)	12.20 (7.18–15.94)	0.001
30' after hypoglycemia	11.35****(4.5–25.88)	24.3&&&(10.5–67.45)	10.50 (6.21–43.44)	<0.001
Response (absolute levels)	−0.3*(−2.4 – +12.0)	+13.1&&(−5.3 – +54.5)	−1.2 (−4.8 – +30.5)	0.018
Percent change (%)	−2.7*(−24.3 – +184.3)	+101.5&(−33.3 – +596.2)	−12.1 (−36.8 – +247.4)	0.011

Differences between OTS and ATL: *$p < 0.05$; **$p < 0.01$; ***$p < 0.005$; ****$p < 0.001$
Differences between ATL and NPAC: &$p < 0.05$; &&$p < 0.01$; &&&$p < 0.005$; &&&&$p < 0.001$
OTS group of athletes affected by overtraining syndrome; *ATL* group of healthy athletes; *NPAC* non-physically active controls, *CI* 95% confidence interval

Table 4.7 Glucose behavior and symptoms of hypoglycemia during the insulin tolerance test (ITT)

Mean ± SD	OTS – affected athletes ($N = 14$)	ATL – healthy athletes ($N = 25$)	NPAC – sedentary subjects ($N = 12$)
Basal serum glucose (mg/dL)	83.5 ± 8.4	81.4 ± 5.1	84.8 ± 6.4
Serum glucose during hypoglycemia	20.7 ± 11.6	18.0 ± 8.3	26.2 ± 12.5
Capillary glucose during hypoglycemia	39.1 ± 9.2	36.2 ± 8.9	43.0 ± 12.1
Time to hypoglycemia (minutes)	23.6 ± 6.4	25.5 ± 5.0	29.7# ± 6.8
Adrenergic symptoms (0–10)	3.0* ± 2.6	5.4 ± 3.0	6.0# ± 2.6
Neuroglycopenic symptoms (0–10)	4.1 ± 2.9	5.5 ± 3.0	4.3 ± 3.0

Differences between OTS and ATL: *$p < 0.05$
Differences between OTS and NCS: #$p < 0.05$
OTS group of athletes affected by overtraining syndrome, *ATL* group of healthy athletes, *NPAC* non-physically active controls, *SD* standard deviation

controls (N). Basal glycemia and the hypoglycemic nadir were similar between the three groups (Table 4.7), and differences in body composition were adjusted, and did not affect the results.

4. In the ITT, healthy athletes disclosed exacerbated and prolonged GH and cortisol responses compared to non-physically active controls (Table 4.5), while this was the only group to disclose a significant response of prolactin to stimulation (Table 4.6) (N/U).

5. OTS-affected athletes presented fewer adrenergic symptoms during hypoglycemia compared to the two other groups, while neuroglycopenic symptoms were maintained, with similar glycemic nadir (Table 4.7) (N).
6. Conversely, the direct stimulation of the adrenal glands using a synthetic ACTH did not yield any difference between the three groups, showing that adrenals are unlikely affected by exercise or by the presence of OTS (Table 4.8).
7. Testosterone levels were higher in healthy athletes than both sedentary and affected athletes, while levels were similar between sedentary and OTS. The testosterone-to-estradiol ratio, an indirect marker of aromatase activity, was approximately two times lower in OTS-athletes, compared to healthy athletes and to sedentary, showing a probable exacerbation of the aromatase activity in OTS, aiming to decrease an anabolic hormone in response to an energy-depleted condition. All other basal hormones were similar between groups (N). Basal hormones are summarized in Table 4.9.

Table 4.8 Mean (±SD) basal serum cortisol and response to a cosyntropin stimulation test (CST, 250 ug IV bolus)

Cortisol (μg/dL) (Mean ± SD)	OTS – affected athletes ($N = 14$)	ATL – healthy athletes ($N = 25$)	NPAC – sedentary subjects ($N = 12$)
Basal	13.1 ± 4.1	12.1 ± 3.2	12.1 ± 5.7
30' pCST	19.1 ± 1.9	19.7 ± 2.4	19.7 ± 3.2
60' pCST	21.9 ± 2.4	22.2 ± 2.9	22.9 ± 4.4

CST cosyntropin stimulation test, *OTS* group of athletes affected by overtraining syndrome, *ATL* group of healthy athletes, *NPAC* non-physically active controls, *SD* standard deviation

Table 4.9 Mean (± standard deviation) basal hormone levels

Mean ± SD	OTS – affected athletes ($N = 14$)	ATL – healthy athletes ($N = 25$)	NPAC – sedentary subjects ($N = 12$)
Total testosterone (ng/dL)	$422.6^{**} \pm 173.2$	$540.3^{\&} \pm 171.4$	405.9 ± 156.3
Estradiol (pg/mL)	$40.1^{**} \pm 10.8$	29.8 ± 13.9	$25.7^{\#\#} \pm 11.2$
Testosterone-to-estradiol ratio	$10.8^{****} \pm 3.7$	20.8 ± 9.9	$20.3^{\#\#} \pm 13.0$
Testosterone-to-cortisol ratio	39.3 ± 19.8	45.0 ± 15.8	39.6 ± 21.3
Total catecholamines (ug/12 h)	$257^{*} \pm 166$	175 ± 69	$133^{\#\#} \pm 54$
Noradrenaline (ug/12 h)	27.4 ± 10.5	22.3 ± 12.7	$17.4^{\#} \pm 9.1$
Dopamine (ug/12 h)	$227^{*} \pm 159$	149 ± 60	$114^{\#\#} \pm 45$

Differences between OTS and ATL: $^{*}p < 0.05$; $^{**}p < 0.01$; $^{***}p < 0.005$; $^{****}p < 0.001$
Differences between ATL and NCS: $^{\&}p < 0.05$; $^{\&\&}p < 0.01$; $^{\&\&\&}p < 0.005$; $^{\&\&\&\&}p < 0.001$
Differences between OTS and NCS: $^{\#}p < 0.05$; $^{\#\#}p < 0.01$; $^{\#\#\#}p < 0.005$; $^{\#\#\#\#}p < 0.001$
OTS group of athletes affected by overtraining syndrome, *ATL* group of healthy athletes, *NPAC* non-physically active controls, *SD* standard deviation

8. Basal lactate levels were unexpectedly lower in healthy athletes than non-physically active participants and also lower than levels in OTS-affected athletes (N/U).
9. CK was exacerbated in affected athletes, compared to healthy ones, after a similar interval between the last training session and the blood collect, while was naturally lower in sedentary, as described in Table 4.10.
10. Neutrophils were higher in healthy athletes than OTS, while lymphocytes were lower compared to sedentary. The neutrophil-to-lymphocyte ratio, a proposed marker of diseases prognosis, was increased in healthy, but not in affected athletes (N). Results are described in Table 4.11.
11. Catecholamines and the catecholamine-to-metanephrine ratio were exacerbated in OTS, compared to healthy athletes (Table 4.9).
12. Healthy athletes had benefits from training in terms of vigor, fatigue, irritability, humor, tension, and lucidity moods, when compared to non-active participants (Table 4.12). Conversely, athletes affected by OTS did not only apparently lose these psychological benefits, but also had worse fatigue and vigor levels compared to sedentary.

Table 4.10 Muscular parameters

Median (95% CI)	OTS – affected athletes ($N = 14$)	ATL – healthy athletes ($N = 25$)	NPAC – sedentary subjects ($N = 12$)
Creatine kinase (U/L)	569* (126–3012)	347&& (92–780)	105.5#### (80–468)
Lactate (nMol/L)	1.11** (0.79–2.13)	0.78& (0.47–1.42)	1.17 (0.57–1.57)

Differences between NCS and ATL: *$p < 0.05$; **$p < 0.01$; ***$p < 0.005$; ****$p < 0.001$
Differences between ATL and NCS: &$p < 0.05$; &&$p < 0.01$; &&&$p < 0.005$; &&&&$p < 0.001$
Differences between OTS and NCS: #$p < 0.05$; ##$p < 0.01$; ###$p < 0.005$; ####$p < 0.001$
OTS group of athletes affected by overtraining syndrome, *ATL* group of healthy athletes, *NPAC* non-physically active controls, *CI* confidence interval

Table 4.11 Immunologic parameters

Mean ± SD	OTS – affected athletes ($N = 14$)	ATL – Healthy athletes ($N = 25$)	NPAC – Sedentary subjects ($N = 12$)
Neutrophils (/mm3)	2986* ± 761	3809 ± 1431	3186 ± 847
Lymphocytes (/mm3)	2498 ± 487	2154& ± 640	2820 ± 810
Neutrophil-to-lymphocyte ratio	1.23* ± 0.34	2.00& ± 1.28	1.27 ± 0.73
Platelet-to-lymphocyte ratio	104.1 ± 34.2	119.1&&& ± 43.4	82.4 ± 19.5

Differences between NCS and ATL: *$p < 0.05$; **$p < 0.01$; ***$p < 0.005$; ****$p < 0.001$
Differences between ATL and NCS: &$p < 0.05$; &&$p < 0.01$; &&&$p < 0.005$; &&&&$p < 0.001$
Differences between OTS and NCS: #$p < 0.05$; ##$p < 0.01$; ###$p < 0.005$; ####$p < 0.001$
OTS group of athletes affected by overtraining syndrome, *ATL* group of healthy athletes, *NPAC* non-physically active controls, *SD* standard deviation

Table 4.12 Psychological parameters

OMS Mean ± SD Median (95%CI)	OTS (N = 14)	ATL (N = 25)	NPAC (N = 12)	P-value overall	ATL vs OTS	ATL vs NPAC	OTS vs NPAC
Total score (−32 to +120)	+52.1 ± 27.2	−6.2 ± 12.8	+33.8 ± 27.9	<0.001	<0.001	<0.001	0.086
Anger subscale (0 to 48)	2.9 ± 6.6	5.8 ± 5.1	11.2 ± 7.0	0.002	0.003	0.036	n/s
Confusion sub (0 to 28)	6.9 ± 5.7	2.2 ± 2.0	5.9 ± 3.8	0.001	0.001	0.019	n/s
Depression sub (0 to 60)	7.5 (0–21.4)	0.0 (0–5.0)	8.5 (1.6–19.9)	0.001	0.008	0.003	n/s
Fatigue sub (0 to 28)	20.0 (9.3–26.7)	2.0 (0.0–4.8)	8.0 (1.1–18.8)	<0.001	<0.001	<0.001	<0.001
Tension sub (0 to 36)	15.7 ± 6.0	6.7 ± 4.3	18.1 ± 5.5	<0.001	<0.001	<0.001	n/s
Vigor sub (0 to 32)	10.9 ± 5.6	25.2 ± 2.3	19.3 ± 5.8	<0.001	<0.001	0.001	<0.001

POMS profile of mood states, *OTS* group of athletes affected by overtraining syndrome, *ATL* group of healthy athletes, *NPAC* non-physically active controls

Table 4.13 Body composition and metabolism

Mean ± SD	OTS (N = 14)	ATL (N = 25)	NPAC (N = 12)	P-value overall	ATL vs OTS	ATL vs NPAC	OTS vs NPAC
Measured to predicted BMR (%)	102.6 ± 8.3	109.7 ± 9.3	100.2 ± 9.5	0.008	0.013	0.012	n/s
Percentage of fat burning compared to total BMR (%)	33.5 ± 21.0	58.7 ± 18.7	29.3 ± 14.2	<0.001	<0.001	<0.001	n/s
Body fat (%)	17.0 ± 6.0	10.8 ± 4.2	21.3 ± 7.2	<0.001	<0.001	<0.001	n/s
Muscle mass (%)	47.2 ± 3.8	50.5 ± 2.3	44.1 ± 3.9	<0.001	0.008	<0.001	0.038
Body water (%)	59.5 ± 3.9	64.7 ± 2.7	57.0 ± 4.9	<0.001	<0.001	<0.001	n/s

BMR basal metabolic rate, *OTS* group of athletes affected by overtraining syndrome, *ATL* group of healthy athletes, *NPAC* non-physically active controls, *SD* standard deviation

13. Libido was the highest among sedentary, while it was severely affected in OTS (Table 4.2).
14. As observed through an indirect calorimetry, healthy athletes had higher measured-to-expected basal metabolic rate (BMR) ratio and fat oxidation than sedentary, whereas both were reduced in OTS when compared to healthy athletes (N) (Table 4.13).
15. Healthy athletes had lower body fat and higher muscle mass and were better hydrated than OTS-affected athletes, while affected athletes still had slightly better levels than sedentary.

A summary of the findings is in Tables 4.14 and 4.15, for biochemical and clinical markers, respectively.

Table 4.14 Summary of the new biochemical markers of overtraining syndrome uncovered by the EROS study

Marker	Findings (from pairwise comparisons)
Basal GH (µg/L)	Exacerbated in healthy athletes (higher than sedentary) Relatively reduced in overtraining syndrome (normal when compared to sedentary)
Early GH response to an ITT (µg/L)	Exacerbated in healthy athletes (higher than sedentary) Relatively reduced in overtraining syndrome (normal when compared to sedentary)
Late GH response to an ITT (µg/L)	Exacerbated in healthy athletes (higher than sedentary) Relatively reduced in overtraining syndrome (normal when compared to sedentary)
Early cortisol response to an ITT (µg/dL)	Exacerbated in healthy athletes (higher than sedentary) Relatively reduced in overtraining syndrome (normal when compared to sedentary)
Late cortisol response to an ITT (µg/dL)	Exacerbated in healthy athletes (higher than sedentary) Relatively reduced in overtraining syndrome (normal when compared to sedentary)
Late ACTH response to an ITT (µg/dL) (pg/mL)	Relatively reduced in overtraining syndrome (normal when compared to sedentary)
Basal prolactin (ng/mL)	Relatively reduced in overtraining syndrome (normal when compared to sedentary)
Early prolactin response to an ITT (ng/mL)	Exacerbated in healthy athletes (higher than sedentary) Relatively reduced in overtraining syndrome (normal when compared to sedentary)
Late prolactin response to an ITT (ng/mL)	Exacerbated in healthy athletes (higher than sedentary) Relatively reduced in overtraining syndrome (normal when compared to sedentary)
Salivary cortisol 30′ after awakening (ng/dL)	Exacerbated in healthy athletes (higher than sedentary) Relatively reduced in overtraining syndrome (normal when compared to sedentary)
Cortisol awakening response (CAR) (%)	Exacerbated in healthy athletes (higher than sedentary) Absolutely decreased in overtraining syndrome (even lower than sedentary)
Total testosterone (ng/dL)	Exacerbated in healthy athletes (higher than sedentary) Relatively reduced in overtraining syndrome (normal when compared to sedentary)
Estradiol (pg/mL)	Absolutely increased in overtraining syndrome (higher than both healthy athletes and sedentary)
Testosterone-to-estradiol ratio	Absolutely decreased in overtraining syndrome (lower than both healthy athletes and sedentary)
Testosterone-to-cortisol ratio	Normal in healthy athletes Normal in overtraining syndrome
Total nocturnal urinary catecholamines (µg/12 h)	Exacerbated in healthy athletes (higher than sedentary) Absolutely increased in overtraining syndrome (even higher than healthy athletes)
Nocturnal urinary noradrenaline (µg/12 h)	Exacerbated in healthy athletes (higher than sedentary) Normal in overtraining syndrome (compared to healthy athletes)

Table 4.14   (continued)

Marker	Findings (from pairwise comparisons)
Nocturnal urinary dopamine (μg/12 h)	Exacerbated in healthy athletes (higher than sedentary) Absolutely increased in overtraining syndrome (even higher than healthy athletes)
Catecholamine-to-metanephrine ratio	Absolutely increased in overtraining syndrome (higher than both healthy athletes and sedentary)
Neutrophils ($*mm^3$)	Relatively reduced in overtraining syndrome (normal when compared to sedentary)
Lymphocytes ($*mm^3$)	Blunted in healthy athletes (lower than sedentary)
Neutrophil-to-lymphocyte ratio	Exacerbated in healthy athletes (higher than sedentary) Relatively reduced in overtraining syndrome (normal when compared to sedentary)
Platelet-to-lymphocyte ratio	Exacerbated in healthy athletes (higher than sedentary)
Creatine kinase (CK) 48 hours post-training (U/L)	Exacerbated in overtraining syndrome (higher than healthy athletes)
Lactate 48 hours post-training (nMol/L)	Blunted in healthy athletes (lower than sedentary) Relatively increased in overtraining syndrome (normal when compared to sedentary)

Table 4.15 Summary of the new clinical and metabolic markers of overtraining syndrome uncovered by the EROS study

Marker	Findings (from pairwise comparisons)
Measured-to-predicted BMR ratio (%)	Exacerbated in healthy athletes (higher than sedentary) Relatively reduced in overtraining syndrome (normal when compared to sedentary)
Fat oxidation (%)	Exacerbated in healthy athletes (higher than sedentary) Relatively reduced in overtraining syndrome (normal when compared to sedentary)
Body fat (%)	Reduced in healthy athletes (lower than sedentary) Relatively increased in overtraining syndrome (normal when compared to sedentary)
Muscle mass (%)	Exacerbated in healthy athletes (higher than sedentary) Relatively reduced in overtraining syndrome (normal when compared to sedentary)
Level of hydration (% of total body weight)	Exacerbated in healthy athletes (higher than sedentary) Relatively reduced in overtraining syndrome (normal when compared to sedentary)
Caloric intake (kcal/kg/day)	Exacerbated in healthy athletes (higher than sedentary) Relatively reduced in overtraining syndrome (normal when compared to sedentary)
Protein intake (g/kg/day)	Exacerbated in healthy athletes (higher than sedentary) Relatively reduced in overtraining syndrome, but higher than sedentary

(continued)

Table 4.15 (continued)

Marker	Findings (from pairwise comparisons)
Carbohydrate intake (g/kg/day)	Exacerbated in healthy athletes (higher than sedentary) Relatively reduced in overtraining syndrome (normal when compared to sedentary)
Self-reported sleep quality (0–10)	Relatively reduced in overtraining syndrome (normal when compared to sedentary)
Working or studying hours (h/day)	Reduced in healthy athletes (lower than sedentary) Relatively increased in overtraining syndrome (normal when compared to sedentary)
Self-reported libido quality (0–10)	Absolutely reduced in overtraining syndrome (lower than both healthy athletes and sedentary)
Overall mood states	Absolutely decreased in overtraining syndrome (lower than both healthy athletes and sedentary)
Fatigue	Blunted in healthy athletes (lower than sedentary) Absolutely exacerbated in overtraining syndrome (even higher than sedentary)
Vigor	Exacerbated in healthy athletes (higher than sedentary) Absolutely decreased in overtraining syndrome (even lower than sedentary)
Depression	Blunted in healthy athletes (lower than sedentary) Relatively increased in overtraining syndrome (normal when compared to sedentary)
Anger	Blunted in healthy athletes (lower than sedentary) Relatively increased in overtraining syndrome (normal when compared to sedentary)
Confusion	Blunted in healthy athletes (lower than sedentary) Relatively increased in overtraining syndrome (normal when compared to sedentary)
Tension	Blunted in healthy athletes (lower than sedentary) Relatively increased in overtraining syndrome (normal when compared to sedentary)

4.2.2 Analysis of the Results: Mechanisms and Aspects Unveiled

Concurrently to the novel findings on OTS, the simultaneous comparisons with healthy sedentary allowed the detection of novel mechanisms not only in OTS but also in healthy athletes, as some previously undescribed intrinsic hormonal and metabolic conditioning adaptive processes in response to exercises found to occur in response to exercise, in addition to those extensively described in the cardiovascular, musculoskeletal, neuromuscular, and neurological systems. These additional findings in healthy athletes allowed the analysis of the findings on OTS under the context of the maintenance or loss of the adaptive processes that athletes were now shown to undergo. This helped unveil a novel understanding of OTS, which has now been demonstrated to be majorly a mix of losses of the mechanisms that promote improvement of sports performance in athletes.

The most remarkable findings, as well as their meanings, are:

(a) In terms of novel and previously described physiological responses and conditioning adaptations to exercise:

1. Healthy athletes disclosed multiple intrinsic and adaptive optimization of hormonal responses to any sort of stimulation, as independent conditioning process of at least three of the HP axes – somato-, lacto-, and corticotropic axes – independently from physical activity and signaling from other systems. As this is the definition for any conditioning process, this has been termed as "hormonal conditioning." To date, this is the first report of an intrinsic and independent hormonal conditioning process, i.e., the occurrence of an enhancement of the hormonal axes irrespective of musculoskeletal, autonomic, and cardiovascular systems. Figure 4.5 summarizes this process.

2. Despite the physiological pulsatile pattern of GH release, its basal levels were unexpectedly higher in healthy athletes than those in sedentary, even taking into account that none of the participants were collected during a GH peak, as all GH levels were below 1.5 μg/L, and also that mean basal glucose levels were similar between groups. This occurred basically because while in general population GH between peaks tend to be undetectable, almost all athletes disclosed detectable GH levels even when not in peaks. The roles of the GH when not in peak are unknown, and the benefits of having non-undetectable GH levels are yet to be understood.

3. A significant increase of prolactin in response to stimulations was present in healthy athletes, while absent in sedentary. The specific roles for the lacto-trophic response in athletes have not been determined, although the occurrence of prolactin increase under stressful situations has been correlated with better prognosis under specific circumstances [12–14] (Fig. 4.6).

4. The hypothalamic-pituitary-adrenal (HPA) adaptations to normal training claimed to be characterized by increased ACTH/cortisol ratio during exercise recovery [11], possibly resulted from modulation of tissue sensitivity to glucocorticoids, were partially confirmed by the EROS study [1].

5. A testosterone increase of testosterone was identified, in accordance with previous findings [15, 16], as a possible enhancer of muscle anabolism, strength, and function.

6. Lactate levels were lower in healthy athletes compared to sedentary, showing an accelerated lactate clearance. This indirectly suggests a quicker and optimized muscle recovery process, in order to recover faster for further trainings. While lactate was lower in healthy athletes than sedentary, CK levels were three times higher in athletes, despite being collected simultaneously [17, 18]. This is an additional argument for the hypothesis that the clearance of lactate is optimized in athletes.

7. Under unhealthy states, or in the presence of any severe, acute, or chronic disease, an increased neutrophil-to-lymphocyte ratio predicts worse prognosis [19, 20]. Also, this can be an indirect risk factor for cardiovascular and

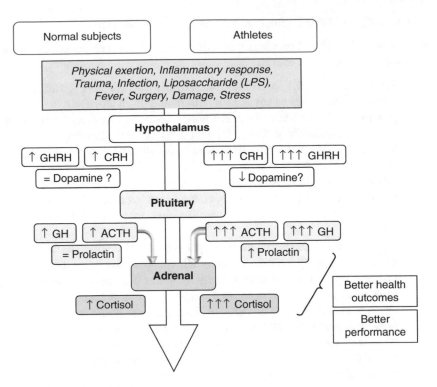

Fig. 4.6 Hormonal conditioning process

inflammatory diseases, once both neutrophils and the neutrophil-to-lymphocyte ratio are increased in obesity and type 2 diabetes mellitus (T2DM). However, healthy athletes also showed an increased ratio, when compared to both sedentary and OTS-affected athletes. Oppositely to the worse prognosis under clinical conditions, the meanings of an increased ratio as a physiological adaptive process have not been explored to date.

8. Catecholamines remained increased in healthy athletes 36 to 48 hours after training, when compared to sedentary, as a possible mechanism of enhancement of the overall metabolic rate, which is corroborated with the increased measured-to-predicted BMR observed in this group. However, whether this finding, which was observed on athletes that practiced both and concurrent endurance and strength sorts of exercises, is also found in pure endurance, pure strength, and explosive (ball games) types of sports is uncertain and unlikely to be similar, as previous literature showed conflicting data [21–23].

9. Both measured-to-expected BMR ratio and percentage of fat oxidation (in relation to total BMR) were both increased which lasted for at least 48 hours after the last training session, which reinforces that training may enhance metabolism not only during and short after exercise. Possibly, the fact that endurance and strength activities were practiced simultaneously and the

inherent irregularity of high intensity functional training (HIFT), which was the prevailing sport modality among athletes evaluated by the EROS study, could also explain the optimization of the metabolism, since the lack of a regularity in training patterns is recognized to exacerbate the already increased metabolism found in athletes.

(b) In terms of novel pathological mechanisms identified to occur in OTS:

1. Hormonal responses to stimulations were slower and blunted in OTS-affected athletes, compared to those in healthy athletes, while indistinguishable from the responses in sedentary controls. This means that while a diffuse exacerbation of the hypothalamic-pituitary (HP) axes was observed in healthy athletes, a loss of the optimization of all hormonal axes was detected in OTS. However, the blunted hormonal responses in OTS was relative (i.e., when compared to the expected responses for athletes), but not overt, as responses were within the normal range for the general population. Figures 4.7 and 4.8 illustrate the differences between OTS and healthy athlete responses to exercise-independent direct hormonal stimulations.

2. The existence of a prolactin response to stimulations, exclusively present in physically active subjects, while absent sedentary, was completely lost in OTS. It means that under OTS, the prolactin response to stimulations become absent and similar to sedentary.

3. The characteristic of non-undetectable GH levels even when not during peaks, which was found in the majority of the healthy athletes, was lost under OTS, as almost all affected athletes showed undetectable levels. The role of the basal GH when not during peaks, as well as the implications of its loss in OTS, is unknown.

4. The increased testosterone levels typically observed in athletes of all sorts of sports, particularly in those that contain strength exercises [24], are completely lost in OTS. The mechanism of the reduction of testosterone levels was due to an exacerbation of the aromatase activity, as estradiol levels were higher in OTS, when compared to both healthy athletes and sedentary. This suggests that the optimization of the gonadotrophic axis (GnRH/LHRH-LH/FSH-gonads) is maintained under OTS, once the "sum" of the sexual hormones (testosterone + estradiol) is similar between OTS and healthy athletes, while these "sums" are higher than that in sedentary.

5. Since testosterone was lower in OTS-affected athletes compared to healthy athletes, and similar to sedentary, while estradiol was higher than in the two control groups, the testosterone-to-estradiol ratio, an indirect marker of the level of aromatase activity, was consequently higher in OTS, more specifically two times higher, when compared to both healthy athletes and sedentary. This demonstrates that the aromatase activity was pathologically exacerbated in OTS, similarly to what is observed in inflammatory and metabolic disorders [25, 26]. This finding reinforces the hypothesis that OTS can be also considered as a dysfunctional metabolic and inflammatory state.

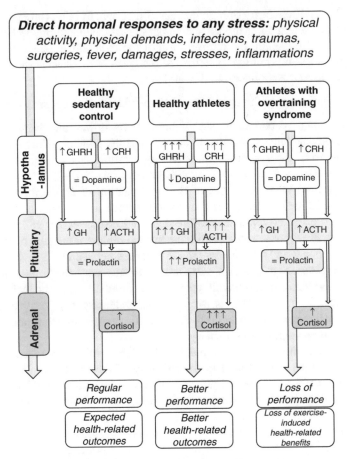

Fig. 4.7 Hormonal conditioning and *deconditioning* (loss of conditioning) processes in healthy athletes and overtraining syndrome, respectively

The dysfunctional increase of the conversion of testosterone into estradiol is likely an additional mechanism to force an anti-anabolic environment, in order to reduce energy expenditure under a chronically energy-deprived environment.

6. Compared to healthy athletes with similar training patterns, OTS-affected athletes presented reduced lactate clearance and the exacerbated CK levels. These two major findings of muscular-related parameters demonstrate that muscle recovery is delayed and compromised under OTS. The impaired

Fig. 4.8 Hormones in healthy athletes and in overtraining syndrome

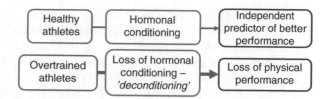

muscle repair prolongs recovery and impairs performance when training is attempted to be maintained at a regular basis, which is a hallmark of OTS. In addition, besides the physiological presence of myokines (muscle-derived cytokines), a pathological increase of non-muscle-derived cytokines inside the muscle tissue reinforces the hypothesis of the existence of a dysfunctional muscular metabolism and functioning in OTS [11, 27].

7. Once an increased neutrophil-to-lymphocyte ratio is considered as a marker of worse prognosis under clinical conditions and inflammatory and metabolic dysfunctions [19, 20], the decreased ratio found in OTS, when compared to healthy sedentary, may have a protective role for this state. However, when analyzed from a different perspective, a reduction of neutrophils found in OTS, compared to the expected levels for athletes, may indicate a mild and non-overt immunologic suppression and imbalance. This is the first report of immunologic changes in OTS that were detected in a simple leukogram, which reinforces the immunological dysfunction hypothesized to be present in OTS, in addition to the previous demonstrations of altered NK cell and T lymphocyte activity, and clusters of differentiation (CDs) in affected athletes [11, 28–30].

8. Unlike other parameters, the increased catecholamines found to occur in healthy athletes were not blunted, but actually exacerbated in OTS, while the inactivation of the catecholamines into metanephrines was shown to be blunted, once the catecholamine-to-metanephrine ratio was increased in OTS [4]. These two characteristics demonstrate attempts to maintain the organism functioning despite the chronic lack of energy and mechanisms of repair. Whether these mechanisms are resulted from an increased motivation that OTS-affected athletes typically tend to present in order to overcome the decrease in sports performance is uncertain. However, although catecholamines may preclude the organism to fade in the short term, its pathological exacerbation will certainly aggravate the already dysfunctional systems and metabolism, as increased catecholamines request metabolic activity from tissues that are already working with *maladaptive* pathways in response to an inhospitable environment, leading to a progression of the multiple dysfunctional *maladptation*-driven processes.

9. The metabolic optimizations that occur in healthy athletes, including increased metabolic rate and fat oxidation, are entirely lost in OTS. The relative reduction of metabolic rate may be an additional adaptation to a chronic depletion of energy. A reduction of fat oxidation means an increased protein catabolism when in the absence of glycogen and glucose availability, which has been hypothesized to be a mechanism to preserve the low energy demanding energy storage of the fat tissue while redirecting the organism toward a proteolysis-preferred energy source of utilization. The protein-structures and muscle tissue are more energy demanding and metabolically active when compared to lipid structures and storage and fat tissue. Thus, the utilization of protein instead of fat as the energy source, through the catabolism of the muscle tissue, will likely lead to further decrease of energy demands, once in this utilization mode, the muscle tissue, a more highly energy demanding tissue, shrank. The hypothesis of the increased muscle catabolism is reinforced by the exacerbation of the indicators of muscle degradation, as seen in the relative increase of lactate and exacerbation of CK levels found in OTS.

10. Despite a higher caloric deficit observed in affected athletes compared to healthy ones, an apparently paradoxical increase of the fat mass was observed in OTS-affected athletes. This finding cannot be fully explained by decrease in metabolic rate and fat oxidation, increased muscle degradation, and altered biochemical and hormonal profile and be an additional dysfunctional consequence of OTS that needs to be further explored.

11. From a total of 44 markers that disclosed differences between OTS, healthy athletes and non-physically active controls, the majority (26 markers; 59.1%) was similar between OTS-affected athletes and sedentary, while different from healthy athletes. The prevailing differences between healthy athletes and sedentary demonstrate that multiple adaptive conditioning processes of different systems occur in athletes, while differences between healthy and OTS-affected athletes and similarities between OTS and sedentary expressed show that the conditioning processes that athletes undergo are lost in OTS. These findings allow the conclusion that OTS results from a sort of a mix of multiple paradoxical losses of conditioning processes, that eventually leads to loss of the physical conditioning. Hence, while metabolic and hormonal progressive conditioning processes help lead to progressive physical conditioning in athletes, the multiple losses of conditioning processes lead to the loss of the physical conditioning. This elucidates the underlying reason for the previously unexplained loss of performance, the hallmark of OTS. Once the multiple *deconditioning* process has now been demonstrated to be the key aspect of OTS, OTS has been proposed to be termed as "paradoxical deconditioning syndrome" (PDS). Table 4.16 describes markers of "deconditioning" and unaltered markers in overtraining syndrome, and Table 4.17 summarizes the "level" of alteration of the markers that were whether relative, overt, or exacerbated.

Table 4.16 Markers of "deconditioning" and unaltered markers in overtraining syndrome

Parameters in which the adaptive changes were completely lost when in OTS	Parameters unaltered during OTS
1. Salivary cortisol 30′ after awakening (ng/dL)	1. Basal cortisol (µg/dL)
2. Cortisol awakening responses (CAR) (%)	2. Early cortisol response to synthetic
3. Early cortisol response to stimulation (µg/dL)	ACTH (cosyntropin) (early direct
4. Late cortisol response to stimulation (µg/dL)	adrenal response to stimulation) (µg/
5. Basal GH (µg/L)	dL)
6. Early GH response to stimulation (µg/L)	3. Late cortisol response to synthetic
7. Late GH response to stimulation (µg/L)	ACTH (cosyntropin) (late direct
8. Basal prolactin (ng/mL)	adrenal response to stimulation) (µg/
9. Early prolactin response to stimulation (ng/mL)	dL)
10. Late prolactin response to stimulation (ng/mL)	4. Basal ACTH (pg/mL)
11. Total testosterone (ng/dL)	5. Early ACTH response to stimulation
12. Estradiol (ng/dL)[a]	(pg/mL)
13. Testosterone-to-estradiol ratio[a]	6. Salivary cortisol at awakening (ng/dL)
14. Neutrophils ([a]mm³)	7. Salivary cortisol at 4 PM (ng/dL)
15. Neutrophil-to-lymphocyte ratio	8. Salivary cortisol at 11 PM (ng/dL)
16. Lactate (nMol/L)	9. IGF-1 (insulin growth-like factor 1
17. Basal metabolic rate (BMR) (% of expected)	(pg/mL)
18. Fat oxidation (% of total BMR)	10. TSH (thyroid-stimulating hormone)
19. Body fat (%)	(µUI/mL)
20. Muscle mass (%)	11. Free T3 (free T3) (pg/ml);
21. Level of hydration (% of body weight)	12. Nocturnal urinary epinephrine
22. Overall mood states	(µg/12 h)
23. Fatigue levels	13. Nocturnal urinary metanephrines
24. Vigor levels[a]	(µg/12 h)
25. Tension levels[a]	14. Nocturnal urinary normetanephrines
26. Depression levels	(µg/12 h)
27. Anger levels	15. Testosterone-to-estradiol ratio
28. Confusion levels	16. ESR (erythrocyte sedimentation rate)
29. Sleep quality	(mm/h)
	17. Ultrassensitive CRP (C-reactive
	protein) (mg/dL)
	18. Vitamin B12 (pg/mL)
	19. Ferritin (ng/mL);
	20. Eusinóphils (*1000/mm³)
	21. Hematocrit (%)
	22. Sleep duration (h/night)

CAR (cortisol awakening response) increase of salivary cortisol levels between awakening and 30 minutes after awakening (%), *GH* growth hormone, *ACTH* Adrenocorticotropic hormone
a = Not only "deconditioned", but also worse even when compared to sedentary controls

12. In addition, seven markers were overtly altered in OTS-affected athletes, as these were different than both healthy athletes and sedentary controls. Thus, although OTS is majorly a relative dysfunction, it also has some characteristics of a frank condition.

Table 4.17 Biochemical markers altered in overtraining syndrome

Altered markers	Level of alteration (overt, relative, or exacerbated*)
Insulin tolerance test (ITT)	
Cortisol during hypoglycemia in response to ITT (μg/dL) Cortisol 30′ after hypoglycemia in response to ITT (μg/dL) Cortisol increase during ITT (μg/dL)	Relative
ACTH 30′ after hypoglycemia in response to ITT (pg/mL) ACTH change during ITT (pg/mL)	Overt
Growth hormone (GH) during hypoglycemia in response to ITT (μg/L) GH 30′ after hypoglycemia in response to ITT (μg/L)	Relative
Prolactin during hypoglycemia in response to ITT (ng/mL) Prolactin 30′ after hypoglycemia in response to ITT (ng/mL) Change in prolactin during ITT (ng/mL)	Relative
Basal hormones	
Basal GH (μg/L)	Relative
Basal prolactin (μg/mL)	Relative
Estradiol (pg/mL)	Overt
Total testosterone (ng/dL)	Relative
Testosterone-to-estradiol ratio	Overt
Other hormones	
Total catecholamines (μg/12 hours)	Exacerbated
Nocturnal urinary norepinephrine (μg/12 hours)	Overt
30′ after awakening salivary cortisol (ng/dL)	Relative
Immunological and muscular markers	
Creatine kinase (U/L)	Exacerbated
Lactate (nmol/L)	Relative
Lymphocytes (/mm^3)	Relative
Neutrophils (/mm^3)	Relative
Platelet to lymphocyte ratio	Relative

*Relative: unaltered when compared with general population, but altered when compared with healthy athletes. Overt: markedly and exclusively altered in overtraining syndrome, when compared with both the healthy athletes and the healthy non-physically active subjects. Exacerbated: changes normally observed in healthy athletes that are exacerbated in the presence of overtraining syndrome

ACTH = adrenocorticotropic hormone

4.2.3   Unexpectedly Remarkable Findings

Some of the findings of both OTS-affected and healthy athletes were unexpectedly present and sometimes opposite to the expected results. Among these surprising findings, some remarkable ones are listed below, due to their potential clinical applications and perspectives for further researches:

1. Since there is an acute increase of lactate levels in response to exercise that tends to be prolonged when physical activity is more intense, the lower lactate found in healthy athletes, when compared to sedentary, was a paradoxical finding. Increased and overcompensated lactate clearance in response to exercise, which eventually leads to decreased levels when compared to non-physically active subjects, may represent an additional mechanism of progressive improvement of performance, as lower lactate is associate with better performance and/ or an indirect marker of better health status, once lactate is directly correlated with poorer outcomes.

2. The non-reduced lactate levels in OTS may be a signal of over-damaged and impaired muscle recovery, while the optimized lactate clearance may still be present, since CK levels were unproportionally higher compared to lactate in OTS-affected athletes.

3. The enhancement of the hypothalamic-pituitary axes as an independent and intrinsic mechanism of hormonal conditioning was highly unexpected, as these occurred irrespectively of exercise, level of performance, or external signaling from other systems. The process of optimization of the endocrine system resembles the extensively demonstrated conditioning of the cardiovascular, autonomic, neuromuscular, and musculoskeletal systems.

4. The demonstration of a clear exposure-response relationship between hormonal responsiveness to stimulations and level of physical performance, since while the optimization of the hormonal responses helps explain the still not fully elucidated mechanisms in athletes of progressive improvement of sports performance in healthy athletes, the loss of these optimized responses may justify the loss of physical performance in OTS.

5. The unveiled data allows the hypothesis of the level of hormonal responsiveness as being at least partially responsible for sports performance, for both explosive capacity, as the hormonal response peaks are amplified in healthy athletes, and for the maintenance of the pace, since hormonal responses were not only increased but also prolonged in healthy athletes. In addition, these hormonal conditioning processes may have an important role and justify the better prognosis, responses, and outcomes in the presence of infections, inflammations, autoimmune disorders, metabolic dysfunctions, and cancers, when these are presented in athletes.

6. The exacerbated increase of catecholamines, as well as its lack of inactivation into metanephrines, found in OTS, is a potential mechanism to maintain a minimal level of functioning in OTS, despite the chronic depletion of energy and mechanisms of repair. Perhaps this may be due to an increase of motivation

among affected athletes, as an attempt to overcome fatigue and loss of performance in OTS.

7. Oppositely to what is largely believed, the cortical layers of the adrenal glands were shown to be unaffected in OTS, at least in the *fasciculata* layer of the adrenal cortex, while they did not undergo any sort of adaptive process to exercise, as its direct stimulation using synthetic ACTH revealed no differences between the two groups of athletes and the group of sedentary control [1].

8. Athletes affected by OTS had fewer adrenergic symptoms when under hypoglycemia, when compared to healthy athletes and sedentary [1, 2]. This may be due to an adaptation of the organism to a lower carbohydrate intake in the long term, which may hypothetically lead to increased frequency, intensity, and prolonged periods of hypoglycemia, particularly during more intense training sessions, in which a higher amount of energy is demanded in a short period of time. This hypothesis is based on the limited speed of glucose generation (gluconeogenesis) from lipids and amino acids, which may be at an insufficient speed to maintain normal serum glucose levels in highly glucose demanding intense trainings, despite the use of ketogenic sources as energy supply, as these may be similarly limited. The chronic exposure to prolonged hypoglycemia, even when mild, leads to adaptive changes, especially the loss of adrenergic sensitivity to lower glycemia, leading to fewer adrenergic symptoms under hypoglycemia. This phenomenon is similar to those patients with insulin-dependent diabetes that constantly present hypoglycemic episodes: they only start presenting symptoms in more severely low glucose, when neuroglycopenic symptoms start to appear, including altered behavior and decreased consciousness.

9. Unexpectedly, excessive physical activity as a trigger of OTS has become unimportant, since athletes and sport coaches learned the consequences of excessive training and not properly periodize training sessions and their intensity and volume.

10. While physical activity does no longer play a so important role as a trigger of OTS, eating patterns, including insufficient caloric, carbohydrate, and protein intake, excessive concurrent cognitive effort, and inadequate sleeping behavior and hygiene become key for the persistently prevailing occurrence of OTS. Possibly, the combination of severely chronic energy deprivation and the inability to replace the mechanisms of repair consumed during the stressful trainings, due to lack of proper recovery, insufficient sleep quality, and concurrent cognitive activity that may be as energy demanding and physically stressful as physical efforts, has become largely responsible for OTS.

11. Overall, athletes affected by OTS disclose a sort of an "energy-saving mode" of functioning, which although controversial shows consistent indicatives of its occurrence, at least in OTS, leading to a global "hypometabolic state," as consistently demonstrated by the concordant findings of reduction of BMR (energy economy) and fat oxidation (for the preservation of energy storage within a tissue that does not demand much energy to be maintained), an anti-anabolic state (indirectly demonstrated by increased conversion of testosterone into estradiol to prevent the construction of high-demanding tissues), and an undermined fight-or-flight response, shown by a relative hypothalamic hyporesponsiveness.

4.3 New Markers of Overtraining Syndrome

New and more accurate markers of OTS have been identified and will likely improve the tools for the early diagnosis of OTS and the level of its understanding and are described in Tables 4.14 and 4.15.

The first new remarkable markers are the flattened GH, cortisol, and prolactin responses to any stimulation, when compared to the expected responses for athletes [1, 2, 5].

Thyroid hormones were shown to be linearly associated with decreased performance, the hallmark of OTS, when were studied longitudinally, although OTS was not necessarily a diagnosis in these cases, owing to the lack of full diagnostic process [31]. Leptin was lower in OTS, even after adjustment for body fat, and IGF-1 had a paradoxical reduction in response to exercise in OTS, although its basal levels were normal [32].

Also compared to healthy athletes, several other parameters for the identification were found, including exacerbated CK and lactate levels, compared with those athletes with similar training patterns and time since last training session, lower lymphocyte-to-neutrophil ratio, measured-to-expected metabolic rate, fat oxidation, and hydration status and lean mass, while increased body fat [4, 5]. Finally, OTS discloses overt compromised fatigue, vigor, and libido, even when compared to sedentary.

In addition, the status of the oxidative stress, i.e., the redox homeostasis, was found to be clearly disrupted, with altered hydroperoxides, plasma antioxidant capacity, red blood cell glutathione, superoxide dismutase, coenzyme Q10, α- and γ-tocopherol, and carotenoids lutein, α-carotene, and β-carotene, when in athletes with unexplained underperformance and normal hematology, biochemistry, thyroid function, immunology, vitamins, and mineral levels [33]. The newly identified redox markers can be correlated with lactate levels as increased oxidative stress together with exacerbated lactate levels can be associated with a forthcoming OTS [34].

Besides the pro- and anti-oxidative balance, oppositely to the classical inflammatory markers, that were unaltered under OTS, specific non-classical cytokines were shown to be affected in OTS. IL-1β was found to increase in response to exercise only under OTS, not in the absence of dysfunctions. Conversely, after exercise IL-10 was lower in affected than healthy athletes, while IL-6 and TNF-α increased at similar extents acutely, but turned out to be exacerbated in the long run in OTS-affected athletes [32]. While when stimulated cytokines can help identify OTS, during resting IL-1β, IL-10, IL-6, and TNF-α were unaltered [32].

When athletes need to be discriminated between non-functional overreaching and overtraining syndrome, a "Training Optimization" (TOP) test has been proposed, and cutoffs were suggested [35] through a retrospective "recalculation," which showed that an increase of ACTH and PRL in response to the TOP 2-test protocol below 200% indicates OTS, whereas increases above 200% are indicative of nun-functional overreaching (NFOR). However, during the same test, mental fatigue seems to be an earlier sign of NFOR or OTS than altered cognitive abilities, reduced physical performance, and biochemical alterations [36].

Additional data reinforced the current proposed biomarkers, including the presence of lower heart ratio variability (HRV), when compared to healthy athletes and to sedentary [37].

Despite the new proposed markers of OTS, more important than the identification of each of these markers is the understanding of how these alterations are usually present in OTS. While none of the described markers are present in all affected athletes, each athlete disclosed a unique combination of dysfunctions. Consequently, a diagnostic guideline for its early identification seems to be more impactful, in order to prevent these athletes from a long-term and career-threatening disease, as will be shown below.

It is noteworthy, however, that some of these newly identified markers need to be compared to those levels expected in athletes, and adapted ranges should be proposed, as normal ranges for the general population will unlikely be helpful, once most of the currently alterations identified in OTS are still within these normal ranges. In Table 4.18 proposal of specific cutoffs for some functional tests are provided and may be helpful as preliminary tools for the diagnosis of OTS.

Table 4.18 Proposed cutoffs and their clinical applicability

Markers	Proposed cutoffs	Applicability for the diagnosis of overtraining syndrome
GH response to an insulin tolerance test (ITT) (µg/l)	>14.5ug/L <5ug/L	High negative predictive value (100%) High accuracy (76.9%)
Prolactin response to an insulin tolerance test (ITT) (ng/mL)	>26.5 ng/mL <10 ng/mL	High negative predictive value (100%) High positive predictive value (100%)
Prolactin increase during an insulin tolerance test (ITT) (ng/mL)	>13 ng/mL	High negative predictive value (100%)
Salivary cortisol (ng/dL) 30 minutes after awakening	>530 ng/dL 370 ng/dL	High negative predictive value (93.9%) High accuracy (80%)
Cortisol response to an insulin tolerance test (ITT) (µg/dL)	>20.5 µg/dL >17.0 µg/dL 19.1 µg/dL	High negative predictive value (100%) High positive predictive value, but non-specific (28.6%) High accuracy (84.6%)
Cortisol increase during an insulin tolerance test (ITT) (ng/mL)	>9.5 ug/dL	High negative predictive value (100%)
ACTH response to an insulin tolerance test (ITT) (pg/mL)	>106 pg/mL	High negative predictive value (92.9%) and high accuracy (80%)
ACTH increase to an insulin tolerance test (ITT) (pg/mL)	<35 pg/mL	High accuracy (80%)
Prolactin response to a training optimization 2-test (TOP)[a]	>200% <200%	93% sensitivity for the diagnosis of OTS 74% sensitivity for the diagnosis of non-functional overreaching
ACTH response to a training optimization 2-test (TOP)[a]	>200% <200%	67% sensitivity for the diagnosis of OTS 74% sensitivity for the diagnosis of non-functional overreaching

aThrough discriminant analysis evaluating psychological changes during the TOP test

4.4 Novel Risk Factors for OTS

Excessive training, although classically described as the main trigger of OTS, has become progressively less important, since training programs are now periodized. Conversely, novel risk factors have been identified and include eating, social, and sleeping patterns. Specific criteria as risk factors for OTS, resulted from the analysis of the differences in the behaviors between healthy and OTS-affected athletes, have been proposed, as listed below:

1. Caloric intake <32 kcal/kg/day.
2. Protein intake <1.6 g/kg/day (irrespective of overall caloric intake).
3. Carbohydrate intake <5.0 g/kg/day (irrespective of overall caloric intake).
4. In a zero to ten self-reported scale for sleep quality, less than six.
5. Work and/or study >8 hours/day.

For the prevention of OTS, in addition to the avoidance of any of these risk factors, changes in any of the POMS questionnaire may be early signs of upcoming OTS, and its detection may be helpful to prevent it.

4.5 Conclusion

Athletes affected by OTS ingested lower carbohydrates, proteins, and overall calories, had worse sleep quality, worked or studied for longer periods, and disclosed blunted and delayed hormonal responses to stimulations, lower testosterone, increased estradiol, increased testosterone-to-estradiol ratio, increased basal lactate levels, exacerbated CK, reduced neutrophils and neutrophil-to-lymphocyte ratio, exacerbated catecholamines and catecholamine-to-metanephrine ratio, reduced vigor, increased fatigue, depression, confusion, tension and anger, loss of libido, lower measured-to-expected basal metabolic rate (BMR) ratio and fat oxidation, higher body fat, lower muscle mass, and reduced hydration compared to healthy athletes while had similar levels of those in sedentary controls. In addition, OTS disclosed disruption of redox homeostasis, including abnormal hydroperoxides, plasma antioxidant capacity, red blood cell glutathione, superoxide dismutase, coenzyme Q10, α- and γ-tocopherol, and carotenoids lutein, α-carotene, and β-carotene; altered non-classical inflammatory pathways, including increased IL-1β, IL-6, and TNF-α and reduced IL-10; and reduced basal leptin and IGF-1 response to exercise.

Conversely, healthy athletes yielded improvements in hormonal responses to stimulations, testosterone, lactate clearance, catecholamines, and increased BMR and % of fat oxidation, which were lost in OTS.

References

1. Cadegiani FA, Kater CE. Hypothalamic-pituitary-adrenal (HPA) axis functioning in overtraining syndrome: findings from endocrine and metabolic responses on overtraining syndrome (EROS) - EROS-HPA axis. Sports Med Open. 2017;3(1):45.
2. Cadegiani FA, Kater CE. Growth hormone (GH) and prolactin responses to a non-exercise stress test in athletes with overtraining syndrome: results from the endocrine and metabolic responses on overtraining syndrome (EROS) – EROS-STRESS. J Sci Med Sport. 2018;21(7):648–53.
3. Cadegiani FA, Kater CE. Body composition, metabolism, sleep, psychological and eating patterns of overtraining syndrome: results of the EROS study (EROS-PROFILE). J Sports Sci. 2018;36(16):1902–10.
4. Cadegiani FA, Kater CE. Basal hormones and biochemical markers as predictors of overtraining syndrome in male athletes: the EROS-BASAL study. J Athl Train. 2019; https://doi.org/10.4085/1062-6050-148-18.
5. Cadegiani FA, Kater CE. Novel insights of overtraining syndrome discovered from the EROS study. BMJ Open Sport Exerc Med. 2019;5(1):e000542. https://doi.org/10.1136/bmjsem-2019-000542.
6. Cadegiani FA, Kater CE, Gazola M. Clinical and biochemical characteristics of high-intensity functional training (HIFT) and overtraining syndrome: findings from the EROS study (the EROS-HIFT). J Sports Sci. 2019;20:1–12. https://doi.org/10.1080/02640414.2018.1555912.
7. Cadegiani FA, Kater CE. Novel causes and consequences of overtraining syndrome: the EROS-DISRUPTORS study. BMC Sports Sci Med Rehabil. 2019;11:21.
8. Cadegiani FA, Kater CE. Eating, sleep, and social patterns as independent predictors of clinical, metabolic, and biochemical behaviors among elite male athletes: the EROS-PREDICTORS study.
9. Cadegiani FA, Kater CE. Inter-correlations among clinical, metabolic, and biochemical parameters and their predictive value in healthy and overtrained male athletes: the EROS-CORRELATIONS study. Front Endocrinol (Lausanne). 2019;10:858.
10. Cadegiani FA, Silva PHL, Abrao TCP, Kater CE. Diagnosis of overtraining syndrome: results of the Endocrine and Metabolic Responses on Overtraining Syndrome (EROS) study —EROS-DIAGNOSIS.
11. Meeusen R, Duclos M, Foster C et al. European College of Sport Science; American College of Sports Medicine. Prevention, diagnosis, and treatment of the overtraining syndrome: joint consensus statement of the european college of sport science and the American College of Sports Medicine. Med Sci Sports Exerc. 2013;45(1):186–205.
12. Eijsbouts AM, van den Hoogen FH, Laan RF, et al. Decreased prolactin response to hypoglycaemia in patients with rheumatoid arthritis: correlation with disease activity. Ann Rheum Dis. 2005;64(3):433–7.
13. Kirk SE, Xie TY, Steyn FJ et al. Restraint stress increases prolactin-mediated phosphorylation of signal transducer and activator of transcription 5 in the hypothalamus and adrenal cortex in the male mouse. J Neuroendocrinol. 2017;29(6):20.
14. Moore KE, Demarest KT, Lookingland KJ. Stress, prolactin and hypothalamic dopaminergic neurons. Neuropharmacology. 1987;26(7B):801–808.21.
15. Hayes LD, Sculthorpe N, Herbert P, et al. Resting steroid hormone concentrations in lifetime exercisers and lifetime sedentary males. Aging Male. 2015;18(1):22–6.
16. Hayes LD, Sculthorpe N, Herbert P, Baker JS, Spagna R, Grace FM. Six weeks of conditioning exercise increases total, but not free testosterone in lifelong sedentary aging men. Aging Male. 2015;18(3):195–200.
17. Kushimoto S, Akaishi S, Sato T, Nomura R, Fujita M, Kudo D, Kawazoe Y, Yoshida Y, Miyagawa N. Lactate, a useful marker for disease mortality and severity but an unreliable marker of tissue hypoxia/hypoperfusion in critically ill patients.

18. Limonciel A, Aschauer L, Wilmes A, Prajczer S, Leonard MO, Pfaller W, Jennings P. Lactate is an ideal non-invasive marker for evaluating temporal alterations in cell stress and toxicity in repeat dose testing regimes. Toxicol In Vitro. 2011;25(8):1855–62.
19. Patrice F, Céline K, Defour JP. What is the normal value of the neutrophil-to-lymphocyte ratio? BMC Res Notes. 2017;10:12.
20. Feng JR, Qiu X, Wang F, et al. Diagnostic value of neutrophil-to-lymphocyte ratio and platelet-to-lymphocyte ratio in Crohn's disease. Gastroenterol Res Pract. 2017:3526460.
21. Kraemer WJ, Gordon SE, Fragala MS, et al. The effects of exercise training programs on plasma concentrations of proenkephalin peptide F and catecholamines. Peptides. 2015;64:74–81.
22. Zouhal H, Jacob C, Delamarche P, Gratas-Delamarche A. Catecholamines and the effects of exercise, training and gender. Sports Med. 2008;38(5):401–23.
23. Jacob C, Zouhal H, Prioux J, Gratas-Delamarche A, Bentué-Ferrer D, Delamarche P. Effect of the intensity of training on catecholamine responses to supramaximal exercise in endurance-trained men. Eur J Appl Physiol. 2004;91(1):35–40.
24. Kraemer WJ, Ratamess NA. Hormonal responses and adaptations to resistance exercise and training. Sports Med. 2005;35(4):339–61.
25. Xu X, Sun M, Ye J. The effect of aromatase on the reproductive function of obese males. Horm Metab Res. 2017;49(8):572–9.
26. Mellouk N, Ramé C, Barbe A, Grandhaye J, Froment P, Dupont J. Chicken is a useful model to investigate the role of Adipokines in metabolic and reproductive diseases. Int J Endocrinol. 2018;2018:4579734.
27. Dressendorfer RH, Wade CE. The muscular overuse syndrome in long-distance runners. Phys Sportsmed. 1983;11(11):116–30.
28. Smith LL. Cytokine hypothesis of overtraining: a physiological adaptation to excessive stress? Med Sci Sports Exerc. 2000;32:317–31.
29. Margonis K, Fatouros IG, Jamurtas AZ, et al. Oxidative stress biomarkers responses to physical overtraining: implications for diagnosis. Free Radic Biol Med. 2007;43(6):901–10.
30. Tanskanen M, Atalay M, Uusitalo A. Altered oxidative stress in overtrained athletes. J Sport Sci. 2010;28(3):309–17.
31. Nicoll JX, Hatfield DL, Melanson KJ, Nasin CS. Thyroid hormones and commonly cited symptoms of overtraining in collegiate female endurance runners. Eur J Appl Physiol. 2018;118(1):65–73.
32. Joro R, Uusitalo A, DeRuisseau KC, Atalay M. Changes in cytokines, leptin, and IGF-1 levels in overtrained athletes during a prolonged recovery phase: a case-control study. J Sports Sci. 2017;35(23):2342–9.
33. Lewis NA, Redgrave A, Homer M, et al. Alterations in redox homeostasis during recovery from unexplained underperformance syndrome in an elite international rower. Int J Sports Physiol Perform. 2018;13(1):107–11.
34. Theofilidis G, Bogdanis GC, Koutedakis Y, Karatzaferi C. Monitoring exercise-induced muscle fatigue and adaptations: making sense of popular or emerging indices and biomarkers. Sports (Basel). 2018;6(4).
35. Buyse L, Decroix L, Timmermans N, Barbé K, Verrelst R, Meeusen R. Improving the diagnosis of nonfunctional overreaching and overtraining syndrome. Med Sci Sports Exerc. 2019.
36. Vrijkotte S, Meeusen R, Vandervaeren C, et al. Mental fatigue and physical and cognitive performance during a 2-bout exercise test. Int J Sports Physiol Perform. 2018;13(4):510–6.
37. Kajaia T, Maskhulia L, Chelidze K, Akhalkatsi V, Kakhabrishvili Z. The effects of nonfunctional overreaching and overtraining on autonomic nervous system function in highly trained athletes. Georgian Med News. 2017;264:97–103.

Chapter 5
Clinical, Metabolic, and Biochemical Behaviors in Overtraining Syndrome and Overall Athletes

Outline

Novel disruptors of OTS, predictors of clinical and biochemical behaviors, and correlations between clinical and metabolic parameters: in this subsection, the author brings new clinical and biochemical changes that disrupt the physiological functioning of the athletes and cause OTS, enlightens how certain hormones, metabolic parameters, and modifiable habits can draw the clinical and biochemical behaviors, and uncovers the newly identified multiple correlations between more than 70 different clinical and biochemical parameters evaluated in OTS.

5.1 Novel Triggers of OTS

Due to its design characteristics, the EROS study allowed the identification of independent predictors of the occurrence of OTS and also of other clinical and biochemical behaviors, as well as linear correlations between the parameters evaluated in the study (those in which causality relationship could not be established) [1–9]. The most remarkable findings regarding disruptors, predictors, and correlations between the tested parameters, both overall and OTS-affected athletes are described below.

From *post hoc* multivariate and logical regression analyses, modifiable factors, mostly habits, that could be independent causes of OTS, i.e., whether a specific habit was solely responsible for the occurrence of some cases of OTS, as well as parameters might be independently modulated by the presence of OTS, irrespective of other characteristics, i.e., even with the same caloric, protein, and carbohydrate intake, the same sleep quality and duration, the same amount of additional sports-related activity, and the same training intensity, volume, frequency, and duration, have been determined [7], since all training patterns were similar between the healthy and OTS-affected athletes in the EROS study.

Although excessive training has traditionally been viewed as the major cause of unexplained reductions in sports performance, the periodization of training sessions did not lead to a decrease in the incidence of OTS, and excessive training is now considered a minor factor in the development of OTS. Conversely, the design of the study and the additional statistic techniques allowed the detection of novel possible etiologies of OTS.

In addition, combinations of different OTS triggers were able to explain all cases of OTS among the participants in the EROS study, i.e., the combination was shown to be "the perfect predictor." Combinations that justified the development of OTS in athletes included dietary and sleep patterns altogether or all dietary patterns together.

Conversely, excessive cognitive demands, each dietary pattern alone, or the combination of two of the three dietary characteristics with other factors were not *sine qua non* conditions to trigger OTS, although not all triggers are necessary to be present in order to develop OTS. Independent triggers of OTS are described in Table 5.1.

Carbohydrate, protein, or overall caloric intake may each, independently disrupt physiological responses to a sport; hence, OTS can be induced without the presence of any of the other risk factors. Noteworthy, OTS may occur after changes in eating, sleeping, and/or social patterns, rather, irrespective of changes in training patterns. However, there is not a specific threshold for each activity or habit that become potentially risky for the occurrence of OTS, as patterns highly depend on the interactions other potential triggers of OTS.

The remaining significant differences in clinical, hormonal, metabolic, psychological, and biochemical behaviors between the ATL and OTS groups, after the adjustments for all baseline characteristics and training, eating, social, and sleep patterns, supported the conclusion that changes in behaviors were inherently due to the presence of OTS, irrespective of any other feature.

Although early hormonal responses to the ITT were predicted independently and positively by carbohydrate intake, the presence of OTS predicted hormonal late responses. Indeed, the commencement of a physical activity at maximum capacity for a short period, which is represented by early responses to stimulation and unaffected by OTS, is not typically observed in athletes with OTS. Conversely, reduced time-to-fatigue, a hallmark of OTS, can be explained by the blunted late hormonal responses independently predicted by the presence of OTS. This indicates an inability to maintain hormonal responses for longer periods in the presence of OTS, which probably explains the impaired pace typical of OTS.

Among the basal hormones, the T:E ratio [4], but not testosterone or estradiol, was disrupted by the presence of OTS, leading to worse metabolic and psychological behaviors. Reduced T:E ratio adds evidence to the hypothesis that OTS induces an anti-anabolic, dysfunctional, and energy-saving environment, irrespective of its triggers, in which reduced testosterone is a protective mechanism against energy expenditure and anabolic activity. However, the underlying mechanisms that lead to

Table 5.1 Independent triggers of OTS

Scenario	Independent variables included	Results (* = positive for independent risk factors and triggers)	Interpretation
Scenario 1 – All modifiable variables	CHO, PROT, CAL, WORK, SLEEP	Perfect separation	Together, modifiable patterns were able to explain all cases of OTS in the athletes studied
Scenario 2 – All modifiable variables, except WORK	CHO, PROT, CAL, SLEEP	Perfect separation	Dietary patterns together with sleep quality were also able to fully explain all cases of OTS in the studied population of athletes
Scenario 3 – All modifiable variables, except CAL	CHO, PROT, WORK, SLEEP	CHO: $p = 0.036$ OR/CL = 1.61 (1.03–2.50) PROT: $p = 0.029$ OR/CL = 16.7 (1.34–208.1) WORK: $p = $ n/s SLEEP: $p = 0.069$ OR/CL = 2.19 (0.94–5.09)	When daily caloric intake is not accounted, not all cases of OTS may be justified. However, in this scenario, both CHO and PROT were shown to be independent triggers of OTS
Scenario 4 – Without specification of each macronutrient	CAL, WORK, SLEEP	CAL: $p = 0.004$ OR/CL = 1.13 (1.04–1.23) WORK: p = n/s SLEEP: p = n/s	When each macronutrient intake is not specified, not all cases of OTS may be justified. However, in this scenario, CAL was enough to independent etiology of OTS
Scenario 5 – Only dietary patterns	CHO, PROT, CAL	CHO: p = n/s PROT: $p = 0.066$ OR/CL = 25.85 (0.81–825.3) CAL: $p = 0.045$ OR/CL = 1.27 (1.01–1.61)	When only dietary patterns are evaluated, we cannot explain all cases of OTS in the studied population. However, in this scenario, overall caloric intake, but not each macronutrient, was able to

CHO Daily carbohydrate intake (g/kg/day), PROT Daily protein intake (g/kg/day), CAL Mean daily caloric intake (kcal/kg/day), WORK Average number of working or studying hours a day, besides training sessions (h/day), SLEEP Self-reported sleep quality (0–10), OTS Overtraining syndrome, OR Odds ratio, CL 95% Confidence Limits, p Level of significance, n/s nonsignificant ($p > 0.1$)

reduced T:E ratio in OTS are unknown. For practical purposes, T:E ratio should be higher than 13.7:1, for total testosterone and estradiol expressed in ng/mL and pg/dL, respectively [4].

The basic immunology panel was also independently affected by the presence of OTS, which supports the theory of involvement of the immune system in the pathophysiology of OTS. Although altered immunology panels (i.e., altered when compared with healthy athletes but similar to those of nonathletes) may be linked to blunted hormonal responses to stress [10, 11], they failed to demonstrate intercorrelations or predictions, at least neutrophils, lymphocytes, and the neutrophil-to-lymphocyte ratio. In addition, other mechanisms, such as an environment with chronic stressors leading to OTS, may directly predict leukocyte composition [12].

Relative dehydration, decrease in muscle mass, and the increase in visceral fat were independently enhanced by OTS, which may reinforce the multiple dysfunctions present in OTS. Possibly, highly oxidative and non-classical inflammatory environment inherently present in OTS may justify the increased visceral fat detected in OTS without concurrent increase in overall body fat.

OTS contributes to severe psychological dysfunctions independently of any other factor. Mood disturbances in OTS are not always fully recoverable. Interestingly, depression was not modulated by OTS, despite previous reports of depression as one of the outcomes of OTS [13–18].

The negative changes in body composition and mood induced by OTS may play underlying roles in the decrease of performance, the key, and the sine qua non characteristic of OTS.

The summary of clinical and biochemical behaviors independently modified by OTS, as well as their estimated level of influence, are disclosed in Table 5.2.

Findings allowed one hypothesize that any type of disruption in eating, sleep, social, or training patterns could lead to a spread of dysfunctional reactions through multiple pathways, as a sort of "domino effect," leading to aberrant changes in hormonal, muscular, immunologic, metabolic, and/or physical behaviors, eventually causing OTS if not promptly addressed.

These findings demonstrated that all dietary and sleep patterns need to be assessed in order to identify athletes at risk for OTS. In clinical practice, dietary characteristics should be assessed prior to other triggers, and whenever they do not indicate the presence of OTS, sleep and social patterns should be investigated.

The use of both healthy and OTS-affected athletes for the analyses is important to predict behavior patterns prior to OTS, as the development of OTS may be understood as a process on a continuum (i.e., the end of an unresolved mixture of attempts to adapt to chronic energy depletion and the mechanisms underlying a recovery-deprived environment) [13–17, 19–21].

Figure 5.1 describes variables included for the analysis of the predictors of OTS, while the summary of the independent predictors of OTS and its disruptions on clinical and biochemical behaviors is illustrated in Fig. 5.2.

5.2   Correlations Between Clinical, Biochemical, and Metabolic Parameters

The unexpected large number of markers recently uncovered in healthy athletes and in OTS allowed the hypothesis of the existence of a web of multiple sorts of interactions between parameters of different natures, which could result in a wide range of benefits and clinical outcomes, as well as the progressively improvement of

Table 5.2 Clinical and biochemical behaviors independently modified by overtraining syndrome (OTS)

Parameters modified by the presence of OTS	p of the influence of OTS*	Level of influence of the presence of OTS* (adjusted R-square)	Other variables that may also influence	Equation for the estimation of the parameter level in male athletes
Late ACTH response to an ITT (30'after hypoglycemia) (pg/mL)	0.002	19.9%	*None*	*n/a-*
Late cortisol response (30'after hypoglycemia) (μg/dL)	0.0005	26.1%	*None*	Cortisol (μg/dL) = 17.86–3.81 (if OTS)
Cortisol response to an ITT (μg/dL)	0.002	22.0%	*None*	*n/a*
Late GH response (30'after hypoglycemia) (μg/L)	0.001	23.0%	*None*	*n/a*
Testosterone-to-oestadiol ratio (T/E)	0.0002	30.7%	*None*	T/E = 14.1 + 12.9 (if OTS)
POMS vigor subscale	<0.0001	83.6%	Sleep quality	POMS vigor subscale = 3.7 + 1.15x (sleep quality) – 11.96 (if OTS)
POMS fatigue subscale	<0.0001	85.7%	Sleep quality	POMS fatigue subscale = 24.5–0.9 x (sleep quality) + 15.3 (if OTS)
POMS tension subscale	<0.0001	42.8%	*None*	
Visceral fat (cm²)	0.002	38.2%	Protein and overall calorie intake	Visceral fat = 47.4–11.9x (protein intake) + 1.3x (calorie intake) +45.1 (if OTS)
Muscle mass (%)	0.028	33.7%	Protein intake	Muscle mass = 47.84 + 1.42x (protein intake) – 3.47 (if OTS)
Body water (%)	0.001	50.5%	Protein and overall calorie intake	Body water = 60.75 + 1.69x (protein intake) – 0.12x (calorie intake) – 5.77 (if OTS)
Neutrophils (/mm³)	0.015	13.8%	Calorie intake	Neutrophils = 4210–60.7x (calorie intake) + 154.4x (CHO intake) – 1724 (if OTS)
Neutrophil-to-lymphocyte ratio	0.015	13.6%	*None*	Ratio = 2.00–1.32 (if OTS)
Parameters modified by the presence of OTS	p of the influence of OTS*	Level of influence of the presence of OTS* (Adjusted R-square)	Other variables that may also influence	Equation for the estimation of the parameter level in male athletes

(continued)

Table 5.2 (continued)

Parameters modified by the presence of OTS	p of the influence of OTS*	Level of influence of the presence of OTS* (adjusted R-square)	Other variables that may also influence	Equation for the estimation of the parameter level in male athletes
Late ACTH response to an ITT (30'after hypoglycemia) (pg/mL)	0.002	19.9%	None	n/a-
Late cortisol response (30'after hypoglycemia) (µg/dL)	0.0005	26.1%	None	Cortisol (µg/dL) = 17.86–3.81 (if OTS)
Cortisol response to an ITT (µg/dL)	0.002	22.0%	None	n/a
Late GH response (30'after hypoglycemia) (µg/L)	0.001	23.0%	None	n/a
Testosterone-to-oestadiol ratio (T/E)	0.0002	30.7%	None	T/E = 14.1 + 12.9 (if OTS)
POMS vigor subscale	<0.0001	83.6%	Sleep quality	POMS vigor subscale = 3.7 + 1.15x (sleep quality) – 11.96 (if OTS)
POMS fatigue subscale	<0.0001	85.7%	Sleep quality	POMS fatigue subscale = 24.5–0.9 x (sleep quality) + 15.3 (if OTS)
POMS tension subscale	<0.0001	42.8%	None	
Visceral fat (cm²)	0.002	38.2%	Protein and overall calorie intake	Visceral fat = 47.4–11.9x (protein intake) + 1.3x (calorie intake) +45.1 (if OTS)
Muscle mass (%)	0.028	33.7%	Protein intake	Muscle mass = 47.84 + 1.42x (protein intake) – 3.47 (if OTS)
Body water (%)	0.001	50.5%	Protein and overall calorie intake	Body water = 60.75 + 1.69x (protein intake) – 0.12x (calorie intake) – 5.77 (if OTS)
Neutrophils (/mm³)	0.015	13.8%	Calorie intake	Neutrophils = 4210–60.7x (calorie intake) + 154.4x (CHO intake) – 1724 (if OTS)
Neutrophil-to-lymphocyte ratio	0.015	13.6%	None	Ratio = 2.00–1.32 (if OTS)

*Other minor influences may also reflect the p-value and the level of influence

CHO Carbohydrate, *ITT* Insulin-tolerant test, *POMS* Profile of mood states, *BMR* Basal metabolic rate, *T/E* Testosterone-to-oestradiol, *OTS* Overtraining syndrome' *n/a* non applicable (non-normal distribution)

Calorie intake = kcal/kg/day; CHO intake = g(CHO)/kg/day; protein intake = g(protein)/kg/day; extra activities = working and/or studying hours besides training; sleep quality = self-reported sleep quality (0 to 10)

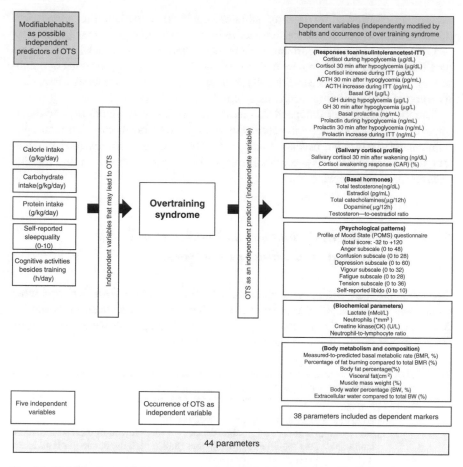

Fig. 5.1 Variables included in the present analysis for the detection of triggers of OTS

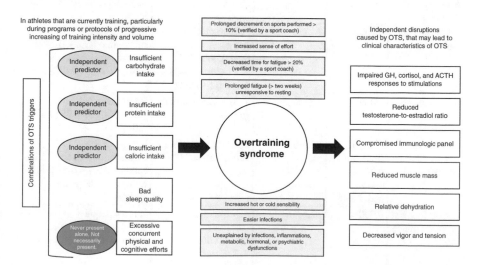

Fig. 5.2 Summary of the predictions of overtraining syndrome (OTS) and its implications

performance observed in healthy athletes, could also explain the paradoxical decrease of sports performance, fatigue, reduced libido, and body changes present in OTS [5, 7]. Potential correlations identified between these newly uncovered parameters could provide a new understanding of the complex processes of conditioning processes that athletes typically undergo and, from this understanding, clarify some of the underlying mechanisms that lead to OTS [5, 7].

The identification of independent predictors and linear correlations, and determine causal relationships and inter-influences, among hormonal, immunologic, inflammatory, muscular, psychological, metabolic, and body composition parameters, that inter-regulate between them irrespective of eating, sleeping, and training patterns, has been revealed [9].

5.2.1   Biochemical Parameters as Predictors of Other Biochemical and Clinical Behaviors

The modulation of clinical and biochemical behaviors from hormones and other parameters in an independent manner from other variables has been proposed but poorly demonstrated. These analyses yielded some findings that were distinct from those reported in the literature and are reported in Table 5.3.

Table 5.3 Clinical or biochemical parameters as independent predictors of other parameters (multivariate linear regression analysis)

Parameter	p of the influence of the modifiable variables	Level of influence by the modifiable variables*	Parameters with significant correlations (and p-value)	Equation for the estimation of the parameter level in male athletes
Total testosterone (ng/dL)	0.0415	8.4%	1. Fat mass (inverse) ($p = 0.0415$)	Testosterone (ng/dL) = 631.77–10.29 × (fat mass)
POMS anger subscore	0.0006	34.7%	1. Estradiol (inverse) ($p = 0.008$). 2. Presence of OTS (direct) ($p = 0.003$)	POMS anger subscore = 25.43–0.24 × (estradiol) − 0.24 (T:E ratio) + 6.97 (if OTS)
Measured-to-predicted BMR (%)	0.026	10.9%	1. T:E ratio (direct) ($p = 0.026$)	BMR ratio (%) = 100.8 + 0.35 (T:E ratio)
Fat oxidation (% of total BMR)	<0.0001 (together with extra-activities)	58.8%	1. Body water (%) (direct) ($p = 0.001$)	Fat oxidation = −66.96 + 2.30x (body water) + 0.51 × (T/E ratio) – 4.99x (extra activities)
Chest-to-waist circumference ratio	0.0003	37.8%	1. Visceral fat (inverse) ($p < 0.0001$) 2. T:E ratio (inverse) ($p = 0.038$)	Ratio = 1.362– 0.012 × (visceral fat) – 0.02 × (T:E ratio)

Although fat mass has been reported to be an independent suppressor of the GH response [22], this was not confirmed, since body fat was not relevant for GH responses, at least in male athletes. Contrariwise, despite the lack of negative effect of body fat on GH release, body fat in males is able to reduce testosterone levels irrespective of other factors, independently from increase of aromatase activity, since T:E ratio was not reduced by increased body fat.

The T:E ratio positively predicted the measured-to-predicted BMR ratio, whereas neither testosterone nor estradiol has the same effect. This ratio may also predict the chest-to-waist circumference ratio, which is the measure that quantifies the "V shape" of the torso. These findings underscore the importance of evaluating the ratios of different hormones for the prediction of metabolic outcomes.

Although an increase in estradiol without a concurrent analysis of testosterone does not necessarily suggest either a beneficial or a harmful outcomes [23–25], in this study, testosterone failed to predict any parameter, including body metabolism or composition characteristics, unless if analyzed concurrently with estradiol. These findings are consistent with previous studies suggesting that a simultaneous increase of testosterone and estradiol has synergistic positive effects, particularly on metabolic parameters [23–27]. Conversely, fat mass reduced testosterone levels, which can be justified by exposure of testosterone to a more intense aromatase enzyme under higher body fat [25, 26].

While estradiol alone did not predict characteristics of body metabolism and composition parameters, it has been shown to reduce anger levels [9]. Although estradiol receptors are widely distributed in the brain [27, 28], their effects on moods in males are still unclear.

In terms of non-biochemical nonhormonal parameters, body water content (in percentage of total body weight) was the only independent predictor of fat oxidation. This finding supports the premise that good hydration status, particularly when properly located within the cells, is key for an adequate occurrence of fat oxidation, as water is a major substract of the fat oxidation pathway [29–31]. Expectedly, dehydration, even when mild, was found to slightly impair fat oxidation.

The graphs demonstrating the most important predictors of behaviors are show in Fig. 5.3.

5.2.2   Linear Correlations

Whereas evaluation for predictions allows the identification of possible causal relationships, linear correlations demonstrate associations, but the determination of causality is unfeasible. However, interesting insights from the identification of linear correlations can be unveiled and different from those uncovered in the analyses of predictions. To avoid random associations, only those that demonstrated moderate to strong correlations ($r > 0.4$; $p < 0.01$) should be considered. Figure 5.4 summarized the strongest correlations that have been identified.

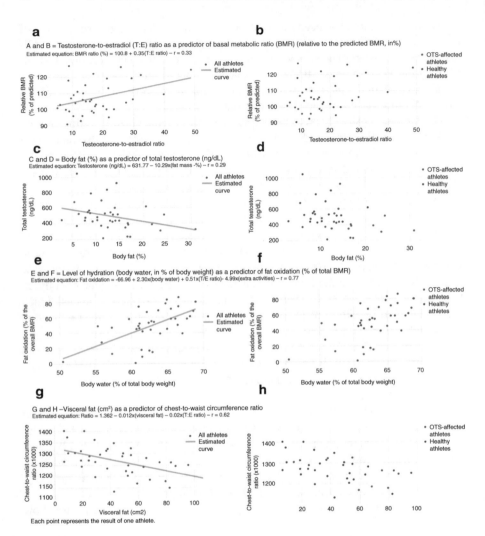

Fig. 5.3 Biological predictors of other clinical, metabolic, and biochemical parameters

Among hormonal responses to stimulation, strict intercorrelations ($r > 0.9$) were found between GH, prolactin, and cortisol for both early and late responses to stimulation tests (Fig. 5.5). Therefore, the levels of responsiveness of the corticotropic, somatotropic, and lactotrophic axes are indistinguishable between them, as these hormones responded equally. The concurrent responses indicate a common, ubiquitous, enhanced, and diffuse hypothalamic responsiveness in athletes, which is detectable when these are compared to sedentary controls. Although ACTH responses to the stimulation test were not strictly correlated with other hormones, its short half-life precluded from drawing conclusions about its correlation with other hormones.

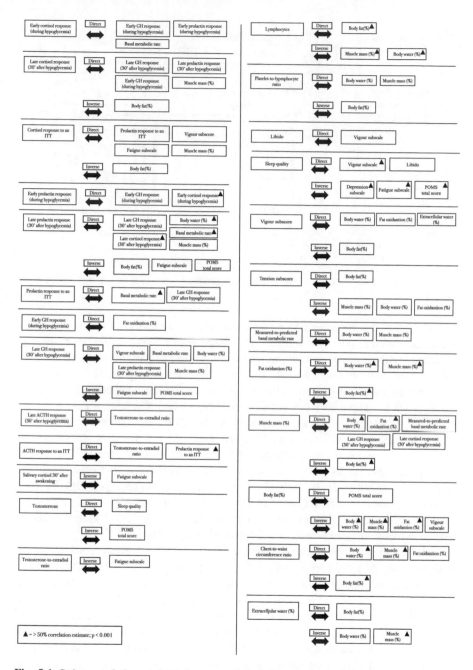

Fig. 5.4 Strict correlations (>0.40) between clinical, hormonal, psychological, and metabolic parameters

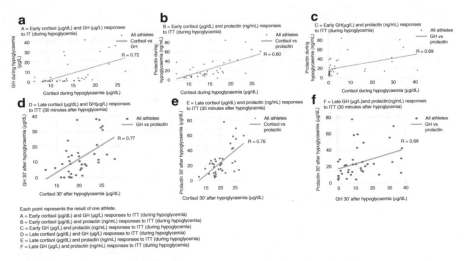

Fig. 5.5 Correlations between cortisol, GH, and prolactin responses to an insulin tolerance test (ITT)

Later hormonal responses to a stressor (in the case of the ITT, late responses were considered as being 30 minutes after the peak of hypoglycemia) were directly correlated with energy levels (higher vigor and lower fatigue), indicating a possible role of sustained hormonal release in response to stress to prevent exhaustion in chronical exposure to stressful circumstances. Prolonged hormonal responses to stimulations have also been correlated with increased muscle mass, lower body fat, and better hydration; however, as mentioned, the distinction between the specific effects of each hormonal response (cortisol, GH, and prolactin) is unfeasible. As extensively described in the literature, the acute biological effects of cortisol and GH are different and in many times opposite from their chronic effects. While the acute release of both GH and cortisol promotes fat oxidation and lipolysis (Fig. 5.6), chronic hypercortisolism may lead to the accumulation of visceral and central fat, even when it is mild, whereas chronic overexposure to GH seems to nullify its fat oxidative function, as observed in patients with acromegaly.

Collectively, the autonomous hormonal conditioning processes, which occur independently of secondary signaling from other systems, such as cardiovascular, musculoskeletal, and autonomic, which has been discovered to occur in athletes, may explain some of the previously unexplained health benefits associated with sports. The concept of hormonal conditioning was based on the identification of enhanced GH, prolactin, and cortisol responses to exercise-independent stimulation tests, when compared to sex-, age-, and BMI-matched healthy sedentary controls, which remained after adjustment for body composition. The optimization of the hormonal behavior may be one of the underlying reasons for decreased extracellular water, decreased anger, fatigue, depression, confusion mood states, and indirect account for reduced fat, increased muscle, and better hydration [32]. Indeed, under OTS, in which the optimization of the hormonal responses is lost, resembling physically inactive subjects [1, 2], the concurrent benefits in response to intense exercise have also become absent.

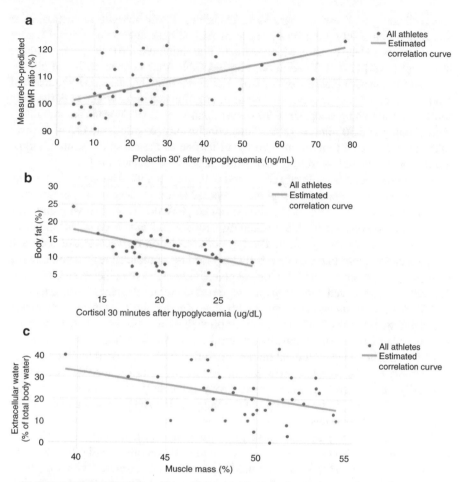

(A) Prolactin 30min after hypoglycemia in response to an insulin tolerance test (ITT) (ng/mL) and relative basal metabolic ratio (% of predicted) (R = 0.71).
(B) Cortisol 30min after hypoglycemia in response to an insulin tolerance test (ITT) (μg/dL) and body fat (%) (R = 0.74).
(C) Muscle mass (%) and extracellular water (% of body water) (R = 0.72).

Fig. 5.6 Other linear correlations

Salivary cortisol collected at 30 minutes after awakening reflects the cortisol response to awakening, which occurs between 15 and 45 minutes after, which is expected to increase between 30% and 70%, is reflected by the cortisol awakening response (CAR), which is the difference (delta) between cortisol at the exact moment of awakening and 30 minutes after. The inverse correlation observed between these parameters and levels of fatigue supported the hypotheses that cortisol response to awakening predicts energy levels, whereas lower cortisol at 30 minutes after awakening and CAR are markers of fatigue, which has been previously demonstrated by different studies [33–40]. Nonetheless, both markers failed to demonstrate the ability to predictor of fatigue in the absence of other factors, when analyzed using multivariate regression. Therefore, lower increase in cortisol after awakening is unlikely the primary cause of fatigue but rather an additional consequence of poor sleep quality, which leads to both impaired CAR [12] and decrease in energy level [41–43]. In addition, blunted CAR after poor sleep is not always resulted from lack of increase of cortisol after awakening, but also a consequence of already increased cortisol levels at the moment of awakening, which is commonly found in unrefreshing sleep, since the determination of CAR depends on both levels at awakening and 30 minutes after. Thus, conclusions regarding the use of cortisol markers and adrenal function for the identification of the cause of chronic fatigue are inappropriate, unless true adrenal insufficiency is suspected [44].

Testosterone level was directly correlated with sleep quality. Although this does not indicate a causal relationship, it supports the possible role of sleep quality in testosterone production, once testosterone physiological peak occurs in the early morning hours, simultaneously with the awakening process and which could be therefore affected by sleep quality [45–47]. The type of association between testosterone and sleep quality is more likely to be a bad sleep leading to reduced testosterone, rather than the opposite, since there is biological plausibility in this hypothesis, and testosterone, although correlated with sleep quality, failed to predict sleeping characteristics.

While testosterone was associated with sleep quality, the T:E ratio was inversely correlated with levels of fatigue, which implies that the ratio was a better indicator of energy level than testosterone or estradiol alone. While estradiol alone was not linked to fatigue, its pathological increase resulted from an exacerbation of aromatase activity may be correlated with increased fatigue, in accordance with the literature [13]. In this case, reduced testosterone secondary to enhanced aromatase activity did not seem to be the origin of the fatigue, since testosterone alone was not connected to energy levels.

In relation to sleep patterns, unlike previous studies [42, 48, 49], in which sleep duration predicted overall clinical and psychological outcomes, in the EROS study, sleep duration in both OTS-affected and healthy was not correlated with any physical, clinical, biochemical, or psychological improvement. Oppositely, sleep quality, even simply measured by a self-report scale, was strongly correlated with better psychological outcomes, including lower depression and fatigue, and higher vigor, indicating that sleep quality led to a global improvement in mood, similarly to what has been unanimously reported [42, 48, 50–52]. We speculate that sleep duration did not demonstrate to be a major factor of mood states or any other characteristic in athletes because whenever sleep quality is of high quality, its duration tends to play a less important role, unless duration is lower than 5 to 6 hours [53, 54].

Hydration and immunity have shown interesting strong correlations, including inverse correlation between body water (% of total body weight) and lymphocyte count and direct correlation between body water and platelet-to-lymphocyte ratio. Although benefits of hydration have been extensively reported in [55–62], its correlations with immunologic system lack [60–62]. Health outcomes related to absolute and subpopulations of lymphocytes have been assessed in athletes and those at high cardiovascular risk [63–68], while correlations between lymphocytes and hydration status have only been described as indirect effects, and findings were inconsistent.

Platelet-to-lymphocyte ratio has been proposed as an inflammatory marker of cardiovascular risk, acute pancreatitis, and sarcopenia [69–73], with some speculations alleging that higher ratio could indicate better hydration status, in a similar manner to the direct linear correlation that has been detected in the EROS study. This perhaps occurs due to a direct correlation between platelet counting and total body water [58, 59, 74], although these associations remain controversial [57–61].A group of associations that have not been considered but have been detected at strong levels was between psychological states and body composition and metabolism. Vigor has been directly correlated with better hydration, higher fat oxidation, less body fat, and increased libido. Since vigor has also been correlated with enhanced hormonal responses to stimulations, as well as better sleep quality [9], which have both in turn also been correlated with improved body metabolism and composition, it has been hypothesized that vigor may be an additional indirect consequence of hormonal responses and sleep quality, similarly to the correlations observed for body composition and metabolism, despite of any established causal relationship.

Conversely, levels of tension have demonstrated opposite correlations than those found for vigor for body composition and metabolism. Correspondingly, tension is likely the only mood correlated with disrupted hormonal responses [43]. Other mood states have not been significantly correlated with body metabolism or composition

While oxidative stress and inflammation are triggers for muscle hypertrophy, chronic stress mediated by the HPA axis leads to the opposite direction [54–78], perhaps due to catabolic effects of exposure to glucocorticoids. Indeed, alterations of the HPA axis seem to disrupt the metabolism of the muscle tissue, toward excessive degradation, eventually leading to muscle mass loss [79–82]. Oppositely, impaired HPA axis leads to independent body fat gain [57], since multiple mechanisms mediated by the HPA axis induce hypertrophy of adipocytes and differentiation of pre-adipocytes and induce a paradoxical pro-inflammatory status including infiltration of macrophages into the adipose tissue, which occurs irrespective of caloric balance, proportion of macronutrient intake, and sleeping patterns. The most probable underlying mechanism that justifies these biological effects of abnormal HPA axis functioning is a major post-receptor conversion from cortisone to cortisol by enhanced activity of 11beta- hydroxysteroid dehydrogenase type 1(11beta-HSD1) [83, 84]. In addition, disruption of the muscle metabolism induced by altered cortisol regulation has also demonstrated direct effects on the metabolism and accumulation of fat [85]. Since the HPA axis is chronically and more severely enhanced by high tension levels, from this perspective, one would expect to find negative correlations between tension levels and body composition. Indeed, besides impaired hydration and far oxidation, higher tension may have indirect effects on muscle catabolism, and increased body fat may through chronic overstimulation of the HPA axis [75–78], besides the direct associations between tension, depression, and anger with increased body fat has [79]. Furthermore, changes in mood states can hypothetically be indirect early manifestations of dehydration.

5.2.3   Hydration as a Key Characteristic for Athletes

The level of hydration is a key characteristic for the overall health, since water is a major substract for multiple reactions of metabolism and thermoregulation [31]. The level of hydration, which has been indirectly measured by the % of body water in relation to total body weight, has been associated with multiple parameters [14, 16, 28], including the predicted-to-measured BMR ratio and fat oxidation, as well as other characteristics of body metabolism.

These correlations seem reasonable since water has been shown to enhance fat oxidation [3, 5, 9, 86–89] and overall metabolic rate [3, 69–72], which can considered as a potential thermogenic [3, 5, 7, 90–92]. A minimum intracellular water necessary for fat oxidation is conceivable in the context of a direct correlation between hydration and fat oxidation [7, 89].

Conversely, the amount of body fat was inversely correlated with fat oxidation, which is expected, once more body fat is a natural consequence of less fat utilization as source of energy, even under glycogen- and glucose-depleted environments, as observed by some studies [93, 94]. In the circumstance of higher body fat under catabolic and glycogen-depleted states may be correlated with proteolysis as the preferred source of energy. Despite the previous reports on the correlation between body fat and fat oxidation, it is still unclear whether a larger fat mass is a result from reduced fat oxidation, or if a greater fat mass impaired fat oxidation through enhanced non-classical inflammatory responses in enlarged, although bidirectional path as a sort of vicious cycle is more likely present. Finally, since dehydration probably leads to lower fat oxidation, body water was expectedly found to be inversely correlated with body fat and chest-to-waist circumference ratio.

Similar to the correlation between fat oxidation and body water, the measured-to-predicted BMR ratio was positively correlated with hydration status and muscle mass, which reinforces the extensive descriptions on the literature that body water content, hydration status, and muscle mass are the major components of the metabolic rate [87, 89–92, 95–97] (Fig. 5.7). Moreover, hydration has been correlated with vigor levels and other mood states at lesser extent, which is supported by the literature [98].

In regard to the location of the water – whether intra- or extracellular – extracellular water accounts for approximately 40% of total body water [99], consistent with the observed in the EROS study for sedentary controls. However, healthy athletes presented lower extracellular water content – of approximately 20% – and consequently higher intracellular content (approx.. 80%), possibly as a mechanism of facilitation for the optimization of overall metabolic pathways, which occurs intracellularly, eventually resulting in optimized metabolism, in the collective sense of the term "metabolism" [100]. Moreover, fat mass has been inversely correlated with relative and absolute intracellular water and consequently lower body water. Some authors speculate that this inverse association is possibly due to the fact that hypertrophic fat cells tend to become more hydrophobic and to consequently contain less water. However, this is still a hypothesis to be further demonstrated.

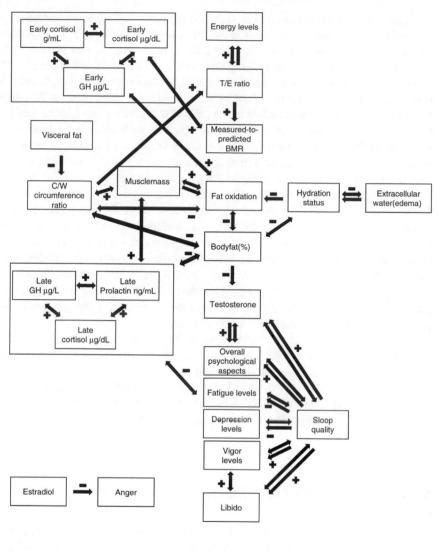

Fig. 5.7 Predictions and correlations between hormonal, psychological, energy, and sleeping patterns

In the top of all other correlations between hydration status, water location, and fat metabolism, an inverse correlation has been detected between hydration status and extracellular water. Considering that extracellular water is that located in the interstice, clinically and radiologically demonstrated as edema, also termed as the "third space," this association means that better hydration could be correlated with less edema, indicating that the amount of water consumed could potentially redirect the placement of water toward inside the cells (the preferred space to be located), rather than interstitially. However, external factors such as disturbed hormonal secretion, excessive sodium intake, and hypercaloric diets may lead to accumulation of extracellular water irrespective of hydration status, resulted from shifts in water compartment redistribution, leading to relative dehydration, i.e., cellular dehydration without low water content [99, 101]. Indeed, besides the dehydration as a result from redistribution of water, overall low water content could also mechanistically induce further dehydration by further enhanced accumulation of extracellular water, as a mechanism of protection from water wasting through vasopressin (ADH) regulation and renin-angiotensin-aldosterone system (RAAS) [102]. In addition, complex interactions between water content, RAAS, HPA axis, direct aldosterone actions and ADH metabolism, and their relationship with chronic stress have been reported, toward a hyperaldosteronism hyperreninemic state with enhanced ADH expression and abnormal HPA regulation [98, 102].

The major influences that have driven water destination were the metabolic environment, sleep quality (worse sleep leads to worse hydration and consequent increase of edema), amount of muscle mass, eating patterns, and mood states. Muscle mass was positively correlated with body water, which suggest that muscle tissue may play an indirect protective role in the prevention of edema, while body fat may have had the opposite effect, in accordance to what has been reported in the literature [79, 81].

5.2.4   Implications of the Findings

The use of additional statistical techniques in the context of more than 100 parameters, 50 participants, and substantial differences, which allowed the statistical power for the performance of these statistical tools, enabled the unprecedented identification of independent predictors and linear correlations between clinical, metabolic, biochemical, and other parameters, which has helped improve the understanding of hormonal and metabolic behaviors and their multiple interactions and influences in athletes.

The summary of the findings and their possible implications is presented in Table 5.4. The fact that hydration status, and, to a lesser extent, muscle mass, were the two major determinants of metabolic rate and fat oxidation [9, 93–99], support the importance of an adequate water intake and to maintain and continuously attempt to build lean muscle for an adequate metabolism and fat oxidation. The lack of body fat effect on GH release, a particular finding in athletes, attenuates the

Table 5.4 Most remarkable findings of the EROS-CORRELATIONS study

Parameter	Markers	Potential implication(s)
Testosterone, estradiol, and T:E ratio		
Total testosterone	1. Decreased body fat (P) 2. Better sleep quality (C)	1. Testosterone is blunted by body fat 2. Better sleep quality may boost testosterone production
Estradiol	1. Lower anger levels (P)	1. Estradiol actions in the male brain improve anger levels
Testosterone-to-estradiol ratio	1. Increased measured-to-predicted basal metabolic rate (P) 2. Increased chest-to-waist circumference ratio (P) 3. Lower fatigue levels (C)	1. The ratio between testosterone and estradiol is more important than testosterone or estradiol alone for body metabolism and composition
Hormonal functional tests		
GH, prolactin, and cortisol responses to an insulin tolerance test	1. Positive (direct) inter-correlations between GH, prolactin, and cortisol in early responses (C) 2. Positive (direct) inter-correlations between GH, prolactin, and cortisol in late responses (C) 3. Lower body fat (C) 4. Higher fat oxidation (C) 5. Higher muscle mass (C) 6. Better hydration (C) 7. Lower fatigue levels (C)	1. Hypothalamic responsiveness to stimulations does not discriminate between different axes 2. Although causality is not confirmed, better hormonal responses are at least linked to more energy and to better body composition
Social and psychological aspects		
Sleep quality	1. Improved overall mood states (C) 2. Lower depression levels (C) 3. Less fatigue levels (C) 4. Higher vigor levels (C)	1. Sleep quality may be more important than hormonal levels or eating patterns for the psychological status of the athletes
Libido	1. Higher vigor levels (C)	
Vigor	1. Lower body fat (C) 2. Higher fat oxidation (C) 3. Better hydration (C) 4. Higher extracellular water (C)	1. Vigor is an indirect marker of less body fat, better fat oxidation, and lower edema
Tension	1. Higher body fat (C) 2. Lower fat oxidation (C) 3. Worse hydration (C) 4. Lower muscle mass (C)	1. Tension is an indirect marker of lower muscle mass, increase of body fat, impaired fat oxidation, and less hydration

(continued)

Table 5.4 (continued)

Parameter	Markers	Potential implication(s)
Body metabolism and composition		
Measured-to-predicted basal metabolic ratio	1. Higher testosterone-to-estradiol ratio (P) 2. Better hydration (P) 3. Higher muscle mass (C)	1. The balance between testosterone and estradiol, more than any hormone alone, is the major predictor of metabolic rate in male athletes 2. Together with the T:E ratio, body water and muscle mass are the two major contributors of the metabolic rate, which means that a minimum content of intracellular water is necessary for a proper metabolism
Fat oxidation	1. Better hydration (C) 2. Higher muscle mass (C) 3. Lower body fat (C)	1. Body water and muscle mass play the most important roles for fat oxidation, the first as part of the pathway for fat oxidation, and the second as a possible signaler for the selective fat catabolism, over protein catabolism 2. Body fat and fat oxidation are inversely correlated; however, whether fat-induced inflammation leads to reduced fat oxidation, or higher body fat is a result from reduced fat oxidation is unknown
Chest-to-waist circumference ratio	1. Higher testosterone-to-estradiol ratio (P) 2. Lower visceral fat (P) 3. Higher muscle mass (C) 4. Higher fat oxidation (C) 5. Better hydration (C) 6. Lower body fat (C)	1. Similarly to other metabolic parameters, the T:E ratio is the most important direct predictor of the W:C ratio, leading to the popular "V shape," highly correlated with an androgen phenotype 2. Once body water is the intracellular water, mostly located within myocytes, rather than adipocytes, this contributes for a higher W/C ratio 3. Muscle mass and body fat are expectedly directly and inversely correlated with W/C ratio, respectively
Muscle mass	1. Late GH response to stimulation (C) 2. Late cortisol response to stimulation (C) 3. Better hydration (C) 4. Higher fat oxidation (C) 5. Lower body fat (C)	1. Although the muscle mass is not the lean mass, i.e., the water within muscles are not accounted, the presence of body water helps provide a muscle anabolic environment and predicts fat oxidation 2. Late hormonal responses, although correlated with muscle mass, are probably two consequences of a same common factor
Fat mass	1. Improved overall mood states (C) 2. Higher vigor levels (C) 3. Decreased hydration (C) 4. Lower muscle mass (C) 5. Decreased fat oxidation (C)	1. Worse psychological moods may be indicators of less healthier environment that naturally tends to save fat storage and catabolize muscle mass 2. All correlated body composition parameters are accordingly

Table 5.4 (continued)

Parameter	Markers	Potential implication(s)
Extracellular water (= edema)	1. Worse hydration (C) 2. Lower muscle mass (C) 3. Increased fat mass (C)	1. The more proper hydration, the less edema; however, what determines the destination of the ingested water is the metabolic environment, not the amount of water intake 2. Fat mass, likely through inflammatory processes, may induce edema, although we did not find prediction relationship

deleterious effects of body fat in physically active subjects, which may explain why some studies showed that some of the multiple harms in obesity may be attenuated with exercise. Conversely, the suppressing effect of body fat on testosterone is maintained regardless of sports activity levels, suggesting that athletes with excessive body fat may have some of the benefits from exercise nullified, which is an additional reason for athletes to maintain a leaner shape.

Since the hormonal responses to an ITT were strictly correlated, *i.e.*, the extent of the increases of GH, cortisol, ACTH, and prolactin was similar within each athlete, the level of hypothalamic responsiveness to stimulation seemed to be diffuse, rather than specific for certain axes. Energy levels were strongly correlated with hormonal status, including prolonged optimization of hormonal responses and better cortisol physiological increase after awakening, although a causal relationship has not been determined. In addition to energy levels, better hormonal responses have been correlated with better body composition, of which causal relationship remains uncertain.

Sleep quality seemed to be the most important determinant of mood states, rather than hormonal levels or eating patterns, which has been demonstrated to play an essential role on overall cognitive and psychological functions. Hydration has now demonstrated to be even more important, particularly for athletes, which should not only be addressed with sufficient water intake but also with healthy eating and sleeping habits. Since better hydration was associated with less edema, which is regulated by other external factors than the amount of water intake.

5.2.5 Summary of the Findings

In summary, testosterone was predicted by fat mass, estradiol predicted anger, and the T:E ratio predicted the measured-to-predicted BMR ratio and chest-to-waist circumference, while hydration status predicted fat oxidation. Early and late somatotropic, corticotropic, and lactotropic responses were strong and strongly correlated, showing a diffuse hypothalamicm rather than axis-specific response to stimulations. Late hormonal responses to stimulations, increased cortisol after awakening, and the T:E ratio were correlated with energy levels. Sleep quality was the major factor correlated with psychological states, while fat oxidation, hydration, muscle mass, and body fat were highly intercorrelated. Water content in the third space has been inversely correlated with hydration and muscle mass and directly correlated with fat mass. The most remarkable correlations and predictors of behaviors are illustrated in Fig. 5.7.

5.3   Predictors of Clinical and Biochemical Behaviors

The EROS study identified more than 50 novel potential biomarkers for OTS, in addition to those recently detected [1–9, 32–36]. Despite the new and the learnings from the EROS-CORRELATIONS arm, in which clinical, biochemical, and metabolic parameters disclosed multiple sorts of interactions, and the EROS-DISRUPTORS arm, which demonstrated novel triggers of OTS, as well as dysfunctions caused by OTS per se, the identification of how modifiable habits can inflect the behaviors of basal and stimulated hormonal, biochemical, muscular, inflammatory, immunologic psychological, body metabolism, and composition parameters in athletes, with and without the presence of OTS, has been performed in the EROS-DISRUPTORS arm, using complementing statistical tools.

For the occurrence of multiple beneficial adaptations that have been detected in athletes, a balance between training, resting, and nutrition is likely crucial, whereas slight changes in any of these patterns should be able to modulate the extension of these benefits. In addition, it has been hypothesized that there would be specific cutoffs from which these habit patterns start to modify clinical, biochemical, and metabolic behaviors in a dysfunction way, eventually leading to OTS.

From the identification of eating, social, and sleep patterns as independent predictors of beneficial or harmful outcomes, one could make more specific recommendations for eating, sleeping, and social patterns, aiming a continuous improvement process in athletes in relation to performance and overall well-being. Expected and observed predictions of each modifiable habit are described in Fig. 5.8, while the relation of habit patterns as independent factors of hormonal responses to stimulations, basal biochemical profile, mood states, body composition, and metabolism is described in Tables 5.5, 5.6, 5.7 and 5.8, respectively.

Other modifiable factors, such as the use of drugs, hormones, smoking, drinking alcohol, and other social behaviors, should also be assessed, since all these characteristics modify overall responses in sports. However, for research purposes, these factors may be employed as exclusion criteria, since it is assumed that athletes were fully aware of the need to avoid drugs, anabolic steroids (unless clinically needed), smoking, alcohol intake (except during special social events), sleep deprivation, and other characteristics of an excessive hedonic living.

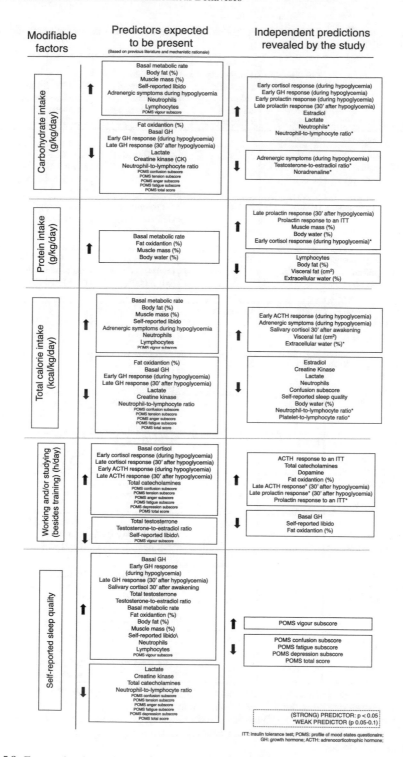

Fig. 5.8 Expected and actual predictions of each modifiable factor

Table 5.5 Modifiable patterns as independent predictors of hormonal responses to stimulations (multivariate linear regression analysis)

Parameter	p of the influence of the modifiable variables	Level of influence by the modifiable variables (adjusted R-Square)	Modifiable variables with significant correlations (and p-value)	Equation for the estimation of the parameter level in male athletes
Hormonal responses to stimulations				
Early cortisol response to an ITT (during hypoglycemia) (µg/dL)	0.029	23.8%	1. CHO intake (direct) ($p = 0.025$)	Cortisol (µg/dL) = 8.33 + 0.5x (CHO intake) + 1.36x (protein intake)
Late cortisol response (30'after hypoglycemia) (µg/dL)	0.0005	26.1%	1. Presence of OTS (inverse) ($p = 0.0005$)	Cortisol (µg/dL) = 17.86–3.81 (if OTS)
Early ACTH response to an ITT (during hypoglycemia) (pg/mL)	0.012	17.5%	1. Calorie intake (direct) ($p = 0.0035$)	ACTH = −67.74 + 2.83x (calorie intake) + 0.92x (Total POMS)
Late ACTH response to an ITT (30'after hypoglycemia) (pg/mL)	0.007	19.9%	1.Presence of OTS (inverse) ($p = 0.002$)	–
Cortisol response to an ITT (µg/dL)	0.004	22.0%	1. Presence of OTS (inverse) ($p = 0.0017$)	–
Basal GH (µg/L)	0.033	9.3%	1. Extra-activities (inverse) ($p = 0.033$)	GH ((µg/L) = 0.97–0.08x (extra activities)
Early GH response to an ITT (during hypoglycemia) (µg/L)	0.017	12.0%	1. CHO intake (direct) ($p = 0.017$)	GH ((µg/L) = −0.78 + 1.29x (CHO intake)
Late GH response (30'after hypoglycemia) (µg/L)	0.0012	23.0%	1. Presence of OTS (inverse) ($p = 0.0012$)	–
Early prolactin response to an ITT (during hypoglycemia) (ng/mL)	0.009	15.0%	1. CHO intake (direct) ($p = 0.009$)	Prolactin (ng/mL) = 8.36 + 2.43x (CHO intake)
Late prolactin response (30'after hypoglycemia) (ng/mL)	0.0002	37.8%	1.Protein intake (direct) ($p = 0.0004$) 2. CHO intake (direct) ($p = 0.038$)	Prolactin (ng/mL) = −28,0.49 + 1.60x (CHO intake) + 10.64x (protein intake) + 2.46x (extra activities)
Prolactin response to an ITT (ng/mL)	0.0133	17.0%	1. Protein intake (direct) ($p = 0.0036$)	Prolactin (ng/mL) = −356,25 + 108.6x (protein intake) + 30.57x (extra activities)

CHO Carbohydrate, *ITT* Insulin-tolerant test, *POMS* Profile of mood states, *BMR* Basal metabolic rate, *T/E* Testosterone-to-oestradiol, *OTS* Overtraining syndrome
Calorie intake = kcal/kg/day; CHO intake = g(CHO)/kg/day; protein intake = g(protein)/kg/day; extra activities = working and/or studying hours besides training; sleep quality = self-reported sleep quality (0 to 10)

Table 5.6 Modifiable patterns as independent predictors of basal hormones and biochemical parameters (multivariate linear regression analysis)

Parameter	p of the influence of the modifiable variables	Level of influence by the modifiable variables (adjusted R-Square)	Modifiable variables with significant correlations (and p-value)	Equation for the estimation of the parameter level in male athletes
Basal hormones				
Oestradiol (pg/mL)	0.008	20.3%	1. Calorie intake (inverse) ($p = 0.002$) 2. CHO intake (direct) ($p = 0.013$)	Oestradiol (pg/mL) = 50.28–0.68x (calorie intake) + 2.32x (CHO intake)
Testosterone-to-oestadiol ratio (T/E)	0.0007	30.7%	1. Presence of OTS (inverse) ($p = 0.0002$)	T/E = 14.1–0.86x (CHO intake) + 12.9 (in case of OTS)
Total nocturnal urinary catecholamines (mg/12 h)	0.0187	11.7%	1. Extra activities (direct) ($p = 0.0187$)	Total NUC = 49.5 + 20.6x (extra activities)
Dopamine (mg/12 h)	0.0136	13.1%	1. Extra activities (direct) ($p = 0.0136$)	Dopamine = 25.7 + 20.1x (extra activities)
Basal biochemistry				
Creatine kinase (CK)	0.02	11.3%	1. Calorie intake (inverse) ($p = 0.02$)	CK = 1488–20.5x (calorie intake)
Lactate	0.0035	22.9%	1. Calorie intake (inverse) ($p = 0.001$)	Lactate = 1.62–0.02x (calorie intake)
Neutrophils (/mm^3)	0.045	13.8%	1. Calorie intake (inverse) ($p = 0.044$) 2. Presence of OTS (inverse) ($p = 0.015$)	Neutrophils = 4210–60.7x (calorie intake) + 154.4x (CHO intake) – 1724 (if OTS)
Lymphocytes (/mm^3)	0.025	10.8%	1. Protein intake (inverse) ($p = 0.025$)	Lymphocytes = 2767 – 207x (protein intake)

CHO Carbohydrate, *T/E* Testosterone-to-oestradiol, *OTS* Overtraining syndrome
Calorie intake = kcal/kg/day; CHO intake = g(CHO)/kg/day; protein intake = g(protein)/kg/day; extra activities = working and/or studying hours besides training; sleep quality = self-reported sleep quality (0 to 10)

Table 5.7 Modifiable patterns as independent predictors of moods and feelings (multivariate linear regression analysis)

Parameter	p of the influence of the modifiable variables	Level of influence by the modifiable variables (Adjusted R-Square)	Modifiable variables with significant correlations (and p-value)	Equation for the estimation of the parameter level in male athletes
Psychology				
POMS confusion subscale	0.0002	33.7%	1. Sleep quality (inverse) ($p = 0.002$) 2. Calorie intake (inverse) ($p = 0.019$)	POMS confusion subscale = 15.25–0.92x (sleep quality) – 0.1x (calorie intake)
POMS depression subscale	0.0001	30.8%	1. Sleep quality (inverse) ($p = 0.0001$)	POMS depression subscale = 17.22–1.66x (sleep quality)
POMS vigor subscale	<0.0001	83.6%	1. Sleep quality (direct) ($p = 0.0002$) 2. Presence of OTS (inverse) ($p < 0.0001$)	POMS vigor subscale = 3.7 + 1.15x (sleep quality) – 11.96 (if OTS)
POMS fatigue subscale	<0.0001	85.7%	1. Sleep quality (direct) ($p = 0.0059$) 2. Presence of OTS (direct) ($p < 0.0001$)	POMS fatigue subscale = 24.5– 0.9 x (sleep quality) + 15.3 (if OTS)
POMS tension subscale	<0.0001	42.8%	1. Presence of OTS (direct) ($p < 0.0001$)	
Adrenergic symptoms (0–10)	0.003	23.7%	1. Calorie intake (direct) ($p = 0.0008$) 2. CHO intake (inverse) ($p = 0.023$)	Symptoms = −0.09 + 0.16x (calorie intake) – 0.45x (CHO intake)
Libido (0–10)	0.018	11.9%	1. Extra activities (inverse) ($p = 0.018$)	Libido = 10.3–0.4x (extra activities)

CHO Carbohydrate, *POMS* Profile of mood states, *OTS* Overtraining syndrome
Calorie intake = kcal/kg/day; CHO intake = g(CHO)/kg/day; protein intake = g(protein)/kg/day; extra activities = working and/or studying hours besides training; sleep quality = self-reported sleep quality (0 to 10)

Table 5.8 Modifiable patterns as independent predictors of body metabolism and composition (multivariate linear regression analysis)

Parameter	p of the influence of the modifiable variables	Level of influence by the modifiable variables (adjusted R-Square)	Modifiable variables with significant correlations (and p-value)	Equation for the estimation of the parameter level in male athletes
Body metabolism and composition				
Fat oxidation (% of total BMR)	<0.0001 (together with body water and T/E ratio)	58.8%	1. Extra activities (inverse) ($p = 0.0001$)	Fat oxidation = −66.96 + 2.30x (body water) + 0.51x (T/E ratio)- 4.99x (extra activities)
Fat mass (%)	0.0001	31.0%	1. Protein intake (inverse) ($p = 0.0001$)	Fat mass = 20.35–3.1x (protein intake)
Muscle mass (%)	0.0006	33.7%	1. Protein intake (direct) ($p = 0.0135$) 2. Presence of OTS (inverse) ($p = 0.0282$)	Muscle mass = 47.84 + 1.42x (protein intake) – 3.47 (if OTS)
Body water (%)	<0.0001	50.5%	1. Protein intake (direct) ($p = 0.0061$) 2. Calorie intake (inverse) ($p = 0.021$) 3. Presence of OTS (inverse) ($p = 0.001$)	Body water = 60.75 + 1.69x (protein intake) – 0.12x (calorie intake) – 5.77 (if OTS)
Visceral fat (cm²)	0.0002	38.2%	1. Calorie intake (direct) ($p = 0.0076$) 2. Protein intake (inverse) ($p = 0.023$) 3. Presence of OTS (direct) ($p = 0.0026$)	Visceral fat = 47.4–11.9x (protein intake) + 1.3x (calorie intake) +45.1 (if OTS)

CHO Carbohydrate, *BMR* Basal metabolic rate, *OTS* Overtraining syndrome
Calorie intake = kcal/kg/day; CHO intake = g(CHO)/kg/day; protein intake = g(protein)/kg/day; extra activities = working and/or studying hours besides training; sleep quality = self-reported sleep quality (0 to 10)

5.3.1 Carbohydrate Intake

Carbohydrate intake had multiple effects on the behavior of hormones and other biochemical parameters. It has been shown to be an independent predictor of overall early hormone responses to an ITT, accounting for up to 24% of responses. It is presumed that early hormonal responses to an ITT can predict explosive and sudden responses in sports performance, which is particularly beneficial for explosive, stop-and-go, ball games, and short-intense exercises and sports. This could explained by the fact that improved responses require greater prompt availability of energy, and carbohydrate, even when storaged as glycogen, are the preferred and easy-to-recruit source of energy for short demands, which in turn may justify why sufficient carbohydrate intake, and consequent prompt availability has been demonstrated to be an independent predictor of early hormone responses to stimulations. Accordingly, carbohydrate deprivation may lead to decreased and delayed hormonal responses, which would indirectly impair performance, which has been indeed identified [1, 2].

Particularly, for GH release, in contrast to the suppressive effect of acute carbohydrate intake on GH release [5, 10, 11], chronic carbohydrate intake had a stimulating effect on the GH response, showing a dual effect of carbohydrate intake on GH release patterns.

Similar to the dual effects on GH release, carbohydrate intake which has also demonstrated an apparent dual effect on aromatase activity (i.e., conversion from testosterone to estradiol). While a very low carbohydrate intake may be related to a pathological increase in aromatase activity [4], corroborating previous similar findings [103–105], excessive carbohydrate intake may also increase aromatase activity, causing increased estradiol and a decreased testosterone-to-estradiol (T/E) ratio, as observed in our previous findings [4–6, 16, 17, 106]. This finding may justify the not fully elucidated finding of higher estradiol levels in obese males, which cannot be not fully explained by the hypertrophy of adipocytes [107–109]. Despite the protective role of overall caloric intake for elite athletes, excessive carbohydrate intake may become inflammatory [4, 103–106], once typical markers of unspecific subclinical metabolic inflammatory states have been correlated with carbohydrate overload, including increased aromatase activity, increased lactate without concurrent increase in CK (unrelated to muscle stimulation), and slight nonsignificant increase in neutrophils. Neutrophils are independently associated with inflammatory status and cardiovascular and neoplastic diseases, in the absence of clinical infections or the use of glucocorticoids [20, 21, 110].

In the particular case of elite athletes, despite some reports suggesting that lower carbohydrate intake does not impair performance, even for elite athletes [104], higher carbohydrate intake has been shown to have positive effects on hormonal profile, and potentially in performance, whereas carbohydrate intake below 5.0 g/kg/day for at least 3 months has been shown to induce harmful effects on hormonal responses and performance [4–7]. The limit above which carbohydrate intake would be considered excessive with consequent pathological increase in aromatase and

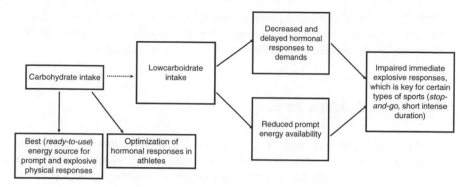

Fig. 5.9 Hypothesized mechanisms for the impaired performance observed in prolonged low carbohydrate

inflammatory activity is unlikely even at very high intake (at least until 10–12 g/kg/day of carbohydrate for elite athletes, no harms should be present). The underlying mechanisms proposed for potential harms of chronic low carbohydrate intake are summarized in Fig. 5.9.

5.3.2   Protein Intake

Protein intake was found to predict the most important parameters of body metabolism and composition in a positive and independent manner, including increased BMR, fat oxidation, muscle mass, and hydration while protects against overall and visceral fat accumulation, accounting for 30%–50% of the variation of body fat. Protein intake has been demonstrated to predict ($p = 0.029$) extracellular water negatively, i.e., ingestion of protein seems to protect from loss of intracellular water for the interstice, thereby preventing edema. All these findings point to the conclusion that protein is a major determinant of body characteristics [3, 5, 103, 107].

The fact that 88% of the athletes of the EROS study ingested whey protein at a daily basis may have contributed to the independent benefits found in the present study, since whey protein intake has been demonstrated to be independently associated with decreased body fat [111], reduced inflammatory parameters [112], and the prevention of fat weight gain [113].

Overall, higher protein intake had beneficial effects on metabolism and body composition in athletes. Previous concerns on kidney and liver safety of higher protein intake have been demonstrated to be unsubstantiated, whereas benefits of additional protein intake have not shown concurrent risks of kidney or liver dysfunctions [111–113]. The EROS study detected that protein intake should be at least 1.6 g/kg/day [3, 5, 6], which is highly consistent with the latest sports nutrition guideline for athletes [104] and previous researches [113].

Indeed, a higher ("unlimited") protein intake among male athletes would have a protective role in the body metabolism and composition, without a plateau or inverse effect at any point, at least when protein intake is up to 4.5 g/kg/day.

5.3.3   Overall Caloric Intake

Overall caloric intake, independent of the proportion between macronutrients, has demonstrated four major influences: positively predicts salivary cortisol 30 minutes after awakening, enhances the clearance of markers of muscle recovery (CK and lactate), prevents aberrant exacerbations of aromatase activity, and prevents a pathological increase of neutrophils in the absence of apparent infections.

These findings mean that higher caloric intake, regardless of the macronutrient content, may increase the level of alertness in the morning, the speed and quality of muscle recovery, may help maintain normal hormonal levels, and supports immune function.

Unlike the predictions for other outcomes, for which each macronutrient has specific roles, muscle recovery higher caloric intake seems to be more important than the intake of each macronutrient.

Despite the positive findings associated with overall caloric intake, this was detected as an independent and direct predictor of visceral fat (although not for total fat), and it also was a predictor of lower muscle mass, when not accompanied by increase of protein intake. Indeed, carbohydrate abuse is frequently associated with low and insufficient protein intake, leading to sarcopenic obesity [107]. Thus, for body composition, the source of calories is at least as important as the total caloric intake, once the effects of higher caloric intake may lead to opposite effects, i.e., higher caloric intake could either induce increase of muscle mass and decrease of body fat, or the opposite, depending on the percentage and absolute amount of protein intake.

In conclusion, increase of caloric intake in elite athletes improved the quality of muscle recovery, hormonal environment, and sports performance. Athletes should have a minimum of 35 kcal/kg/day [3, 5, 6] to achieve these benefits. As a change in the paradigm, any macronutrient (i.e., protein, carbohydrate, or fat) can be added to the diet, even if the amount exceeds the athletes daily caloric needs, without major risks, with likely more benefits and risks, even for carbohydrate in case of intense and arduous training programs.

5.3.4   Other Activities

For elite athletes, excessive concurrent physical and cognitive efforts may lead to harmful effects, although different from those related to insufficient caloric, protein, and carbohydrate intake. The number of hours of studying and/or working was an independent enhancer of ACTH response to stimulation, albeit without concomitant-enhanced cortisol release, as would be expected in response to

ACTH. The lack of cortisol response to enhance ACTH release can be hypothesized to be a sort of hyporesponsiveness of the adrenals to ACTH stimulation. Conversely, direct adrenal stimulation in the same participants did not disclose differences in cortisol responsiveness, irrespective of any characteristic, which does not support the hypothesis of changes in the sensitivity of the *fasciculata* layer of the adrenal cortex to ACTH.

Excessive mental activity seems to blunt basal GH levels, which suggests that more studying or working may lead to lower GH levels when these are not in release peak. However, unlike basal GH, cognitive demands have not been reflected in tGH responses to stimulations.

The duration of working and/or studying among elite athletes directly predicted urinary catecholamines. Nonetheless, despite the acute biological actions of catecholamines on fat oxidation and increase of metabolic rate, paradoxical reductions in fat oxidation and BMR have been correlated with increased cognitive activity.

Indeed, although catecholamines acutely increase fat oxidation, chronic exposure may have the opposite effect, in a similar manner that happens in hypercortisolism states. The *fight-or-flight* response, a prompt reaction to severe stressors, when present chronically, typically observed in psychological stress, can paradoxically lead to fat weight gain and decreased BMR [8], despite the similar increase of cortisol and catecholamines as seen in acute responses. Possibly, a decreased sensitivity of the fat tissue to catecholamines is in accordance with a lack of fat loss to be expected under excessive catecholamines [36, 37]. To reinforce this hypothesis, an unexpected lack of fat loss occurs in patients with pheochromocytoma (catecholamine-producing tumors), who are chronically exposed to higher catecholamine levels [8].

Considering all the findings, it is speculated that when physical and cognitive demands are concurrently present, fat oxidation and BMR get impaired, which is resulted from an environment exposure to multiple sources of chronic stress [6–14, 35]. The impaired metabolism associated with excessive cognitive and physical demands may be due to insufficient resting and recovery, as cognitive stress precludes appropriate physical recovery. Athletes should avoid excessive cognitive activities during periods of higher training demands and avoid intensification of trainings when under high cognitive effort.

5.3.5   Sleeping

While duration of sleep did not predict any marker or outcome, sleep quality was the most important predictor of psychological outcomes and the only modifiable factor that was able to modulate overall mood states.

Also, sleep quality was an independent and inverse predictor of total caloric intake, i.e., better sleep quality could be able to reduce overall caloric intake, irrespective of other factors, such as training characteristics. However, further increase of sleep quality does not lead to additional reduction and consequent insufficient caloric intake.

5.3.6 Summary of How Habit Patterns May Drive Clinical and Biochemical Behaviors

The current manuscript presented some remarkable and highly practical novel findings, which may contribute to the unclear correlations between behaviors, hormone profile, and metabolism, which are described in Fig. 5.10. Some of the most remarkable findings were:

(a) Carbohydrate intake predicts quick hormonal responses to stress and improves explosion responses during exercises, when above 5.0 g/kg/day, as carbohydrate predicted the early GH, cortisol, and prolactin responses (during hypoglycemia) to an insulin tolerance test (ITT). Opposite to the suppressive effect of carbohydrate intake on the acute GH release, higher carbohydrate intake stimulates chronic GH release.

(b) Together, carbohydrate and protein intake predicted the late prolactin response (30 minutes after hypoglycemia) to an ITT.

(c) Estradiol levels in male were directly influenced by carbohydrate intake whereas inversely predicted by overall caloric intake.

(d) Muscle recovery speed was directly predicted by overall calorie intake, regardless of the proportion of macronutrients.

(e) Protein intake protects against body fat accumulation, specifically the visceral fat, independently from caloric intake, energy expenditure, and metabolic rate, which is in accordance with previous studies. In addition, protein intake above 1.6 g/kg/day tends to provide further improvements in body composition, metabolism, and hydration. Increase of protein intake could be used as a strategy for prevention of fat gain under several different circumstances.

(f) Visceral fat was inversely predicted by protein intake and directly predicted overall caloric intake.

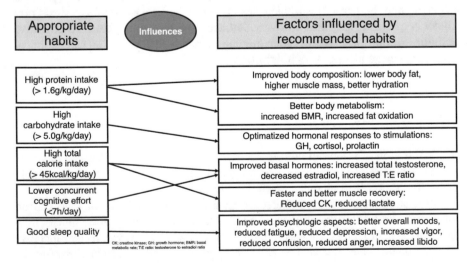

Fig. 5.10 Summary of the influences of modifiable habits on clinical and biochemical behaviors

1. Good sleep quality is the major determinant of mood states, which has been demonstrated to independent predict depression, confusion, anger, vigor, fatigue, and tension.

(g) Excessive concurrent physical and cognitive effort in athletes undergoing intense training independently decreases fat oxidation, increases muscle catabolism, and impairs libido. For those who practice intense physical activity for more than 300 minutes a week for at least four to five times a week, a concurrent routine of work or study for at least 8 hours may be harmful for several clinical, biochemical, hormonal, and psychological responses, regardless of the presence of OTS. It means that periods of more intense physical activity should be accompanied by reduction of working or studying hours, while periods that require more cognitive demands should not be associated with intense physical activity.

Healthy athletes that follow adequate sleeping, training, nutrition, and social patterns experiment multiple benefits resulted from the expected behaviors predicted by the modifiable habits above cited, as summarized in Table 5.9.

Table 5.9 Most remarkable findings of the EROS study in healthy athletes

Study/tests	Remarkable findings in healthy athletes
EROS-HPA axis	
Basal ACTH and cortisol and their response to an insulin tolerance test (ITT)	(1) Prompter cortisol response (compared to nonathletes and OTS-affected athletes) (2) Optimized cortisol response (compared to nonathletes and OTS-affected athletes)
Salivary cortisol rhythm (SCR)	(3) Higher salivary cortisol 30 minutes after awakening (compared to nonathletes and OTS-affected athletes)
EROS-STRESS	
GH response to an ITT	(4) Higher basal GH (compared to non-athletes and OTS-affected athletes) (5) Prompter GH response (compared to non-athletes and OTS-affected athletes) (6) Optimized GH response (compared to nonathletes and OTS-affected athletes)
Prolactin response to an ITT	(7) Prompter prolactin response (compared to nonathletes and OTS-affected athletes) (8) Optimized prolactin response (compared to nonathletes and OTS-affected athletes)
EROS-BASAL	
Hormonal markers	(9) Higher total testosterone (ng/dL) (compared to nonathletes and OTS-affected athletes) (10) Higher total catecholamines and noradrenaline (compared to nonathletes)

(continued)

Table 5.9 (continued)

Study/tests	Remarkable findings in healthy athletes
Biochemical markers	(11) Lower lactate (compared to nonathletes and OTS-affected athletes) (12) Lower neutrophils (compared to nonathletes and OTS-affected athletes) (13) Higher lymphocytes (compared to nonathletes and OTS-affected athletes)
Ratios	(13) Lower neutrophil-to-lymphocyte (compared to nonathletes and OTS-affected athletes)
EROS-PROFILE	
General patterns	(14) Better self-reported sleep quality (compared to nonathletes and OTS-affected athletes)
Psychological patterns	(15) Better overall moods, anger, confusion, vigor, depression, tension, and fatigue subscales (compared to nonathletes and OTS-affected athletes)
Body metabolism analysis	(16) Higher measured-to-predicted basal metabolic rate (BMR) (compared to nonathletes and OTS-affected athletes) (17) Higher percentage of fat burning compared to total BMR (compared to nonathletes and OTS-affected athletes)
Body composition	(18) Lower body fat percentage (compared to nonathletes and OTS-affected athletes) (19) Higher muscle mass weight (compared to nonathletes and OTS-affected athletes) (20) Higher body water percentage (compared to nonathletes and OTS-affected athletes) (21) Extracellular water compared to total BW (compared to nonathletes) (22) Lower visceral fat (compared to nonathletes and OTS-affected athletes)

5.4   Conclusion

Excessive training has become a minor factor for the development of OTS after the periodization of training sessions, whereas dietary and sleep patterns together and dietary characteristics alone were able to detect all cases of OTS, without the presence of training overload.

Hypothalamic-pituitary responses to stimulation were diffuse and indistinguishable between the different axes. A late hormonal response to stimulation, increased cortisol after awakening, and the T:E ratio were correlated with vigor and fatigue. The T:E ratio was also correlated with body metabolism and composition, testosterone was predicted by fat mass, and estradiol predicted anger. Hydration status was inversely correlated with edema, and inter-correlations were found among fat oxidation, hydration, and body fat.

With regard to modifiable habit patterns, carbohydrate intake seems to predict quick hormonal responses to stress and improves explosive responses during exercise, protein intake seems to improve body composition and metabolism, overall caloric intake, independent of the its source, has been demonstrated to predict muscle recovery, sleep quality seemed to improve mood states, and excessive concurrent cognitive effort in athletes participating in intense training may impair metabolism and libido. These results support the premise that eating, sleep, and social patterns affect metabolic, hormonal, and clinical behaviors in athletes and should be addressed to prevent dysfunctions.

Correlations and predictions between clinical, hormonal, biochemical markers, and their respective behaviors occur as a comprehensive web of influences through multiple and multidirectional chain reactions that work synergistically between them. This allows the hypothesis of the existence of several novel metabolic mechanisms of conditioning processes that are induced by exercise, From these findings, it can be deduced that sports performance results from a combination of hormonal and metabolic environment, energy, water availability, psychological, and muscular status. Of these novel endocrine and metabolic adaptations identified in athletes, at least 40–50% are nullified under OTS [5, 7, 9]. The multiple correlations, predictions, and interactions that have been revealed show that further studies should not evaluate hormonal, biochemical, metabolic, clinical, and psychological aspects separately from each other, as this is unlikely to provide answers to important questions.

References

1. Cadegiani FA, Kater CE. Hypothalamic-pituitary-adrenal (HPA) axis functioning in overtraining syndrome: findings from endocrine and metabolic responses on overtraining syndrome (EROS) - EROS-HPA axis. Sports Med Open Sports Med Open. 2017;3(1):45.
2. Cadegiani FA, Kater CE. Growth hormone (GH) and prolactin responses to a non-exercise stress test in athletes with overtraining syndrome: results from the endocrine and metabolic responses on overtraining syndrome (EROS) - EROS-STRESS. J Sci Med Sport. 2018;21(7):648–53.
3. Cadegiani FA, Kater CE. Body composition, metabolism, sleep, psychological and eating patterns of overtraining syndrome: results of the EROS study (EROS-PROFILE). J Sports Sci. 2018;36(16):1902–10.
4. Cadegiani FA, Kater CE. Basal hormones and biochemical markers as predictors of overtraining syndrome in male athletes: the EROS-BASAL study. J Athl Train. 2019; https://doi.org/10.4085/1062-6050-148-18.
5. Cadegiani FA, Kater CE. Novel insights of overtraining syndrome discovered from the EROS study. BMJ Open Sport Exerc Med. 2019;5(1):e000542. https://doi.org/10.1136/bmjsem-2019-000542.
6. Cadegiani FA, Kater CE, Gazola M. Clinical and biochemical characteristics of high-intensity functional training (HIFT) and overtraining syndrome: findings from the EROS study (the EROS-HIFT). J Sports Sci. 2019;20:1–12. https://doi.org/10.1080/02640414.2018.1555912.
7. Cadegiani FA, Kater CE. Novel causes and consequences of overtraining syndrome: the EROS-DISRUPTORS study. BMC Sports Sci Med Rehabil. 2019;11:21.

8. Cadegiani FA, Kater CE. Eating, sleep, and social patterns as independent predictors of clinical, metabolic, and biochemical behaviors among elite male athletes: the EROS-PREDICTORS study.

9. Cadegiani FA, Kater CE. Inter-correlations among clinical, metabolic, and biochemical parameters and their predictive value in healthy and overtrained male athletes: the EROS-CORRELATIONS study. Front Endocrinol (Lausanne). 2019;10:858.

10. Farhat K, Bodart G, Charlet-Renard C, et al. Growth hormone (GH) deficient mice with GHRH gene ablation are severely deficient in vaccine and immune responses against *Streptococcus pneumoniae*. Front Immunol. 2018;9:2175.

11. Bodart G, Farhat K, Renard-Charlet C, et al. The severe deficiency of the somatotrope GH-releasing hormone/growth hormone/insulin-like growth factor 1 axis of $Ghrh^{-/-}$ mice is associated with an important splenic atrophy and relative B lymphopenia. Front Endocrinol (Lausanne). 2018;9:296.

12. Wilhelm I, Born J, Kudielka BM, Schlotz W, Wust S. Is the cortisol awakening rise a response to awakening? Psychoneuroendocrinology. 2007;32:358–66. https://doi.org/10.1016/j.psyneuen.2007.01.008.

13. Meeusen R, Duclos M, Foster C et al. European College of Sport Science; American College of Sports Medicine. Prevention, diagnosis, and treatment of the overtraining syndrome: joint consensus statement of the European College of Sport Science and the American College of Sports Medicine. Med Sci Sports Exerc. 2013;45(1):186–205.

14. Penz M, Kirschbaum C, Buske-Kirschbaum A, Wekenborg MK, Miller R. Stressful life events predict one-year change of leukocyte composition in peripheral blood. Psychoneuroendocrinology. 2018;94:17–24.

15. Nederhof E, Zwerver J, Brink M, Meeusen R, Lemmink K. Different diagnostic tools in nonfunctional overreaching. Int J Sports Med. 2008;29(7):590–7.

16. Lehmann M, Foster C, Keul J. Overtraining in endurance athletes: a brief review. Med Sci Sports Exerc. 1993;25(7):854–62.

17. Rietjens GJ, Kuipers H, Adam JJ, et al. Physiological, biochemical and psychological markers of strenuous training-induced fatigue. Int J Sports Med. 2005;26(1):16–26.

18. Smit PJ. Sports medicine and 'overtraining'. S Afr Med J. 1978;54(1):4.

19. Slivka DR, Hailes WS, Cuddy JS, Ruby BC. Effects of 21 days of intensified training on markers of overtraining. J Strength Cond Res. 2010;24(10):2604–12.

20. Angeli A, Minetto M, Dovio A, Paccotti P. The overtraining syndrome in athletes: a stress-related disorder. J Endocrinol Invest. 2004;27(6):603–12.

21. Budgett R. Fatigue and underperformance in athletes: the overtraining syndrome. Br J Sports Med. 1998;32:107–10.

22. Rahim A, O'Neill P, Shalet SM. The effect of body composition on hexarelin- induced growth hormone release in normal elderly subjects. Clin Endocrinol. 1998;49:659–64. https://doi.org/10.1046/j.1365-2265.1998.00586.x.

23. van Koeverden ID, de Bakker M, Haitjema S, van der Laan SW, de Vries JPM, Hoefer IE, et al. Testosterone to oestradiol ratio reflects systemic and plaque inflammation and predicts future cardiovascular events in men with severe atherosclerosis. Cardiovasc Res. 2019;115:453–62. https://doi.org/10.1093/cvr/cvy188.

24. Chan YX, Knuiman MW, Hung J, Divitini ML, Handelsman DJ, Beilby JP, et al. Testosterone, dihydrotestosterone and estradiol are differentially associated with carotid intima-media thickness and the presence of carotid plaque in men with and without coronary artery disease. Endocr J. 2015;62:777–86. https://doi.org/10.1507/endocrj.EJ15-0196.

25. Aguirre LE, Colleluori G, Fowler KE, Jan IZ, Villareal K, Qualls C, et al. High aromatase activity in hypogonadal men is associated with higher spine bone mineral density, increased truncal fat and reduced lean mass. Eur J Endocrinol. 2015;173:167–74. https://doi.org/10.1530/EJE-14-1103.

26. Xu X, Wang L, Luo D, Zhang M, Chen S, Wang Y, et al. Effect of testosterone synthesis and conversion on serum testosterone levels in obese men. Horm Metab Res. 2018;50:661–70. https://doi.org/10.1055/a-0658-7712.

27. Arevalo MA, Azcoitia I, Garcia-Segura LM. The neuroprotective actions of oestradiol and oestrogen receptors. Nat Rev Neurosci. 2015;16:17–29. https://doi.org/10.1038/nrn3856.
28. Russell N, Grossmann M. Mechanisms in endocrinology: estradiol as a male hormone. Eur J Endocrinol. 2019:EJE-18–1000.R2. https://doi.org/10.1530/EJE-18-1000.
29. Charrière N, Miles-Chan JL, Montani JP, Dulloo AG. Water-induced thermogenesis and fat oxidation: a reassessment. Nutr Diabetes. 2015;5:e190. https://doi.org/10.1038/nutd.2015.41.
30. Purdom T, Kravitz L, Dokladny K, Mermier C. Understanding the factors that effect maximal fat oxidation. J Int Soc Sports Nutr. 2018;15:3. https://doi.org/10.1186/s12970-018-0207-1.
31. Jequier E, Constant F. Water as an essential nutrient: the physiological basis of hydration. Eur J Clin Nutr. 2018;64:115–23. https://doi.org/10.1038/ejcn.2009.111.
32. Cadegiani FA, Kater CE. Enhancement of hypothalamic-pituitary activity in male athletes: evidence of a novel hormonal mechanism of physical conditioning. BMC Endoc Dis. 2019;1:117. https://doi.org/10.1186/s12902-019-0443-7.
33. Oosterholt BG, Maes JH, Van der Linden D, Verbraak MJ, Kompier MA. Burnout and cortisol: evidence for a lower cortisol awakening response in both clinical and non-clinical burnout. J Psychosom Res. 2015;78:445–51. https://doi.org/10.1016/j.jpsychores.2014.11.003.
34. Grossi G, Perski A, Ekstedt M, Johansson T, Lindström M, Holm K. The morning salivary cortisol response in burnout. J Psychosom Res. 2005;59:103–11. https://doi.org/10.1016/j.jpsychores.2005.02.009.
35. Sjörs A, Ljung T, Jonsdottir IH. Long-term follow-up of cortisol awakening response in patients treated for stress-related exhaustion. BMJ Open. 2012;2:e001091. https://doi.org/10.1136/bmjopen-2012-001091.
36. Nater UM, Maloney E, Boneva RS, Gurbaxani BM, Lin JM, Jones JF, et al. Attenuated morning salivary cortisol concentrations in a population based study of persons with chronic fatigue syndrome and well controls. J Clin Endocrinol Metab. 2008;93:703–9. https://doi.org/10.1210/jc.2007-1747.
37. Stalder T, Kirschbaum C, Kudielka BM, Adam EK, Pruessner JC, Wust S, et al. Assessment of the cortisol awakening response: expert consensus guidelines. Psychoneuroendocrinology. 2016;63:414–32. https://doi.org/10.1016/j.psyneuen.2015.10.010.
38. Elder GJ, Wetherell MA, Barclay NL, Ellis JG. The cortisol awakening response–applications and implications for sleep medicine. Sleep Med Rev. 2014;18:215–24. https://doi.org/10.1016/j.smrv.2013.05.001.
39. Clow A, Hucklebridge F, Stalder T, Evans P, Thorn L. The cortisol awakening response: more than a measure of HPA axis function. Neurosci Biobehav Rev. 2010;35:97–103. https://doi.org/10.1016/j.neubiorev.2009.12.011.
40. Smyth N, Thorn L, Hucklebridge F, Evans P, Clow A. Detailed time course of the cortisol awakening response in healthy participants. Psychoneuroendocrinology. 2015;62:200–3. https://doi.org/10.1016/j.psyneuen.2015.08.011.
41. Zhang J, Ma RC, Kong AP, So WY, Li AM, Lam SP, et al. Relationship of sleep quantity and quality with 24-h urinary catecholamines and salivary awakening cortisol in healthy middle-aged adults. Sleep. 2011;34:225–33. https://doi.org/10.1093/sleep/34.2.225.
42. Pires GN, Bezerra AG, Tufik S, Andersen ML. Effects of acute sleep deprivation on state anxiety levels: a systematic review and meta-analysis. Sleep Med. 2016;24:109–18. https://doi.org/10.1016/j.sleep.2016.07.019.
43. Obasi EM, Chen TA, Cavanagh L, Smith BK, Wilborn KA, McNeill LH, et al. Depression, perceived social control, and hypothalamic-pituitary- adrenal axis function in African-American adults. Health Psychol. 2019; https://doi.org/10.1037/hea0000812. [Epub ahead of print].
44. Cadegiani FA, Kater CE. Adrenal fatigue does not exist: a systematic review. BMC Endocr Disord. 2016;16:48. https://doi.org/10.1186/s12902-016-0128-4.
45. Abu-Samak MS, Mohammad BA, Abu-Taha MI, Hasoun LZ, Awwad SH. Associations between sleep deprivation and salivary testosterone levels in male university students: a prospective cohort study. Am J Mens Health. 2018;12:411–9. https://doi.org/10.1177/1557988317735412.

46. Lee DS, Choi JB, Sohn DW. Impact of sleep deprivation on the hypothalamic- pituitary-gonadal axis and erectile tissue. J Sex Med. 2019;16:5–16. https://doi.org/10.1016/j.jsxm.2018.10.014.
47. Arnal PJ, Drogou C, Sauvet F, Regnauld J, Dispersyn G, Faraut B, et al. Effect of sleep extension on the subsequent testosterone, cortisol and prolactin responses to total sleep deprivation and recovery. J Neuroendocrinol. 2016;28:12346. https://doi.org/10.1111/jne.12346.
48. Banks S, Dinges DF. Behavioral and physiological consequences of sleep restriction. J Clin Sleep Med. 2007;3:519–28. https://doi.org/10.1080/15402000701244445.
49. Pérez-Fuentes MDC, Molero Jurado MDM, Simón Márquez MDM, Barragán Martín AB, Gázquez Linares JJ. Emotional effects of the duration, efficiency, and subjective quality of sleep in healthcare personnel. Int J Environ Res Public Health. 2019;16:E3512. https://doi.org/10.3390/ijerph16193512.
50. Park YK, Kim JH, Choi SJ, Kim ST, Joo EY. Altered regional cerebral blood flow associated with mood and sleep in shift workers: cerebral perfusion magnetic resonance imaging study. J Clin Neurol. 2019;15:438–47. https://doi.org/10.3988/jcn.2019.15.4.438.
51. Schwarz J, Axelsson J, Gerhardsson A, Tamm S, Fischer H, Kecklund G, et al. Mood impairment is stronger in young than in older adults after sleep deprivation. J Sleep Res. 2019;28:e12801. https://doi.org/10.1111/jsr.12801.
52. Lo JC, Ong JL, Leong RL, Gooley JJ, Chee MW. Cognitive performance, sleepiness, and mood in partially sleep deprived adolescents: the need for sleep study. Sleep. 2016;39:687–98. https://doi.org/10.5665/sleep.5552.
53. Chaput JP, Dutil C, Sampasa-Kanyinga H. Sleeping hours: what is the ideal number and how does age impact this? Nat Sci Sleep. 2018;10:421–30. https://doi.org/10.2147/NSS.S163071.
54. Grandner MA, Drummond SP. Who are the long sleepers? Towards an understanding of the mortality relationship. Sleep Med Rev. 2007;11:341–60. https://doi.org/10.1016/j.smrv.2007.03.010.
55. Popkin BM, D'Anci KE, Rosenberg IH. Water, hydration, and health. Nutr Rev. 2010;68:439–58. https://doi.org/10.1111/j.1753-4887.2010.00304.x.
56. Armstrong LE. Challenges of linking chronic dehydration and fluid consumption to health outcomes. Nutr Rev. 2012;70(Suppl 2):S121–7. https://doi.org/10.1111/j.1753-4887.2012.00539.x.
57. Liska D, Mah E, Brisbois T, Barrios PL, Baker LB, Spriet LL. Narrative review of hydration and selected health outcomes in the general population. Nutrients. 2019;11:E70. https://doi.org/10.3390/nu11010070.
58. El-Sharkawy AM, Sahota O, Lobo DN. Acute and chronic effects of hydration status on health. Nutr Rev. 2015;73(Suppl 2):97–109l. https://doi.org/10.1093/nutrit/nuv038.
59. Shirreffs SM. Markers of hydration status. Eur J Clin Nutr. 2003;57:S6–9. https://doi.org/10.1038/sj.ejcn.1601895.
60. Svendsen IS, Killer SC, Gleeson M. Influence of hydration status on changes in plasma cortisol, leukocytes, and antigen-stimulated cytokine production by whole blood culture following prolonged exercise. ISRN Nutr. 2014;2014:561401. https://doi.org/10.1155/2014/561401.
61. Penkman MA, Field CJ, Sellar CM, Harber VJ, Bell GJ. Effect of hydration status on high-intensity rowing performance and immune function. Int J Sports Physiol Perform. 2008;3:531–46. https://doi.org/10.1123/ijspp.3.4.531.
62. Borgman MA, Zaar M, Aden JK, Schlader ZJ, Gagnon D, Rivas E, et al. Hemostatic responses to exercise, dehydration, and simulated bleeding in heat-stressed humans. Am J Physiol Regul Integr Comp Physiol. 2019;316:R145–56. https://doi.org/10.1152/ajpregu.00223.2018.
63. Roh HT, Cho SY, So WY, Paik IY, Suh SH. Effects of different fluid replacements on serum HSP70 and lymphocyte DNA damage in college athletes during exercise at high ambient temperatures. J Sport Health Sci. 2016;5:448–55. https://doi.org/10.1016/j.jshs.2015.09.007.
64. Hom LL, Lee EC, Apicella JM, Wallace SD, Emmanuel H, Klau JF, et al. Eleven days of moderate exercise and heat exposure induces acclimation without significant HSP70 and apopto-

sis responses of lymphocytes in college-aged males. Cell Stress Chaperones. 2012;17:29–39. https://doi.org/10.1007/s12192-011-0283-5.

65. Kim YN, Shin HS. Relationships of total lymphocyte count and subpopulation lymphocyte counts with the nutritional status in patients undergoing hemodialysis/peritoneal dialysis. Kosin Med J. 2017;32:58–71. https://doi.org/10.7180/kmj.2017.32.1.58.

66. Altunayoglu Cakmak V, Ozsu S, Gulsoy A, Akpinar R, Bulbul Y. The significance of the relative lymphocyte count as an independent predictor of cardiovascular disease in patients with obstructive sleep apnea syndrome. Med Princ Pract. 2016;25:455–60. https://doi.org/10.1159/000447697.

67. Dmitrieva NI, Burg MB. Elevated sodium and dehydration stimulate inflammatory signaling in endothelial cells and promote atherosclerosis. PLoS One. 2015;10:e0128870. https://doi.org/10.1371/journal.pone.0128870.

68. Mitchell JB, Dugas JP, McFarlin BK, Nelson MJ. Effect of exercise, heat stress, and hydration on immune cell number and function. Med Sci Sports Exerc. 2002;34:1941–50. https://doi.org/10.1097/00005768-200212000-00013.

69. Balta S, Ozturk C. The platelet-lymphocyte ratio: a simple, inexpensive and rapid prognostic marker for cardiovascular events. Platelets. 2015;26:680–1. https://doi.org/10.3109/09537104.2014.979340.

70. Ye GL, Chen Q, Chen X, Liu YY, Yin TT, Meng QH, et al. The prognostic role of platelet-to-lymphocyte ratio in patients with acute heart failure: a cohort study. Sci Rep. 2019;9:10639. https://doi.org/10.1038/s41598-019-47143-2.

71. Akboga MK, Canpolat U, Yayla C, Ozcan F, Ozeke O, Topaloglu S, et al. Association of platelet to lymphocyte ratio with inflammation and severity of coronary atherosclerosis in patients with stable coronary artery disease. Angiology. 2016;67:89–95. https://doi.org/10.1177/0003319715583186.

72. Liaw FY, Huang CF, Chen WL, Wu LW, Peng TC, Chang YW, et al. Higher platelet-to-lymphocyte ratio increased the risk of sarcopenia in the community-dwelling older adults. Sci Rep. 2017;7:16609. https://doi.org/10.1038/s41598-017-16924-y.

73. Gasparyan AY, Ayvazyan L, Mukanova U, Yessirkepov M, Kitas GD. The platelet-to-lymphocyte ratio as an inflammatory marker in rheumatic diseases. Ann Lab Med. 2019;39:345–57. https://doi.org/10.3343/alm.2019.39.4.345.

74. Eccles R, Mallefet P. Observational study of the effects of upper respiratory tract infection on hydrationstatus. Multidiscip Respir Med. 2019;14:36. https://doi.org/10.1186/s40248-019-0200-9.

75. Yang DF, Shen YL, Wu C, Huang YS, Lee PY, Er NX, et al. Sleep deprivation reduces the recovery of muscle injury induced by high-intensity exercise in a mouse model. Life Sci. 2019;235:116835. https://doi.org/10.1016/j.lfs.2019.116835.

76. Wang JP, Chi RF, Wang K, Ma T, Guo XF, Zhang XL, et al. Exp oxidative stress impairs myocyte autophagy, resulting in myocyte hypertrophy. Physiology. 2018;103:461–72. https://doi.org/10.1113/EP086650.

77. Ng TP, Lu Y, Choo RWM, Tan CTY, Nyunt MSZ, Gao Q, et al. Dysregulated homeostatic pathways in sarcopenia among frail older adults. Aging Cell. 2018;17:e12842. https://doi.org/10.1111/acel.12842.

78. Berr CM, Stieg MR, Deutschbein T, Quinkler M, Schmidmaier R, Osswald A, et al. Persistence of myopathy in Cushing's syndrome: evaluation of the German Cushing's registry. Eur J Endocrinol. 2017;176:737–46. https://doi.org/10.1530/EJE-16-0689.

79. Solomon AM, Bouloux PM. Modifying muscle mass—the endocrine perspective. J Endocrinol. 2006;191:349–60. https://doi.org/10.1677/joe.1.06837.

80. Pedersen BK, Febbraio MA. Muscles, exercise and obesity: skeletal muscle as a secretory organ. Nat Rev Endocrinol. 2012;8:457–65. https://doi.org/10.1038/nrendo.2012.49.

81. Lemche E, Chaban OS, Lemche AV. Neuroendorine and epigenetic mechanisms subserving autonomic imbalance and HPA dysfunction in the metabolic syndrome. Front Neurosci. 2016;10:142. https://doi.org/10.3389/fnins.2016.00142.

82. Romanello V, Sandri M. Mitochondrial quality control and muscle mass maintenance. Front Physiol. 2015;6:422. https://doi.org/10.3389/fphys.2015.00422.

83. Peng K, Pan Y, Li J, Khan Z, Fan M, Yin H, et al. 11beta-hydroxysteroid dehydrogenase type 1(11beta-HSD1) mediates insulin resistance through JNK activation in adipocytes. Sci Rep. 2016;6:37160. https://doi.org/10.1038/srep37160.

84. Chapman K, Holmes M, Seckl J. 11beta-hydroxysteroid dehydrogenases: intracellular gate-keepers of tissue glucocorticoid action. Physiol Rev. 2013;93:1139–206. https://doi.org/10.1152/physrev.00020.2012.

85. Hu F, Liu F. Mitochondrial stress: a bridge between mitochondrial dysfunction and metabolic diseases? Cell Signal. 2011;23:1528–33. https://doi.org/10.1016/j.cellsig.2011.05.008.

86. Brennan K, Gallo S, Slavin M, Herrick J, Jonge LD. Water consumption increases resting fat oxidation. New Orleans, LA: Poster presentation at Obesity Week 2016; 2016.

87. Stookey JJD. Negative, null and beneficial effects of drinking water on energy intake, energy expenditure, fat oxidation and weight change in randomized trials: a qualitative review. Nutrients. 2016;8:19. https://doi.org/10.3390/nu8010019.

88. Keller U, Szinnai G, Bilz S, Berneis K. Effects of changes in hydration on protein, glucose and lipid metabolism in man: impact on health. Eur J Clin Nutr. 2003;57(Suppl 2):S69–74. https://doi.org/10.1038/sj.ejcn.1601904.

89. Boschmann M, Steiniger J, Franke G, Birkenfeld AL, Luft FC, Jordan J. Water drinking induces thermogenesis through osmosensitive mechanisms. J Clin Endocrinol Metab. 2007;92:3334–7. https://doi.org/10.1210/jc.2006-1438.

90. Boschmann M, Steiniger J, Hille U, Tank J, Adams F, Sharma AM, et al. Water-induced thermogenesis. J Clin Endocrinol Metab. 2003;88:6015–9. https://doi.org/10.1210/jc.2003-030780.

91. Dennis EA, Dengo AL, Comber DL, Flack KD, Savla J, Davy KP, et al. Water consumption increases weight loss during a hypocaloric diet intervention in middle- aged and older adults. Obesity. 2010;18:300–7. https://doi.org/10.1038/oby.2009.235.

92. González-Alonso J, Calbet JA, Nielsen B. Muscle blood flow is reduced with dehydration during prolonged exercise in humans. J Physiol. 1998;513(Pt 3):895–905. https://doi.org/10.1111/j.1469-7793.1998.895ba.x.

93. Klaas RW, Smeets A, Lejeune PM, Wouters-Adriaens MPE, Margriet S, Westerterp P, et al. Dietary fat oxidation as a function of body fat. Am J Clin Nutr. 2008;87:132–5. https://doi.org/10.1093/ajcn/87.1.132.

94. Zurlo F, Lillioja S, Del Puente AE, Nyomba BL, Raz I, Saad MF, et al. Low ratio of fat to carbohydrate oxidation as predictor of weight gain: study of 24-h RQ. Am J Phys. 1990;259:E650–7. https://doi.org/10.1152/ajpendo.1990.259.5.E650.

95. MacKenzie-Shalders KL, Byrne NM, King NA, Slater GJ. Are increases in skeletal muscle mass accompanied by changes to resting metabolic rate in rugby athletes over a pre-season training period? Eur J Sport Sci. 2019;19:885–92. https://doi.org/10.1080/17461391.2018.1561951.

96. McPherron AC, Guo T, Bond ND, Gavrilova O. Increasing muscle mass to improve metabolism. Adipocytes. 2013;2:92–8. https://doi.org/10.4161/adip.22500.

97. Müller MJ, Langemann D, Gehrke I, Later W, Heller M, Glüer CC, et al. Effect of constitution on mass of individual organs and their association with metabolic rate in humans–a detailed view on allometric scaling. PLoS One. 2011;6:e22732. https://doi.org/10.1371/journal.pone.0022732.

98. Moyen NE, Ganio MS, Wiersma LD, Kavouras SA, Gray M, McDermott BP, et al. Hydration status affects mood state and pain sensation during ultra-endurance cycling. J Sports Sci. 2015;33:1962–9. https://doi.org/10.1080/02640414.2015.1021275.

99. Horowitz M, Samueloff S. Plasma water shifts during thermal dehydration. J Appl Physiol Respir Environ Exerc Physiol. 1979;47:738–44. https://doi.org/10.1152/jappl.1979.47.4.738.

100. Stefanaki C, Pervanidou P, Boschiero D, Chrousos GP. Chronic stress and body composition disorders: implications for health and disease. Hormones. 2018;17:33–43. https://doi.org/10.1007/s42000-018-0023-7.
101. Ritchie RF, Ledue TB, Craig WY. Patient hydration: a major source of laboratory uncertainty. Clin Chem Lab Med. 2007;45:158–66. https://doi.org/10.1515/CCLM.2007.052.
102. Murck H, Schussler P, Steiger A. Renin-angiotensin-aldosterone system: the forgotten stress hormone system: relationship to depression and sleep. Pharmacopsychiatry. 2012;45:83–95. https://doi.org/10.1055/s-0031-1291346.
103. Ludwig DS, Hu FB, Tappy L, Brand-Miller J. Dietary carbohydrates: role of quality and quantity in chronic disease. BMJ. 2018;361:k2340.
104. Jäger R, Kerksick CM, Campbell BI, et al. International society of sports nutrition position stand: protein and exercise. J Int Soc Sports Nutr. 2017;14:20.
105. Noakes T, Volek JS, Phinney SD. Low-carbohydrate diets for athletes: what evidence? Br J Sports Med. 2014;48(14):1077–8.
106. Shaner AA, Vingren JL, Hatfield DL, Budnar RG Jr, Duplanty AA, Hill DW. The acute hormonal response to free weight and machine weight resistance exercise. J Strength Cond Res. 2014;28(4):1032–40.
107. Batsis JA, Villareal DT. Sarcopenic obesity in older adults: aetiology, epidemiology and treatment strategies. Nat Rev Endocrinol. 2018;14(9):513–37.
108. Ferrand C, Redonnet A, Prévot D, Carpéné C, Atgié C. Prolonged treatment with the beta3-adrenergic agonist CL 316243 induces adipose tissue remodeling in rat but not in Guinea pig: (1) fat store depletion and desensitization of beta-adrenergic responses. J Physiol Biochem. 2006;62(2):89–99.
109. Frontini A, Vitali A, Perugini J, et al. White-to-brown transdifferentiation of omental adipocytes in patients affected by pheochromocytoma. Biochim Biophys Acta. 2013;1831(5):950–9.
110. Suh B, Shin DW, Kwon HM, et al. Elevated neutrophil to lymphocyte ratio and ischemic stroke risk in generally healthy adults. PLoS One. 2017;12(8):e0183706.
111. McAdam JS, McGinnis KD, Beck DT et al. Effect of whey protein supplementation on physical performance and body composition in army initial entry training soldiers. Nutrients. 2018;10(9):E1248.
112. Nabuco HCG, Tomeleri CM, Sugihara Junior P et al. Effects of pre- or post-exercise whey protein supplementation on body fat and metabolic and inflammatory profile in preconditioned older women: A randomized, double-blind, placebo-controlled trial. Nutr Metab Cardiovasc Dis. 2018: S0939-4753(18)30335-1.
113. Pezeshki A, Fahim A, Chelikani PK. Dietary whey and casein differentially affect energy balance, gut hormones, glucose metabolism, and taste preference in diet-induced obese *Rats*. J Nutr. 2015;145(10):2236–44.

Chapter 6
New Understanding, Concepts, and Special Topics on Overtraining Syndrome

Outline

From an overtraining-centered to a multifactorial understanding of OTS: the author summarizes the major change in this paradigm of the understanding of OTS, in an illustrative fashion. This is a key subsection of the book, since it brings OTS and burnout together for future researches.

    The pivotal sequence of dysfunctions that leads to OTS: the author demonstrates the logical dysfunctional pathways that lead to OTS, from the collection of findings of the EROS and all the other studies. The rationale of OTS should be easily learned from this perspective. Basically, a combination of chronic deprivations, including insufficient caloric intake, lack of good sleep quality, insufficient rest, and concurrent physical and cognitive extreme demands, leads to chronic deprivation of energy and mechanisms of repair at a cell level. The organism needs to adapt in a way that it maintains survival and functioning. Multiple adaptations do occur, but as a "price" to maintain "normal" functioning under such an extreme inhospitable environment, these adaptations are highly dysfunctional and toxic. When these adaptations combine and interact, they work in a synergistically manner boosting the level of dysfunctions. The result is OTS. And a quite similar sequence of facts occurs in burnout syndrome.

    Overtraining syndrome: an overt or a relative dysfunction? When findings in OTS are analyzed from the perspective of normal ranges and general populations, they do not show any abnormality. However, when compared to sex-, age-, and BMI-matched healthy athletes, several alterations emerge. A paradoxical? No. The answer is on the fact that healthy athletes also differ from the general population and do show some "alterations" that are actually physiological. This is explored by the author in this section.

F. Cadegiani, *Overtraining Syndrome in Athletes*, https://doi.org/10.1007/978-3-030-52628-3_6

Unexpected serendipitous findings of novel conditioning processes in healthy athletes: Since a two-group control started to be employed to compare with OTS-affected athletes, highly unexpected differences between healthy athletes and sedentary have been detected. Among these, a true process of intrinsic hormonal conditioning has been described in athletes, similarly to those that occur in the cardiovascular and musculoskeletal systems, as well as several other metabolic benefits.

The novel key pathophysiological characteristic of overtraining syndrome: The key change in the understanding of OTS led to a redefinition of its pathophysiology, which is described by the author, using highly illustrative figures.

New concepts on overtraining syndrome: Similarly to the changes in the pathophysiology of OTS, new concepts have arisen from the multiple novel findings, and they are thoroughly explained, one-by-one, by the author in this section.

6.1   Introduction

The classic theories and hypotheses for OTS proposed partial explanations for its pathophysiology and underlying mechanisms for its occurrence, but could not fully elucidate the major aspects of OTS, even when they were joined and unified. Also, most of these theories and hypotheses failed to be demonstrated in clinical studies.

The new findings on OTS allowed a more precise and comprehensive view of OTS, through the improvement of the classic theories, proposal of new concepts, and the generation of a new understanding of OTS, from the combination of the new concepts and the classical theories revisited.

6.2   From an Overtraining-Centered to a Multifactorial Understanding of OTS

The excessive training-centered view of OTS, in which excessive training is a sine qua non condition for its occurrence, has been weakened since the learning and the avoidance of excessive training did not prevent or decrease the incidence of OTS.

Instead, the understanding that OTS occurs as a result from the combination and interactions between multiple chronic stressors, including lack of good sleep quality; excessive concurrent cognitive effort; insufficient carbohydrate, protein, or overall caloric intake; and possibly, but not obligatorily, excessive training overload, better explains the true etiology of OTS [1–4]. Indeed, concurrent mental fatigue can be an exacerbating factor under the presence of overreaching or OTS, although mental exhaustion is only able to impair sports performance if any overreaching or overtraining state was already installed [5]. Figure 6.1 illustrates the change in the understanding of OTS.

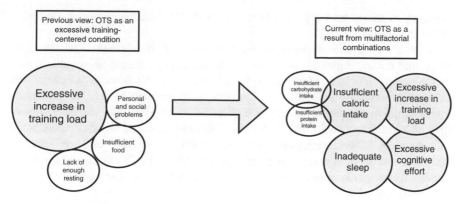

Fig. 6.1   Previous versus current view on overtraining syndrome

The additional triggers that have been identified were:

1. Chronic energy deprivation with overall caloric insufficiency (not necessarily due to hypocaloric diet)
2. Specific protein deficit
3. Specific carbohydrate insufficient intake
4. Unrefreshing sleep
5. Lack of sleep hygiene
6. Excessive cognitive activity in high-demanding jobs
7. Excessive studying at the same time of intensification of the trainings
8. Exposure to highly stressful situations

which can be combined, and which also depend on the severity of each trigger, for the development of OTS. Hence, the concept of *sum and interactions of factors* better describes the triggering process of OTS, rather than excessive training alone.

6.3   The Pivotal Sequence of Dysfunctions that Leads to OTS

From the identification of multiple different aspects correlated with OTS, it can be inferred that OTS is now understood as resulted from a chronic exposure to a hostile environment, fully depleted of energy availability and mechanisms of repair, due to lack of proper recovery and excessive exposure to oxidative stress without sufficient repairing mechanisms (anti-oxidative stress system) that forces organism to get adapted to maintain survival and functionality, as thoroughly explained previously (Chaps. 4 and 5) [1]. The only viable adaptations that are able to maintain survival and functionality under a hostile environment are highly dysfunctional, termed as *maladaptations*, which occurs in most tissues and organs.

The multiple sorts of *maladaptive* processes lead to multiple dysfunctions, diffusely, which eventually lead to OTS.

Indeed, although oxidative stress may play a positive role during the process of the beneficial adaptive processes, when excessive and without appropriate repair, oxidative stress can lead to pathological impaired muscle recovery and exacerbation of muscle catabolism [6, 7]

In summary, a mix of deprivations, including caloric, protein, carbohydrate, sleep, and resting deprivations, leads to chronic energy depletion, excessive oxidative stress, and lack of mechanisms of repair, which force adaptations and changes to maintain survival and functioning in an inhospitable environment. These adaptive processes are highly dysfunctional and harmful and lead to multiple dysfunctions, including metabolic, hormonal, immunologic, non-classic inflammation, muscular, psychiatric, neurological, autonomic, and cardiovascular. The combinations and interactions between these dysfunctions and the unfavorable tissue environment that act in a negatively synergistic manner are what eventually lead to OTS.

6.4   Overtraining Syndrome: An Overt or a Relative Dysfunction?

One of the most important questions regarding OTS is whether dysfunctions of OTS are actual, overt, relative, or inexistent. Previous studies showed inconsistent results, not only due to the challenges regarding the assessment of the diagnosis and evaluation of OTS, but also because these studies yielded conflicting findings, which depended on the sort of control group. If healthy athletes were the control group, differences tended to be observed, and therefore OTS was considered as a true dysfunctional condition. Conversely, if sedentary control or normal ranges for general population were used for the analysis of the findings on OTS, normal results were observed, and OTS was demonstrated to be a non-actual disorder.

This apparent contradiction occurred because athletes affected by OTS had similar levels to general population, which would classify OTS as a "non-abnormal condition," at the same time parameters in these athletes were altered when compared to the expected for healthy athletes. The simultaneous use of both control groups, of healthy athletes and of non-athletes, allowed the understanding of the context of OTS and confirmed the fact that parameters in OTS are altered when compared to healthy athletes, but normal when compared to age-, BMI-, and sex-matched healthy non-physically active controls.

From this perspective, OTS is indeed is a combination of actual but relative dysfunctions, and a truly existent, but not overt, syndrome, with multiple patterns of clinical, metabolic, and biochemical behaviors present in OTS, which shows that only 4.5% of markers are overtly altered, whereas the majority of the parameters different from healthy athletes, but similar to sedentary.

Furthermore, OTS is not a frank dysfunction, neither clinically nor biochemically, but is characterized by a combination of losses of the multiple adaptive process that athletes were shown to undergo, which eventually leads to a loss of performance, the key feature of OTS.

The sequence of Figs. 6.2, 6.3, 6.4, 6.5, and 6.6 depicts in an illustrative manner how clinical, metabolic, and biochemical parameters behave in the presence of

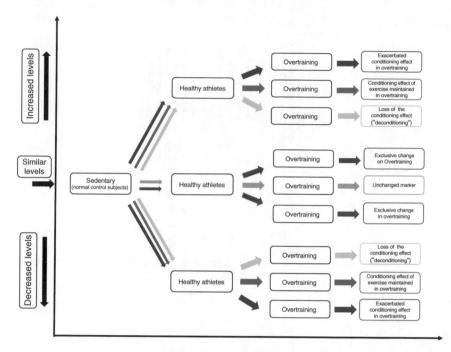

Fig. 6.2 Patterns of clinical, metabolic, and biochemical behaviors that can be observed in overtraining syndrome (OTS) when compared to levels in healthy athletes and healthy sedentary

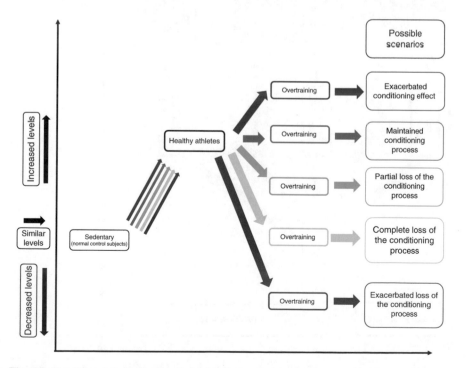

Fig. 6.3 Possible scenarios for the parameters in overtraining syndrome (OTS) according to the comparisons with parameters that disclosed conditioning adaptive changes in athletes

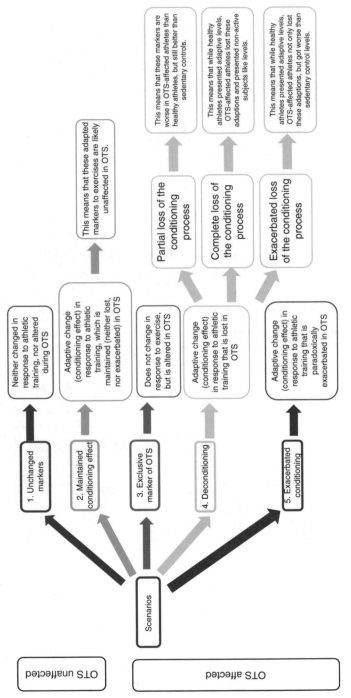

Fig. 6.4 Conditioning and *deconditioning scenarios* depicted in OTS

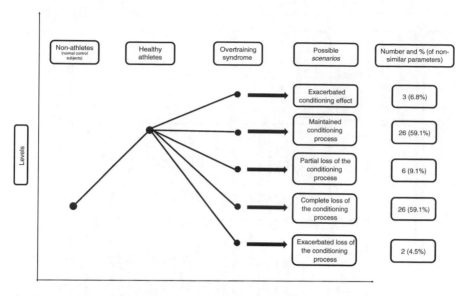

Fig. 6.5 Behaviors of parameters in overtraining syndrome (OTS) among those that demonstrated adaptive (conditioning) changes in response to athletic training

OTS, from the perspective of levels typically found in healthy athletes and healthy non-physically subjects.

Figure 6.2 summarizes the possible behaviors that can be possibly detected in OTS, from the perspective of both healthy athletes and general population, for all parameters, including those that are increased, similar, or decreased in healthy athletes compared to the second control group, of healthy sedentary. Figure 6.3 presents the possible scenarios for the clinical, metabolic, and biochemical parameters in OTS, when compared to parameters in healthy athletes that disclosed adaptive conditioning processes, when compared to healthy sedentary. Figure 6.4 depicts the *scenarios* present in OTS. Figure 6.5 summarizes the percentage of overtly, relative, and absent parameters, while Fig. 6.6 depicts which markers are affected overtly, relatively, or not-affected.

The characteristic of OTS as being a condition in which the recognition of its dysfunctions is heavily based on a substantial level of relativeness challenges its diagnosis and may also explain why it has taken a long time to finally identify its biomarkers and mechanisms. Besides, as the identification of the dysfunctions on OTS is basically dependent on the expected levels for athletes, and there are no adapted ranges for athletes for any parameter, a proper diagnostic assessment of OTS becomes almost unfeasible.

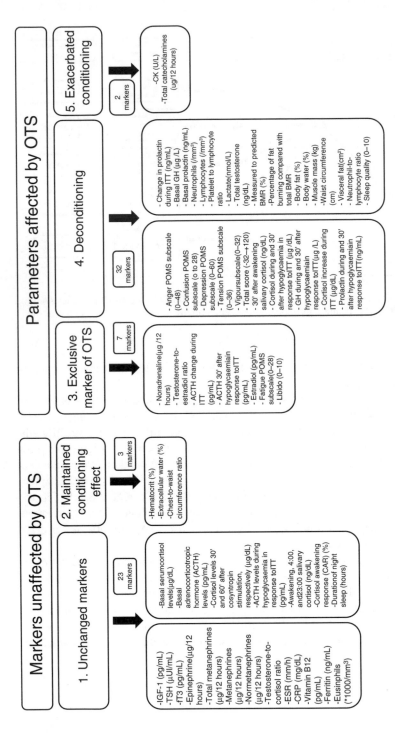

Fig. 6.6 Overtly or relatively affected, and non-affected markers of OTS

BMR = Basal metabolic rate; CK =Creatine kinase; CRP = Creactive protein; ESR = Erythrocyte sedimentation rate; ITT = Insulin tolerance test;; POMS = Profile of Mood States.

6.5   Unexpected Serendipitous Findings of Novel Conditioning Processes in Healthy Athletes

Since parameters were different between OTS-affected and healthy athletes, while similar between OTS and sedentary controls, it is predictable that levels in athletes would be different from those in sedentary, which indeed happened when these groups were compared.

However, many of the differences between athletes and sedentary that had similar baseline characteristics, including BMI and absence of diseases, in parameters that hypothetically would not depend on sports performance or would not be influenced by any conditioning process or adaptation, were unexpected, in the light of an absence of data supporting or reporting these differences.

Hence, the unforeseen yet serendipitous findings on healthy athletes, when compared to sedentary, revealed the existence of additional independent conditioning processes, besides those previously described in the cardiovascular, musculoskeletal, neuromuscular, autonomic, and neurologic systems.

True and intrinsic novel hormonal and metabolic conditioning processes were identified in athletes [8]. These newly described phenomena were shown to occur in an independent matter from any external stimulation, including signaling including other systems and physical effort, which means that although chronic, intense, and regular physical activity induced the occurrence of hormonal and metabolic conditioning process, these optimizations become autonomous and independent from exercising. Hormonal conditionings include optimization of GH and cortisol, responses, not only physical-dependent, but also to non-exercise stimulations, that yield faster, exacerbated, and prolonged responses; the existence of a prolactin response to any sort of stimulation, which is absent in non-physically active subjects; chronic increase of basal testosterone levels; increased and non-undetectable basal GH levels; increase of catecholamines even after long periods of resting; optimized and overcompensated lactate; increased basal metabolic rate and increased proportion of fat oxidation (in relation to overall metabolic rate) during resting, and increased intracellular hydration. These newly identified mechanisms were termed as "hormonal conditioning" and "metabolic conditioning," respectively.

6.6   Novel Concepts on Overtraining Syndrome

In addition to the classic hypotheses and new view on OTS, other new concepts on OTS have been proposed to provide a more thorough and full understanding of OTS. Altogether, these mechanisms may fully explain the pathogenesis of OTS. Table 6.1 summarizes the new concepts on OTS, and Fig. 6.7 compares, drives, and puts into context from the classical theories to the new concepts, as well as how they relate.

Table 6.1 New concepts on overtraining syndrome

New concept	Background	Evidence
Sum, intensity, and interactions of multiple factors as the underlying mechanism of overtraining syndrome	After the correction of excessive training, OTS unexpectedly persisted. As complex as its pathophysiology, its occurrence depends on the number of risk factors present, as well as the severity of each risk factor.	OTS-affected athletes disclosed a markedly larger number and more severely presentation of risk factors, compared to healthy athletes. Although each affected athlete disclosed a unique combination of risk factors, the number and intensity of risk factors were similar between all affected cases, while substantially higher than healthy athletes, with no overlapping results
Multiple chronic *maladaptations*	Under a hostile and roughly inhospitable environment, organism needs to adapt in order to maintain survival and a minimum functioning level. The multiple forced adaptations are however highly dysfunctional, leading to long-term destructive consequences. These were termed as *maladaptations*	Multiple unexpected findings were unveiled, among which some are strong demonstrations of the existence of *maladaptive* changes in OTS. These *maladaptations* may also justify the underlying mechanisms and harmful consequences of OTS
Unrepaired damage	The excessive exposure to large amount of reactive oxygen species (ROS) depletes the protection (anti-oxidant) and repair capacity, which are not properly regenerated, due to insufficient recovery. Further exposition to large amounts of ROS occurs without any protection and repair and leads to substantial and diffuse damages and inability to be repaired	Increased cytokines, relatively increased lactate and creatine kinase (CK), impaired muscle recovery, blunted hydration, possible long-term exposure to mild hypoglycemia during training sessions, and prolonged mental and physical exhaustion that were identified in OTS are large causes and indicatives of excessive damage without appropriate defense and repairing
Hypometabolism (hypometabolic state)	Since overtraining syndrome results from a chronic deprivation of energy, mechanisms toward a sort of "forced energy-saving mode" functioning system are theoretically essential to maintain survival	Indicators of reduced metabolism were extensively detected in overtraining syndrome, allowing the hypothesis of OTS as being a "hypometabolic state." Demonstrations for these findings include decreased hormonal responses to simulations, blunted basal anabolic hormones, increased conversion from anabolic to non-anabolic hormones, decreased metabolic rate and fat oxidation, decreased of the highly metabolically active muscle tissue, and cognitive activity suppression, in OTS

Table 6.1 (continued)

New concept	Background	Evidence
Relative diffuse hypothalamic *hyporesponsiveness* or "functional relative panhypopituitarism of hypothalamic origin state"	The "fight-or-flight" response, a primitive and sturdy reaction to stressors, is exacerbated in healthy athletes, as an adaptive process to chronic and regular exposure to a positive stressor (exercising), and may eventually contribute to a progressive improvement of performance. In a sick and energy-depleted environment, including OTS, non-essential energy demands are suppressed, including the "fight-or-flight" reaction. It means that the exacerbated reaction, as a positive adaptation to exercise, is lost under OTS, which may explain the loss of physical performance found in this syndrome	Enhanced diffuse hypothalamic responses to stressful stimulations were demonstrated to be present in healthy athletes when compared to similar healthy sedentary control and include quicker, exacerbated, and prolonged GH, prolactin, and cortisol responses. These positive adaptations are blunted in OTS, demonstrating a loss of positively optimized "fight-or-flight" responses expected for athletes
"Chronic mild neuroglycopenic state"	The energy demands from intense trainings are unlikely to be completely supplied by non-glycogenic non-direct glucose sources of energy, due to a limited capacity speed of generation of ATPs from fatty acids and amino acid pathways, possibly leading to mild chronic decrease of glucose levels. Chronic mild hypoglycemia has particular important consequences in the brain, including flattened central hormonal responses, psychological disturbances, and mental fatigue and exhaustion	During the hypoglycemic episode induced by an insulin tolerance test (ITT), OTS-affected athletes had fewer adrenergic symptoms, showing that these athletes are chronically adapted to hypoglycemia, similarly to what is typically identified in patients with uncontrolled diabetes and multiple episodes of hypoglycemia. The reduced response to hypoglycemia may also indicate a chronic exposure to lower glucose levels
"Chronic severe physical and mental exhaustion"	The chronic mental and physical exhaustion that is unresponsive to resting in OTS may be a protective mechanism to avoid the expenditure of energy in all non-essential demands. In short, unresponsive exhaustion should be the most clinically relevant manifestation of the chronic shortage and need for energy	Both mental and physical fatigue were strictly correlated with markers of reduced metabolism and impaired responses to stimulations, showing that exhaustion is indeed a probable mechanism to reduce activity and energy demands

(continued)

Table 6.1 (continued)

New concept	Background	Evidence
"Burnout syndrome of the athlete"	The characterization of the psychological disturbances of athletes affected by overtraining syndrome resembles the hallmark features of burnout syndrome, originally related to workplace. Moreover, analog characteristics between an unfavorable workplace and unfavorable training site seem to be similar	The overall scales and subscale of the Profile of Mood States (POMS) questionnaire as well as questions from specific burnout questionnaires yielded similar results between typical burnout syndrome and OTS states
"Actual yet relative syndrome"	Researchers struggled to find markers of OTS because parameters showed inconsistent results. Basically, at the same time that differences were found for a variety of markers between affected and healthy athletes, these were unaltered when compared to normal ranges for general population. Since athletes undergo several different adaptive processes, this population should have particular adapted ranges of many of these parameters. In this context, OTS is hypothesized to be a relative syndrome, as dysfunctions may be detected when compared to the expected for athletes, but are not overt, once levels tend to be within the ranges for the general references	When OTS-affected athletes were simultaneously compared with both healthy athletes and healthy sedentary controls, for a large number of parameters (more than 60), approximately 60% of the markers were shown to be relatively altered, i.e., altered when compared to healthy athletes, while similar to sedentary. Hence, there is sufficient data to conclude that OTS is indeed a relative syndrome, what justifies the previous unsuccessful attempts to detect biomarkers. At the same time, these comparisons revealed the existence of multiple additional conditioning processes in response to exercise
Unifying theory of the "paradoxical deconditioning syndrome"	Once OTS has been shown to be triggered by other factors than excessive exercise, the expression "overtraining" no longer reflects the key feature of OTS. The majority of the findings were found to be relatively altered in OTS, as a sort of mix of losses of biochemical, metabolic, and hormonal conditioning processes, which are likely the underlying dysfunction that leads to the loss of physical conditioning, the hallmark of the syndrome. These reactions, as well as the loss of physical performance, are paradoxical when in the context of a progressive training, which is expected to lead to opposite results	Athletes disclosed a wide variety of "deconditioning" process in a variety of tissues, organs, and systems, paradoxically from the expected, as demonstrated and explained in the other new concepts. Hence, the expression "paradoxical deconditioning process" better describes and defines the key underlying characteristic that leads to decreased physical performance, the hallmark of the syndrome, which has been suggested therefore to be reframed to "paradoxical deconditioning syndrome" (PDS)

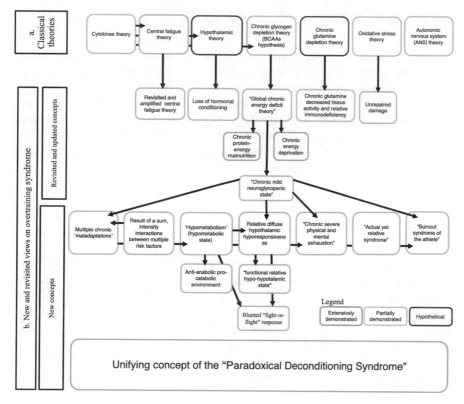

Fig. 6.7   Classical theories and new concepts on overtraining syndrome

6.6.1   Unrepaired Damage

A full depletion of mechanisms of repair typically occurs in elite athletes, particularly when under extreme efforts. In OTS, there is an inability to regenerate these mechanisms due to insufficient resting (unrefreshing sleep and concurrent cognitive effort) and insufficient energy availability (chronic glycogen depletion) for the recovery and regeneration process. The complete absence of mechanisms of repair, as these have not been regenerated, leads to an exponential increase of reactive oxygen species (ROS) with consequent generalized damage. This diffuse and severe damage is mostly unrepaired, due to the persistence of the shortage of mechanisms of repair. This is the concept of "unrepaired damage," which is the extensive presence of highly harmful, degenerative, and reactive damage in a broad number of structures, present in OTS.

The generalized unrepaired damage leads to a systemic increase of pro-inflammatory cytokines, relative immunodeficiency, and additional mechanisms to prevent further energy spending and worsening of this damage.

6.6.2   Multiple Chronic "Maladaptations"

The combination of multiple and unrepaired damage, associated with chronic energy deprivation, increased inflammation, and altered immunologic function forces the occurrence of adaptations, in order to maintain survival and functionality, despite the highly adverse cell and tissue environment. Even though these chronic adaptations occur, they are vastly dysfunctional and include and cause diffuse aberrant reactions and pathways and can be properly termed as "maladaptations." This is the concept of "multiple chronic *maladaptations*," a major characteristic of OTS, and probably one of its root causes.

6.6.3   "Energy Saving" Mode

The presence of simultaneous anti-anabolic and pro-catabolic metabolic environment, impaired muscle recovery and increased muscle degradation, preference for protein instead of fat utilization as energy, and suppressed responses to stimulations that demands energy demonstrate the existence of energy saving-driven functioning mode, termed as "energy saving mode." Although controversial, this concept may be an actual additional strategy to maintain survival under the shortage of energy availability. In the "saving mode," the catabolic preference for proteolysis over lipolysis as the primary source of non-glucose energy aims the preservation of the less metabolically active storage.

6.6.4   "Hypometabolism"

In complement to the understanding of the metabolism as functioning on "energy saving mode" under OTS, the overall metabolism is conclusively reduced in OTS, which is evidenced by the presence of decreased hormonal responses to simulations, blunted basal anabolic hormones, increased conversion from anabolic to non-anabolic hormones, decreased metabolic rate and fat oxidation, decreased of the highly metabolically active muscle tissue, and cognitive activity suppression. Combined, all these alterations provide sufficient support for the existence of a diffuse reduction in the metabolism, concepted as "hypometabolism," as a typical characteristic of OTS.

6.6.5   Relative Hypothalamic Hyporesponsiveness

Another important novel mechanism identified in OTS is diffuse non-specific blunted central responses to stimulations in comparison to what is expected for athletes, at least for all tested hormones, including GH, prolactin, and cortisol responses to direct stimulations of the hypothalamic-pituitary axes, although these responses are still within normal range when compared to the general population. The multiple hormonal under-responses in OTS are actually a loss of the optimization of the responses typically found in healthy athletes, i.e., collectively, this is a relative, non-overt dysfunction. This dysfunction is better described now demonstrated concept of "relative hypothalamic hyporesponsiveness," a concept, now demonstrated, that contributes to an impaired "fight-or-flight" primitive reaction, largely present in OTS, as an attempt to save the high energy demands that this reaction requires. Since the optimized hypothalamic responses present in athletes occurs as a sort of positive channeling of the "fight-or-flight" response that leads to improved performance, the loss of this optimization is the main underlying reason that OTS disclosed a loss of performance.

6.6.6 Singular Sum of the Number, Intensity, Combinations, and Interactions Between Environmental Stressors

The sum, interaction, intensity, and level of "aberration" of the alterations and factors are resulted from a dose-, intensity-, volume-, and time-dependent manner that will lead to overreaching. If not properly addressed, OTS, a prolonged, more dysfunctional, and less reversible condition than overreaching, installs. Hence, OTS does not result from only one single stressor. Instead, a chronic exposure to a minimum number, intensity, combinations, and interactions between different environmental triggering factors is key for its occurrence. This is better understood if one considers that although healthy athletes may present some of the stressors that lead to OTS, the sum and intensity of these factors are substantially lower than in affected athletes. In addition, it is unlikely that two affected athletes disclose the exact same risk factors, as each case of OTS was found to be triggered from one-of-a-kind combination of stressors. Consequently, neither none of the triggering factors and pathological alterations is ubiquitously present, nor there is a pre-determined combination of these aspects.

Thus, the concept of "singular sum of the number, intensity, combinations and interactions between environmental stressors" seems to be essential for the understanding of OTS.

6.6.7 Reinforcing the Theory of the Autonomic Dysfunction

Attempts to maintain functioning go beyond dysfunctional adaptations, as the increased catecholamines observed in athletes are exacerbated under OTS. The predominance of sympathetic over parasympathetic system under OTS has also been identified in studies that directly evaluated heart rate variability (HRV), with reduced vagal stimulation and heart rate variability [9]. The exacerbated catecholamines and the disturbed autonomic system have already been proposed among the classic theories and have been now reinforced [9].

6.6.8 Burnout Syndrome of the Athlete

Burnout syndrome is a disease recently recognized by the World Health Organization (WHO) and is characterized by symptoms of depression, anxiety, and depersonalization. From a collective perspective, burnout may be resulted from the increasing commitment to perform at high levels in all aspects of life at the same time: professionally, financially, socially, intellectually, physically, and sexually. In this scenario, the "high human performance" neglects allowances and indulgences, which are crucial for mental health. The toxic excessive motivation precludes from noticing signs of fatigue and exhaustion, until the point that burnout occurs. Analogically, burnout can be understood as a car that has consumed all the gas (the source of energy), including the extra gas for emergencies, and start to destroy its own pieces to maintain survival and functioning, which in turn makes the recovery process highly challenging. The consumption of the pieces of the car can be compared to the *maladaptations* that occur in energy-deprived environments in athletes, which eventually leads to OTS.

In addition, the characterization of the psychological disturbances of athletes affected by overtraining syndrome resembles the hallmark features of burnout syndrome, originally related to workplace. Indeed, the overall scale and subscales of the Profile of Mood States (POMS) questionnaire as well as questions from specific burnout questionnaires yielded similar results between typical burnout syndrome and OTS states.

Moreover, analog characteristics between an unfavorable workplace and unfavorable training site seem to be similar, which has been described as the burnout syndrome of the athlete (BSA) has been described and is depicted in a further chapter, but which is perfectly corresponding with OTS.

6.6.9 "Chronic Mild Neuroglycopenic State"

Athletes suspected of OTS experiment change in personality and important mood disturbances. These abnormalities are similar to those present under mild hypoglycemia.

Indeed, it has been detected that during hypoglycemic episodes, OTS-affected athletes presented fewer adrenergic symptoms compared to healthy athletes and sedentary [1, 2], which is typically identified in patients with uncontrolled diabetes and multiple episodes of hypoglycemia. It seems that the chronic exposure to hypoglycemia decreases adrenergic sensitivity, leading to reduction of adrenergic symptoms. This finding suggests that OTS athletes are chronically adapted to hypoglycemia, which is in accordance with the finding that these same athletes had lower carbohydrate intake, while they maintained intense training.

In this scenario, the energy demands from intense trainings are unlikely supplied by non-glycogenic non-direct glucose sources of energy in the speed required, due to limited capacity to generate ATPs from fatty acids and amino acid pathways. The combination of low glycogenic storage, low carbohydrate supply, and intense trainings that demand large amounts of prompt energy availability leads to overconsumption of plasma glucose, eventually leading to hypoglycemia. While adrenergic symptoms are progressively lost with the increasing number and duration of hypoglycemic episodes, neuroglycopenic manifestations do not decrease accordingly, possibly because central neurological system (CNS) is not as adaptable as the autonomic system. The same athletes that had fewer adrenergic effects of hypoglycemia did not present lower neuroglycopenic symptoms during hypoglycemia.

Altogether, these findings support the concept that OTS is also a chronic mild neuroglycopenic state. The chronic mild hypoglycemia has particular important consequences in the brain, including flattened central hormonal responses, psychological disturbances, and mental fatigue and exhaustion, which accurately matches with OTS manifestations.

6.6.10 Actual Yet Relative Syndrome

The historical struggle to detect biomarkers of OTS is based on the inconsistency and sometimes contradictory findings from studies. At the same time that OTS disclosed important different results compared to healthy athletes, OTS-affected athletes had no overt abnormal levels when analyzed through general reference ranges.

However, the premise that athletes undergo multiple different adaptations, and may therefore disclose significant differences when compared to general population, and which has been further confirmed, had not been considered.

When one realizes that differences between OTS and healthy athletes do not necessarily mean overt alterations because healthy athletes may also differ from sedentary, even for non-exercise-induced parameters, the understanding of OTS may become feasible.

When OTS-affected athletes were simultaneously compared with both healthy athletes and healthy sedentary controls for a large number of parameters, approximately 60% of the markers were shown to be relatively altered, i.e., altered when compared to healthy athletes, albeit similar to sedentary and within reference ranges.

Hence, there is sufficient data to conclude that OTS is actually a relative, non-overt syndrome, what justifies the previous unsuccessful attempts to detect biomarkers.

From this perspective, OTS should be comprehended as a loss of the adaptations found to occur in athletes, making athletes affected by OTS regress to sedentary biochemical behaviors.

6.6.11 Chronic Severe Physical and Mental Exhaustion

There is an interesting profile of athletes that further develop OTS that these athletes tend to avoid usual signs of exhaustion in a conscious or unconscious pretending manner and artificially overcome the exhaustion with more "motivation" and "focus." Since overcoming barriers is worshiped as a characteristic of superiority, this provides additional "motivation" for further barriers. This "motivation" eventually becomes pathological and toxic.

In response, chronic mental and physical severe exhaustion appears as a protective mechanism to avoid the expenditure of energy in all non-essential demands and is unresponsive to resting and improvements in dietary and sleeping patterns. This unresponsive exhaustion should be the most clinically relevant manifestation of the chronic shortage and need for energy among the manifestations of OTS.

To reinforce this concept, both mental and physical fatigue have been strictly correlated with markers of reduced metabolism and impaired responses to stimulations, showing that exhaustion is indeed a probable mechanism to induce reduction of overall activities and energy demands.

6.6.12 Unifying Theory of the "Paradoxical Deconditioning Syndrome"

None of the novel concepts and previous theories alone is able to fully explain the pathophysiology of OTS. However, when unified, there is sufficient substantiation to claim that all major underlying mechanisms that lead to OTS have been uncovered.

Collectively, the majority of the findings were found to be relatively altered in OTS, as a sort of mix of losses of biochemical, metabolic, and hormonal conditioning processes, which are likely the underlying dysfunction that leads to the loss of physical conditioning, the hallmark of the syndrome. Besides, once OTS has been shown to be triggered by other factors than excessive exercise, the expression "overtraining" no longer reflects the key feature of OTS.

These reactions, as well as the loss of physical performance, are seen as paradoxical when in the context of a progressive training, which is expected to lead to opposite results, as demonstrated in athletes that have been shown to disclose a wide variety of loss of conditioning ("deconditioning") processes in a variety of tissues, organs, and systems, paradoxically from the expected.

Hence, the expression "paradoxical deconditioning process" better describes and defines the key underlying characteristic that leads to decreased physical performance, the hallmark of OTS. Hence, it has been suggested that overtraining syndrome, which does not encompass the current major characteristics of the syndrome within the name, should be reframed for "paradoxical deconditioning syndrome" (PDS).

6.7   The Novel Key Pathophysiological Characteristic of Overtraining Syndrome as a "Paradoxical Deconditioning Syndrome" (PDS)

While in healthy athletes many endocrine and metabolic processes of optimization were identified, most of these adaptive mechanisms are lost when OTS occurs. The loss of the quicker, increased, and prolonged hormonal responses, which are partly responsible for a faster and more intense explosive reaction, and increased maintenance of the pace throughout a training session, respectively, coincided with the loss of these characteristics of physical performance, the hallmark of OTS.

The correlations found between healthy sedentary, healthy athletes, and then OTS-affected athletes, regarding the hormonal responsiveness and physical performance, reinforce the existence of an exposure-effect relationship between hormonal responses and sports performance.

This is an additional argument for the statement that the multiple conditioning processes play among the most important roles in the control of the physical performance.

The exposure-effect relationship has a likely causal relationship, in which the loss of the increased hormonal responsiveness to stimulation, alongside with the multiple losses of the adaptive processes that athletes undergo, which can be collectively described as multiple losses of conditioning, or multiple *deconditioning* processes, is probably the underlying reasons for the loss of physical conditioning, or physical *deconditioning*.

Hence, OTS athletes disclose a paradoxical response to the expected, since a progressive gain, not loss, of performance should occur progressively with the training program, resulting from a paradoxical mix of losses, instead of progressive improvements, of many of the conditioning processes that athletes undergo.

Finally, OTS can be considered as a paradoxical phenomenon – of both biochemical and hormonal responses and physical performance – from what would be expected from this population. Therefore, OTS is better conceived and described by its fundamental dysfunction as a "sort of mix of multiple *deconditioning* processes."

Although highly inserted into the scientific and general communities, the term "overtraining syndrome" considers that excessive training is the major and only cause for this underperformance syndrome, which no longer accounts for its most cases, since the learning of the importance of the periodization of the training sessions, intensity, volume, and speed of progression. Moreover, the term "overtraining syndrome" does not describe or synthetize the key aspect of the syndrome. For these reasons, this term should be reconsidered since despite its unequivocal historical importance, it can now be considered as a misname.

Instead, the proposed term *paradoxical deconditioning syndrome (PDS)* better describes the hallmark of the syndrome, once the multiple biochemical and metabolic *deconditioning* processes lead to the physical *deconditioning*, the sine qua non characteristic of OTS, which means that *deconditioning* is within one expression what better represents the central aspect of OTS. Also, unlike OTS that limits the triggers to the excessive training, PDS does not limit which triggers may cause OTS. Figure 6.8 summarizes the major aspects of the new understanding of OTS as actually and eventually being a paradoxical deconditioning syndrome (PDS).

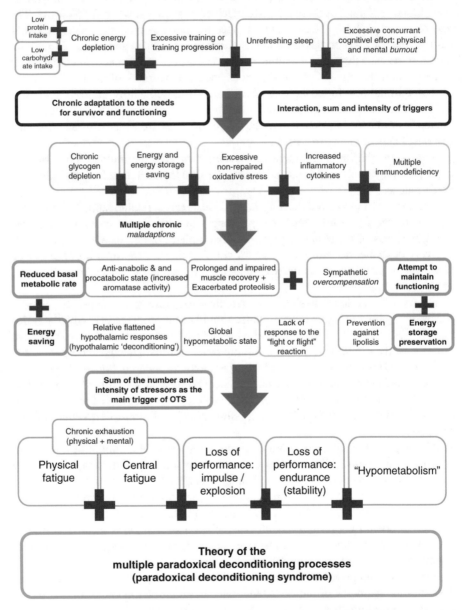

Fig. 6.8 Summary of new understanding of overtraining syndrome (OTS), as being a paradoxical deconditioning syndrome (PDS)

6.8 Conclusions

Instead of excessive training, the combination of multiple chronic stressors, including lack of good sleep quality, excessive concurrent cognitive effort, and insufficient carbohydrate, protein, or overall caloric intake, leads to multiple types of *maladaptive* processes forced to occur to maintain survival and functioning under chronic energy depletion and shortage of mechanisms of repair lead to multiple dysfunctions, diffusely, which eventually lead to OTS. The struggle to identify biomarkers on OTS has historically occurred because OTS has been demonstrated to be a dysfunction with neither overt nor absent abnormalities, but relative instead.

The novel findings in overtraining syndrome allowed multiple new concepts to be suggested as pieces of its pathophysiology. When combined altogether, the unifying theory of "paradoxical deconditioning syndrome" has been proposed, which may finally clarify the pathogenesis of OTS.

The term *paradoxical deconditioning syndrome (PDS)* better describes the hallmark of the syndrome, of multiple *deconditioning* processes that lead to the physical *deconditioning*, the hallmark of OTS.

References

1. Cadegiani FA, Kater CE. Novel insights of overtraining syndrome discovered from the EROS study. BMJ Open Sport Exerc Med. 2019;5(1):e000542. https://doi.org/10.1136/bmjsem-2019-000542.
2. Cadegiani FA, Kater CE. Novel causes and consequences of overtraining syndrome: the EROS-DISRUPTORS study. BMC Sports Sci Med Rehabil. 2019;11:21.
3. Cadegiani FA, Kater CE. Eating, Sleep, and Social Patterns as Independent Predictors of Clinical, Metabolic, and Biochemical Behaviors Among Elite Male Athletes: The EROS-PREDICTORS Study. Front Endocrinol (Lausanne). 2020;11:414. Published 2020 Jun 26. https://doi.org/10.3389/fendo.2020.00414.
4. Cadegiani FA, Kater CE. Inter-correlations among clinical, metabolic, and biochemical parameters and their predictive value in healthy and overtrained male athletes: the EROS-CORRELATIONS study. Front Endocrinol (Lausanne). 2019;10:858.
5. Buyse L, Decroix L, Timmermans N, Barbé K, Verrelst R, Meeusen R. Improving the diagnosis of nonfunctional overreaching and overtraining syndrome. Med Sci Sports Exerc. 2019;51:2524.
6. Lewis NA, Redgrave A, Homer M, et al. Alterations in redox homeostasis during recovery from unexplained underperformance syndrome in an elite international rower. Int J Sports Physiol Perform. 2018;13(1):107–11.
7. Theofilidis G, Bogdanis GC, Koutedakis Y, Karatzaferi C. Monitoring exercise-induced muscle fatigue and adaptations: making sense of popular or emerging indices and biomarkers. Sports (Basel). 2018;6(4):153.
8. Cadegiani FA, Kater CE. Enhancement of hypothalamic-pituitary activity in male athletes: evidence of a novel hormonal mechanism of physical conditioning. BMC Endocr Disord. 2019;1:117. https://doi.org/10.1186/s12902-019-0443-7.
9. Kajaia T, Maskhulia L, Chelidze K, Akhalkatsi V, Kakhabrishvili Z. The effects of non-functional overreaching and overtraining on autonomic nervous system function in highly trained athletes. Georgian Med News. 2017;264:97–103.

Chapter 7
The Underappreciated Athlete: Overtraining Syndrome in Resistance Training, High-Intensity Functional Training (HIFT), and Female Athletes

Outline

Overtraining syndrome in resistance training and high-intensity functional training (HIFT): in researches on OTS, these training regimens have been less assessed compared to endurance sports. Although findings are inconsistent, peculiar alterations and triggers are likely found in these populations of athletes, besides additional particularities in HIFT regarding training patterns and biomarkers in both healthy and OTS athletes. In the present chapter, the author will explore specific pathophysiological mechanisms and propose biomarkers of OTS for athletes that practice sports with resistance, including weight-lifting and HIFT. as well as depict the particularities that can be detected in these populations. The author will also provide specific recommendations for the growing number of HIFT athletes.

Particularities of the diagnosis of overtraining syndrome in female athletes: OTS has been predominantly researched in male athletes. However, some particularities in females have been identified, which is highlighted by the author. Also, peculiarities in the research in female athletes are brought in the book.

Sports on the extremes: are resulted from the continuous seek for more difficult tasks to overcome and are represented by bodybuilders during the preparation for the competitions in the stage, fighters that need extreme weight loss to change category, and ultramarathons and triathlons. The author depicts the potential of these circumstances to induce OTS, as well as the potential long-term consequences.

© The Editor(s) (if applicable) and The Author(s), under exclusive license
to Springer Nature Switzerland AG 2020
F. Cadegiani, *Overtraining Syndrome in Athletes*,
https://doi.org/10.1007/978-3-030-52628-3_7

7.1   Overtraining Syndrome in Resistance Training

Unlike endurance and explosive sports, overtraining syndrome has been poorly explored in resistance training. A previous theory hypothesized that on the opposite autonomic manifestation of OTS in endurance athletes, in which the parasympathetic system prevails over sympathetic, leading to reduced heart rate (HR), in weight lifting and other strength athletes affected by OTS, increased HR and predominant sympathetic autonomic system are found.

In a systematic review, from 22 studies initially selected, 8 reported or resulted in decreased performance, whereas 4 described athletes with underperformance who failed to recover during follow-up [1]. Similarly to studies on endurance, explosive, and mixed sports, methodology lacked standardization for the detection of OTS and identification of biomarkers [2].

One of the major contributors for the field of OTS on resistance training is Fry, who has published among the most important studies with weight lifters [3–5]. Indeed, among the 12 studies identified by the systematic review as true FOR, NFOR, or OTS, 7 were published by him as the first author. He identified a sort of hyposensitivity and decreased beta2-adrenergic receptor in the skeletal muscle tissue, with an insufficient compensatory increase of circulating catecholamines. However, more recently Sterczala (co-authored by Fry) demonstrated the opposite effect on the beta2-AR density on the muscle [6].

Only one study, from Fry, evaluated hormonal levels and identified inconclusive findings on testosterone and cortisol levels [3], since total and free testosterone disclosed different results, as well as testosterone/cortisol and free-testosterone/cortisol ratios, while in the study conducted by Sterczala, none of these parameters were altered [6]. In another study using the same sample, Fry identified unaltered catecholamines, but exacerbated responses to exercise-stress stimulation tests [4], whereas in a third study reduced maximum lactate and increased CK levels were detected in OTS in resistance training [5].

Besides Fry and Sterczala, only four other authors (Warren, Margonis, Warren, Nicoli, and Hecksteden) published researches with resistance athletes that disclosed true performance decrement [7–10], whereas only Fry and Margonis evaluated actual OTS (with duration of decreased performance of more than 1 month) [3, 7].

Correspondingly to the complex clinical manifestations and multifactorial nature of OTS, symptoms of potential *maladaptations* in resistance training also include increased and persistent fatigue and muscle soreness, and in almost all cases, additional stressors besides training contributed to the unexplained underperformance state, although continuous training to muscular failure was associated with more severe *maladaptations* (>4 months).

7.2 Overtraining Syndrome in High-Intensity Functional Training (HIFT)

7.2.1 Peculiarities of the High-Intensity Functional Training (HIFT)

High-intensity functional training *(HIFT)*, popularly termed as its patented version (CrossFit), is alleged to improve overall physical conditions through gymnastics, athletics, weight lifting, cycling, running, and rowing abilities [11, 12]. "*In HIFT athletes* exercise in a manner that continuous changes in the combination and patterns of exercises, in order to enhance the ten currently recognized domains of physical conditioning: strength; agility; balance; flexibility; cardiovascular and respiratory resistance; muscular resistance; potency; speed; coordination; and precision [13, 14].

The increasing popularity of such regimens is a response to the monotony of exercise performed alone and repeatedly; moreover, the competitiveness intrinsic to group-based high-intensity functional training is highly entertaining [11, 15, 16].

Despite the popularity of these regimens, there are very few published studies on the metabolic and hormonal aspects of HIFT and similar modalities [17, 18]. Additionally, although overtraining syndrome (OTS) has been clinically observed in HIFT athletes, its incidence in HIFT has only been recently reported [19].

This profusion of healthy and OTS-affected *HIFT athletes* enabled a post hoc analysis of the specific metabolic and clinical aspects of CF. Our aim was to investigate specific endocrine and metabolic responses and eating, social, psychological, and body characteristics of healthy and OTS-affected *HIFT athletes*.

7.2.2 Findings on High-Intensity Functional Training (HIFT) and Related Overtraining Syndrome (OTS)

Results are presented in Tables 7.1, 7.2, 7.3, and 7.4, for results of the hypothalamus-pituitary-adrenal (HPA) axis, hormonal responses to an insulin tolerance test (ITT), basal biochemical and hormonal parameters, and psychological, eating, and social patterns, and body metabolism and composition markers, respectively.

In HIFT, direct stimulation of the adrenal glands through the cortisol response to synthetic ACTH did not show differences (Table 7.1). Conversely, cortisol response to an insulin tolerance test (ITT) was prompter and exacerbated in healthy athletes, compared to OTS and sedentary, while ACTH was lower in OTS compared to healthy HIFT athletes (Table 7.2). GH response to an ITT was also earlier and

Table 7.1 Specific features of high-intensity functional training (HIFT) in healthy and overtraining syndrome (OTS): findings of the hypothalamus-pituitary-adrenal (HPA) axis

Mean (±SD) or median (95% CI)	CF-OTS (n = 9) p vs CF-ATL	CF-ATL (n = 22) p vs NPAC	NPAC (n = 12) p vs CF-OTS	Overall p-value	Pairwise p-value
Cortisol response to CST					
Basal (µg/dL)	14.2 ± 4.8	11.9 ± 3.2	12.1 ± 5.7	n/s	
30' cortisol	19.1 ± 1.9	19.8 ± 2.4	19.7 ± 3.2	n/s	
60' cortisol	22.1 ± 2.3	22.5 ± 2.9	22.9 ± 4.4	n/s	
Cortisol response to ITT					
Basal (µg/dL)	11.8 ± 3.0	12.2 ± 2.9	10.9 ± 2.8	n/s	
During hypoglycemia	12.7 ± 3.3	16.2 ± 5.5*	11.8 ± 3.1	0.01	*0.02 vs NPAC
30' after hypoglycemia	18.8 ± 2.8&	22.0 ± 2.9*	16.9 ± 4.1	0.005	*<0.001 vs NPAC &<0.041 vs CF-ATL
Δ increase: basal to 30'	7.0 ± 1.6	9.8 ± 3.5*	5.9 ± 3.9	0.019	*0.002 vs NPAC
ACTH response to ITT					
Basal (pg/mL)	19.4 (11.9–36.0)	17.9 (6.3–39.2)	21.4 (8.7–37.8)	n/s	
During hypoglycemia	27.9 (7.4–219.2)	54.0 (8.8–169.7)	29.5 (14.8–191.7)	n/s	
30' after hypoglycemia	32.4 (9.9–80.4)&	74.8 (21.9–196.8)	51.4 (22.7–137.5)	0.014	&<0.004 vs CF-ATL
Absolute increase	10.6 (−15.0 to 50.9)&	79.3 (7.2–189.5)	38.0 (0.5–108.8)	0.023	&<0.002 vs CF-ATL
Δ % increase	79%	299%	200%	–	
Salivary cortisol rhythm					
Awakening (pg/mL)	367 ± 254	342 ± 136	266 ± 149	n/s	
30' after awakening	314 ± 133&	522 ± 162*	393 ± 149	0.004	*<0.03 vs NPAC &<0.002 vs CF-ATL
4 PM	186 ± 135	157 ± 85.1	130 ± 57	n/s	
11 PM	103 ± 47	98 ± 39	83 ± 11	n/s	
Cortisol awakening response (CAR)	19%	69%	79%		

OTS overtraining syndrome, *CF-OTS* subgroup of CrossFit athletes affected by OTS, *CF-ATL* subgroup of healthy CrossFit athletes, *NPAC* non-physically active controls, *CST* Cosyntropin stimulation test, *ITT* insulin tolerance test, *SD* standard deviation, *CI* confidence interval

optimized by three to four times in healthy HIFT athletes, compared to HIFT-OTS and sedentary, while basal GH was higher in healthy HIFT athletes compared to sedentary (Table 7.2). Prolactin was increased by two times in response to an ITT in healthy HIFT athletes, while absent in OTS (Table 7.2). Similarly, basal prolactin levels were higher in healthy HIFT athletes than other groups. During ITT, time to

Table 7.2 Specific features of high-intensity functional training (HIFT) in healthy and overtraining syndrome (OTS): hormonal responses to an insulin tolerance test (ITT)

Mean (±SD) or median (95% CI)	CF-OTS (n = 9) p vs CF-ATL	CF-ATL (n = 22) p vs NPAC	NPAC (n = 12) p vs CF-OTS	Overall p-value	Pairwise p-value
GH response to ITT					
Basal (µg/L)	0.12 (0.05–1.03)	0.24 (0.1–1.29)*	0.06 (0.03–0.47)	0.007	*0.002 vs NPAC
During hypoglycemia	0.40 (0.05–4.16)&	3.20 (0.17–34.64)*	0.16 (0.05–8.13)	0.005	*0.005 vs NPAC &<0.016 vs CF-ATL
30' after hypoglycemia	8.41 (0.22–14.35)&	14.20 (2.36–37.63)*	4.80 (0.33–27.36)	0.017	*0.012 vs NPAC &<0.037 vs CF-ATL
Prolactin response to ITT					
Basal (ng/mL)	6.9 (5.3–13.8)&	12.1 (7.1–23.3)	10.6 (7.9–15.7)	0.047	&<0.014 vs CF-ATL
During hypoglycemia	8.9 (4.5–56.3)&	17.0 (12.2–63.8)*	12.20 (7.2–15.9)	0.001	*<0.01 vs NPAC &<0.001 vs CF-ATL
30' after hypoglycemia	15.1 (5.1–24.4)&	23.2 (10.4–68.4)*	10.5 (6.2–43.4)	0.004	*<0.003 vs NPAC &<0.024 vs CF-ATL
Δ response	+3.8 (−0.6 to +12.8)	+11.4 (−5.4 to +55.2)	−1.2 (−4.8 to +30.5)	0.053	
Δ% response	+35%	+91%*	−12%	0.10	*<0.032 vs NPAC
Glucose during ITT					
Basal (mg/dL)	79.1 ± 5.5	81.0 ± 4.8	84.8 ± 6.4	n/s	
During hypoglycemia	17.6 ± 11.6	19.1 ± 8.1	26.2 ± 12.5	n/s	
Time to hypoglycemia (min)	21.0 ± 2.6&	26.2 ± 4.9	29.7 ± 6.8#	0.003	&<0.0003 vs CF-ATL #<0.004 vs CF-OTS
Adrenergic symptoms (0–10)	1.9 ± 2.1&	5.6 ± 3.0	6.0 ± 2.6#	0.002	&<0.003 vs CF-ATL #<0.003 vs CF-OTS
Neuroglycopenic symptoms (0–10)	3.8 ± 3.5	5.2 ± 3.0	4.3 ± 3.0	n/s	

OTS overtraining syndrome, *CF-OTS* subgroup of CrossFit athletes affected by OTS, *CF-ATL* subgroup of healthy CrossFit athletes, *NPAC* non-physically active controls, *ITT* insulin tolerance test, *SD* standard deviation, *CI* confidence interval

Table 7.3 Specific features of high-intensity functional training (HIFT) in healthy and overtraining syndrome (OTS): basal biochemical and hormonal results

Mean (±SD) or median (95% CI)	CF-OTS ($n = 9$)	CF-ATL ($n = 22$)	NPAC ($n = 12$)	Overall p-values	Pairwise p-values
Biochemical basal levels					
Hematocrit (%)	44.8 ± 2.8	43.8 ± 2.4[*]	46.4 ± 2.4	0.05	[*]<0.028 vs NPAC
MCV (fL)	85.8 ± 2.8	87.5 ± 4.9	88.3 ± 3.9	n/s	
Neutrophils (/mm³)	2898 ± 629[&]	3917 ± 1431	3186 ± 847	0.048	[&]<0.032 vs CF-ATL
Lymphocytes (/mm³)	2502 ± 603	2116 ± 646[*]	2820 ± 810	0.039	[*]<0.016 vs NPAC
Platelets (x 1000/ mm³)	265 ± 40	231 ± 38	225 ± 63	n/s	
LDLc (mg/dL)	109 ± 17	120 ± 70	104 ± 17	n/s	
HDLc (mg/dL)	52.3 ± 6.2	61.1 ± 14.7	51.5 ± 8.7	n/s	
Triglycerides (mg/ dL)	82 ± 28	90 ± 44	152 ± 87	n/s	
Pairwise p-values	n/s	n/s	n/s		
Creatinine (mg/dL)	1.17 ± 0.10	1.17 ± 0.15	1.01 ± 0.10	n/s	
Pairwise p-values	n/s	n/s	n/s		
Vitamin B12 (pg/ mL)	529 ± 278	552 ± 197	442 ± 155	n/s	
Neutrophil to lymphocyte ratio	1.21 ± 0.33	2.10 ± 1.32	1.27 ± 0.73	0.006	
Pairwise p-values	0.013	<0.001	n/s		
Platelet to lymphocyte ratio	112.1 ± 34.2	120.1 ± 45.6	82.4 ± 19.5	0.015	
Pairwise p-values	n/s	0.004	n/s		
Eosinophils (/mm³)	132 (55–459)	91 (29–375)	193 (51–549)	n/s (0.08)	
Pairwise p-values	n/s	0.026	n/s		
CRP (mg/dL)	0.12 (0.04–1.79)	0.06 (0.02–0.49)	0.08 (0.02–0.23)	n/s	
ESR (mm/1 h)	4.0 (2.0–12.0)	2.0 (2.0–12.9)	2.0 (1.5–6.0)	n/s	
Creatine kinase (U/L)	324 (122–1650)	359 (91–789)	105.5 (80–468)	0.005	
Pairwise p-values	n/s	0.006	0.005		
Ferritin (ng/mL)	190.1 (48.7–380.8)	171.0 (83.2–384.2)	229.4 (90.5–540.4)	n/s	
Lactate (nMol/L)	1.10 (0.79–2.00)	0.82 (0.44–1.44)	1.17 (0.57–1.57)	0.044	
Pairwise p-values	0.025	0.048	n/s		
Basal hormone levels					
Total testosterone (ng/dL)	412.3 ± 210.5	541.8 ± 177.0	405.9 ± 156.3	0.023	

Table 7.3 (continued)

Mean (±SD) or median (95% CI)	CF-OTS (n = 9)	CF-ATL (n = 22)	NPAC (n = 12)	Overall p-values	Pairwise p-values
Pairwise p-values	0.038	0.042	n/s		
Estradiol (pg/mL)	39.5 ± 13.1	28.6 ± 14.4	25.7 ± 11.2	0.062	
Pairwise p-values	0.038	n/s	0.03		
IGF-1 (ng/mL)	181 ± 52	175 ± 54	184 ± 59	n/s	
Free T3 (pg/mL)	3.2 ± 0.6	3.2 ± 0.5	3.3 ± 0.5	n/s	
TSH (μUI/mL)	2.4 ± 1.0	1.8 ± 0.8	1.8 ± 0.9	n/s	
Total catecholamines (μg/12 h)	273 ± 203	172 ± 66	133 ± 54	0.016	
Pairwise p-values	0.05	n/s	0.014		
Noradrenaline (μg/12 h)	31.6 ± 10.3	22.3 ± 12.2	17.4 ± 9.1	0.018	
Pairwise p-values	0.035	n/s	0.005		
Epinephrine (μg/12 h)	3.0 (1.4–6.8)	2.0 (1.0–10.7)	2.0 (0.6–7.9)	n/s	
Dopamine (μg/12 h)	237 ± 196	147 ± 58	114 ± 45	n/s	
Pairwise p-values	n/s	n/s	n/s		
Total metanephrines (μg/12 h)	284 ± 176	223 ± 79	221 ± 107	n/s	
Metanephrine (μg/12 h)	53.3 ± 37.2	45.0 ± 26.9	41.7 ± 25.4	n/s	
Normetanephrine (μg/12 h)	95.2 ± 59.1	86.6 ± 39.7	90.8 ± 48.2	n/s	
Urinary volume (mL/12 h)	1125 ± 466	1177 ± 776	1045 ± 656	n/s	
Pairwise p-values	n/s	n/s	n/s		
Testosterone to estradiol ratio	10.9 ± 4.6	21.8 ± 10.1	20.3 ± 13.0	0.008	
Pairwise p-values	0.002	n/s	0.037		
Testosterone to cortisol ratio	38.7 ± 24.3	45.8 ± 15.9	39.6 ± 21.3	n/s	
Pairwise p-values	n/s	n/s	n/s		

OTS overtraining syndrome, *CF-OTS* subgroup of CrossFit athletes affected by OTS, *CF-ATL* subgroup of healthy CrossFit athletes, *NPAC* non-physically active controls, *SD* standard deviation, *CI* confidence interval, *CRP* C-reactive protein, *ESR* erythrocyte sedimentation rate, *IGF-1* insulin-like growth factor 1, *TSH* thyroid stimulating hormone, *MCV* median corpuscular volume, *LDL* low density lipoprotein, *HDL* high density lipoprotein

Table 7.4 Specific features of high-intensity functional training (HIFT) in healthy and overtraining syndrome (OTS): psychological, eating, and social patterns, and body metabolism and composition analyses

Mean (±SD) or median (95% CI)	CF-OTS (n = 9)	CF-ATL (n = 22)	NPAC (n = 12)	
Eating patterns				
>0.5 g/kg of CHO intake after training (% of affected subjects)	33%	55%	n/a	–
Daily use of whey protein (% of affected subjects)	20%	55%	0%	–
Follow a diet plan? (% of affected subjects)	89%	86%	0%	–
Estimated daily calorie intake per weight (kcal/kg/day)	26.7 ± 3.3	52.5 ± 6.4	54.7 ± 3.9	<0.001
Pairwise p-values	<0.001	n/s	<0.001	
Daily protein calorie intake (g/kg/day)	1.64 ± 0.62	2.82 ± 0.86	0.92 ± 0.49	<0.001
Pairwise p-values	0.005	<0.001	n/s	
Daily CHO calorie intake (g/kg/day)	3.09 ± 1.53	6.95 ± 3.14	9.85 ± 4.49	<0.001
Pairwise p-values	0.008	0.039	<0.001	
Daily fat calorie intake (g/kg/day)	0.87 ± 0.51	1.49 ± 1.27	1.29 ± 0.91	0.072
Pairwise p-values	n/s	n/s	n/s	
Sleeping and social patterns				
Sleep quality (0–10)	6.6 ± 1.7	8.2 ± 1.7	7.0 ± 0.9	0.028
Pairwise p-values	0.018	0.050	n/s	
Duration of night sleep (hours)	6.0 ± 1.2	6.5 ± 0.8	6.7 ± 0.9	n/s
Pairwise p-values	n/s	n/s	n/s	
Initial insomnia (% of affected subjects)	33%	23%	25%	–
Terminal insomnia (% of affected subjects)	22%	9%	42%	–
Wake up more than two times at night (% of affected subjects)	44%	18%	50%	–
Number of hours of work or study (hours/day)	8.4 ± 2.5	6.9 ± 1.5	9.3 ± 2.9	0.026
Pairwise p-values	n/s	0.009	n/s	
Libido (0–10)	6.2 ± 3.3	7.7 ± 1.6	8.8 ± 1.0	0.047
Pairwise p-values	n/s	n/s	0.017	
Profile of Mood States (POMS) questionnaire (total score and subscales)				
Total score (−32 to +120)	+46.0 (+11.8 to +92.2)	−7.5 (−23.9 to +17.8)	+30.5 (−3.3 to +81.7)	<0.001
Pairwise p-values	<0.001	<0.001	n/s	
Anger (0–48)	17.0 (4.0–20.8)	5.0 (1.0–15.7)	9.0 (4.0–24.5)	0.023
Pairwise p-values	0.019	0.034	n/s	
Confusion (0–28)	4.0 (1.4–17.8)	2.0 (0.0–5.0)	5.5 (1.6–10.9)	0.004
Pairwise p-values	0.003	0.017	n/s	

Table 7.4 (continued)

Depression (0–60)	4.0 (0.4–16.6)	0.5 (0.0–9.9)	8.5 (1.6–19.9)	0.003
Pairwise p-values	0.037	0.001	n/s	
Fatigue (0–28)	17.0 (11.2–27.2)	1.5 (0.0–4.9)	8.0 (1.1–18.8)	<0.001
Pairwise p-values	<0.001	0.002	0.047	
Tension (0–36)	14.0 (5.4–23.6)	6.0 (1.0–14.8)	17.5 (11.0–25.5)	<0.001
Pairwise p-values	0.006	<0.001	n/s	
Vigor (0–32)	10.0 (7.4–21.6)	26.0 (21.1–28.0)	19.5 (11.1–27.0)	<0.001
Pairwise p-values	<0.001	0.007	n/s	
Body composition and metabolism				
Measured to predicted BMR (%)	103.7 ± 9.7	109.5 ± 9.5	100.2 ± 9.5	0.026
Pairwise p-values	n/s	0.015	n/s	
Percentage of fat oxidation compared to total BMR (%)	36.8 ± 19.7	60.0 ± 17.2	29.3 ± 14.2	<0.001
Pairwise p-values	0.003	<0.001	n/s	
Body fat (%)	14.9 ± 4.1	11.2 ± 4.2	21.3 ± 7.2	<0.001
P-value between groups and subgroups	n/s	<0.001	n/s	
Muscle mass (%)	48.8 ± 3.0	50.4 ± 2.1	44.1 ± 3.9	<0.001
Pairwise p-values	0.021	<0.001	n/s	
Body water (%)	60.9 ± 2.8	64.6 ± 2.6	57.0 ± 4.9	<0.001
Pairwise p-values	0.025	<0.001	n/s	
Extracellular water (%)	15.8 ± 9.8	19.5 ± 12.1	31.7 ± 16.7	0.007
Pairwise p-values	n/s	0.023	0.015	
Visceral fat (cm^2)	55.0 ± 24.3	36.7 ± 20.8	83.8 ± 41.8	<0.001
Pairwise p-values	n/s	<0.001	n/s	
Chest-to-waist circumference ratio	1.26 ± 0.06	1.26 ± 0.06	1.16 ± 0.07	<0.001
Pairwise p-values	n/s	<0.001	0.004	

OTS overtraining syndrome, *CF-OTS* subgroup of CrossFit athletes affected by OTS, *CF-ATL* subgroup of healthy CrossFit athletes, *NPAC* non-physically active controls, *SD* standard deviation, *CI* confidence interval, *BMR* basal metabolic rate, *CHO* carbohydrate

hypoglycemia was shorter in OTS than healthy HIFT athletes and sedentary, whereas adrenergic symptoms during hypoglycemia were less noticed in OTS than in healthy HIFT athletes and sedentary (Table 7.2).

In the SCR, salivary cortisol (SC) levels were higher in healthy HIFT than OTS and sedentary when collected 30 minutes after awakening, while similar at awakening, 4 PM and 11 PM (Table 7.1). Cortisol awakening response (CAR), measured through increase (%) between awakening and 30 minutes after awakening, was lowest in HIFT-OTS.

Among basal hormones, testosterone was higher in healthy HIFT athletes than in HIFT-OTS and sedentary, estradiol was lower in HIFT-OTS compared to the other groups, and consequently, the testosterone-to-estradiol ratio was approximately two times lower in HIFT-OTS than in healthy HIFT and sedentary (Table 7.3).

Thyroid-stimulating hormone (TSH), free T3 (fT3), and insulin-like growth factor 1 (IGF-1) were similar between groups.

Total nocturnal urinary catecholamines (NUC), noradrenaline, and dopamine fractions were higher in HIFT-OTS than in healthy HIFT. Conversely, total and fractioned metanephrines were similar between groups (Table 7.3).

In regard to basal biochemical markers, neutrophils were lower in HIFT-OTS than healthy HIFT athletes, lymphocytes were lower in healthy HIFT compared to sedentary, neutrophil-to-lymphocyte ratio was higher in healthy HIFT than in HIFT-OTS and sedentary, eosinophils were lower in healthy HIFT than in sedentary controls, and lactate was higher in OTS than in healthy HIFT athletes (Table 7.3). Platelets, platelet-to-lymphocyte ratio, lipid profile, creatinine, vitamin B12, creatine kinase (CK), C-reactive protein (CRP), and erythrocyte sedimentation rate (ESR) were similar between groups.

HIFT-OTS had worse sleep quality compared to healthy HIFT athletes, while sleep duration was similar between groups (Table 7.4). Healthy HIFT athletes tended to have lower cognitive demands than sedentary and slightly lower than HIFT-OTS, although not significantly. Self-reported libido was lower in HIFT-OTS athletes compared to other groups (Table 7.4).

Although both groups of athletes similarly followed a diet plan, HIFT-OTS athletes ingested less than half the calories than healthy HIFT athletes and sedentary, owing to the carbohydrate daily ingestion of less than three times among OTS-HIFT athletes, compared to healthy HIFT athletes and sedentary, whereas healthy HIFT athletes had higher daily protein intake than HIFT-OTS and sedentary controls (Table 7.4).

On the total and subscales of Profile of Mood States (POMS), HIFT-OTS had worse than healthy HIFT athletes for total score and all sub-scales: depression, fatigue, anger, confusion, tension, and vigor subscales. Compared to sedentary, healthy HIFT athletes had improved moods for all scales, while HIFT-OTS scored similar to sedentary controls, except for worse fatigue and vigor in HIFT-OTS (Table 7.4).

Percentage of fat oxidation compared to total basal metabolic rate (BMR) was significantly higher in healthy HIFT athletes than HIFT-OTS and sedentary, whereas measured-to-predicted BMR ratio was higher in CF-ATL than in sedentary, while not-significantly higher than HIFT-OTS. Body fat was lower in healthy HIFT athletes than in sedentary, while muscle mass and body water were higher in healthy HIFT athletes than in HIFT-OTS and sedentary controls (Table 7.4). Extracellular water was higher in sedentary compared to both healthy and OTS HIFT athletes.

7.2.3   Specific Findings on High-Intensity Functional Training (HIFT)-Related Overtraining Syndrome (OTS)

Particularities of the findings on both healthy HIFT athletes and HIFT-OTS are described in Table 7.5. In regard to the OTS particularities in HIFT, OTS-related *HIFT* tended to show less compromised early GH, cortisol, and prolactin responses and higher overall hormone responses compared to overall OTS.

Table 7.5 Summary of the findings of *CrossFit* in healthy and OTS athletes

Tests	Markers	Adaptive changes in healthy crossfitters	Specific changes in "crossfitters" (not observed in overall athletes)	Features of OTS in crossfitters
Cortisol response to ITT	Cortisol levels during hypoglycemia (μg/dL)	Higher than sedentary Only significantly increased in healthy crossfitters	Both CF-OTS and CF-ATL showed slightly higher levels than OTS and ATL, respectively	Not statistically different compared to healthy crossfitters[a]
	Cortisol levels 30 minutes after hypoglycemia (μg/dL)	Higher than OTS crossfitters and sedentary		Lower differences between OTS and healthy crossfitters, but still statistically different
	Cortisol increase during an ITT (μg/dL)		Higher in CF-OTS than overall OTS athletes	Not statistically different compared to healthy crossfitters[a]
ACTH response to ITT	ACTH levels 30 minutes after hypoglycemia (pg/mL)	–		Lower than healthy crossfitters and sedentary (statistically lower than healthy crossfitters)
	ACTH increase during an ITT (pg/mL)	–		Lower than healthy crossfitters and sedentary (statistically lower than healthy crossfitters)
GH response to ITT	Basal GH (μg/L)	Higher than sedentary		Although two times lower than healthy crossfitters, was not statistically significant[a]
	GH levels during hypoglycemia (μg/L)	Higher than OTS crossfitters and sedentary. Only significantly increased in healthy crossfitters	Slightly higher in CF-ATL (3.2 μg/L) than ATL (2.5 μg/L)	Not significant increase in the early response
	GH levels 30 minutes after hypoglycemia (μg/L)	Higher than OTS crossfitters and sedentary	Higher in CF-OTS (8.4 μg/L) than OTS (3.3 μg/L) and in CF-ATL (14.2 μg/L) than ATL (12.7 μg/L)	Lower differences between OTS and healthy crossfitters, but still substantially and statistically different (almost two times lower)

(continued)

Table 7.5 (continued)

Tests	Markers	Adaptive changes in healthy crossfitters	Specific changes in "crossfitters" (not observed in overall athletes)	Features of OTS in crossfitters
Prolactin response to ITT	Basal prolactin levels (ng/mL)	–		Significantly lower than healthy crossfitters[a]
	Prolactin levels during hypoglycemia (ng/mL)	Higher than OTS crossfitters and sedentary		Non-significant increase
	Prolactin levels 30 minutes after hypoglycemia (ng/mL)	Higher than OTS crossfitters and sedentary	Slightly higher in CF-OTS (15.1 ng/mL) than OTS (11.3 ng/mL), but still did not increase significantly in response to an ITT	Non-significant increase.
Glucose response and related symptoms during ITT	Time to hypoglycemia (minutes)	–		Lower than healthy crossfitters and sedentary[a]
	Adrenergic symptoms (0–10)	–		More important symptoms than healthy crossfitters and sedentary
Salivary cortisol rhythm (SCR)	Cortisol 30' after awakening (pg/mL)	Higher than other groups		
	Cortisol awakening response (CAR) (%)			More prominent differences (almost 3.5 times lower) than healthy crossfitters
Biochemical markers	Creatine kinase (CK) (U/L)			Not different and actually slightly lower levels, when compared to healthy crossfitters[a]
	Lactate (nMol/L)	Lower than OTS crossfitters and sedentary		
	Neutrophils (/mm³)	Higher than OTS crossfitters		
	Neutrophil-to-lymphocyte ratio	Higher than OTS crossfitters and sedentary		
	Platelet-to-lymphocyte ratio	Higher than sedentary		

Table 7.5 (continued)

Tests	Markers	Adaptive changes in healthy crossfitters	Specific changes in "crossfitters" (not observed in overall athletes)	Features of OTS in crossfitters
Hormonal markers	Total catecholamines (µg/12 h)		Not significantly increased in CF-ATL compared to NPAC. Slightly but significantly higher than healthy crossfitters and sedentary	Slightly but significantly higher than healthy crossfitters and sedentary
	Noradrenaline (µg/12 h)		Not significantly increased in CF-ATL compared to sedentary.	Higher than healthy crossfitters and sedentary[a]
	Estradiol (pg/dL)			Higher than healthy crossfitters and sedentary
	Testosterone (ng/dL)	Higher than OTS crossfitters and sedentary		
	Testosterone-to-estradiol ratio			Lower than healthy crossfitters and sedentary
Eating patterns	Calorie intake (kcal/kg/day)	Higher than OTS crossfitters and sedentary		Lower than healthy crossfitters and sedentary
	Protein intake (g/kg/day)	Higher than OTS crossfitters and sedentary		
	Carbohydrate intake (g/kg/day)		Lower in CF-OTS than OTS (3.09 vs 3.76 g/kg/day), while similar between CF-ATL and ATL	Lower than healthy crossfitters and sedentary
	Follow a diet plan? (% of affected subjects)			Equally followed as healthy crossfitters (which was not able to prevent OTS)
	Daily use of whey protein			Less used than healthy crossfitters.
	Post-workout CHO intake (>0.5 g/kg)			Less used than healthy crossfitters

(continued)

Table 7.5 (continued)

Tests	Markers	Adaptive changes in healthy crossfitters	Specific changes in "crossfitters" (not observed in overall athletes)	Features of OTS in crossfitters
Social patterns	Self-reported sleep quality (0–10)	Better sleep than OTS crossfitters and sedentary	Better in CF-OTS than OTS (6.6 vs 6.2), but still worse than ATL	
	Self-reported libido (0–10)			Worse libido than sedentary, but similar to healthy crossfitters[a]
	Work and/or study (h/day)	Worked and/or studied less than sedentary		Worked and/or studied similarly to healthy crossfitters
Psychological patterns	Profile of Mood State (POMS) questionnaire total score (−32 to +120) and depression, tension, vigor, confusion, and anger subscales	Better than OTS crossfitters and sedentary – in all subscales	Slight less fatigues and tension in CF-OTS than overall OTS	–
	POMS fatigue subscale (0–28)			Worse than sedentary
Body metabolism analysis	Measured to predicted BMR ratio (%)	Higher than sedentary		Similar to healthy crossfitters[a]
	Fat oxidation (% of total BMR)	Higher than OTS crossfitters and sedentary		
Body composition	Body fat percentage (%)	Lower than sedentary	Lower in CF-OTS than OTS (14.9 vs 17.0%)	Similar to healthy crossfitters[a]
	Muscle mass weight (%)	Higher than OTS crossfitters and sedentary		
	Body water percentage (%)	Higher than OTS crossfitters and sedentary		
	Visceral fat (cm^2)	Lower than sedentary		Similar to healthy crossfitters[a]
	Extracellular water	Higher than sedentary		Higher than sedentary
	Chest-to-waist circumference	Higher than sedentary		Higher than sedentary

ITT insulin tolerance teste, *OTS* overtraining syndrome, *ITT* insulin tolerance test, *GH* growth hormone
[a]Not found in overall OTS athletes

Among the specific features of OTS in HIFT, a lack of increased CK has been observed [19–22] when compared to healthy "HIFT athletes," possibly showing a less compromised muscle function, and faster recovery, when compared to overall OTS, although lactate levels were still significantly higher than healthy "HIFT athletes."

The higher catecholamine levels in OTS-affected HIFT athletes, when compared to overall OTS athletes, may aid the optimization of an "overcompensation" process to maintain the energy levels of OTS-affected athletes, which may explain why HIFT athletes seem to be clinically and psychologically less affected [23, 24].

In regard to psychological aspects, the increased motivation and reduced monotony associated with HIFT may also explain the less affected moods among OTS "HIFT athletes." In relation to eating patterns, HIFT-OTS had lower carbohydrate intake compared to all OTS, while approximately three times lower than healthy HIFT athletes, despite the similar percentages of OTS and healthy HIFT athletes that followed long-term low-carbohydrate diet plans, although healthy HIFT athletes tended to have larger number of ad libitum free meals with increased carbohydrate intake, compared to HIFT-OTS.

Although OTS "HIFT athletes" presented lower BMR, lower fat oxidation, and higher body fat compared to healthy "HIFT athletes," these differences were not statistically different, showing a less compromised body metabolism and composition in HIFT-specific OTS, compared to overall OTS. These could be consequences of the less affected hormonal optimization and the higher catecholamines found in HIFT-OTS compared to all OTS athletes.

7.2.4 *Discussion on High-Intensity Functioning Training (HIFT)-Related Overtraining Syndrome (OTS)*

Compared to other sport modalities, HIFT has unique patterns, as it requires multiple abilities, given the complexity and irregularity of its activities. These characteristics may lead to unique or enhanced adaptive changes, compared to athletes that exclusively perform endurance sports, such as triathlon or long-distance running; strength (or weightered), such as weight lifting; or anaerobic and explosion sports. Previous studies on HFIT presented data only on some orthopedic and psychological aspects of the HFIT practitioners, but not on metabolic or hormonal aspects.

When analyzed altogether, most of the findings of the EROS study persisted in subgroups of HIFT athletes, despite the smaller number of athletes, which reinforces the large extension and consistency of the differences between healthy and OTS-affected "HIFT athletes" and between healthy "HIFT athletes" and sedentary, eliciting the presence of multiple adaptations to HIFT, while these adaptive changes were lost under OTS. Indeed, the adaptive changes observed in overall healthy athletes [25–28] were observed in healthy HIFT athletes, who also yielded earlier and enhanced cortisol, GH, and prolactin responses to an ITT, increased neutrophils, lower lactate, increased testosterone, improved sleep quality, better psychological states, increased measured-to-predicted basal metabolic rate (BMR) ratio and fat oxidation, less body fat, more muscle mass, and better hydration, when compared to healthy sedentary controls.

However, unlike overall sports, total and fractioned NUC were not increased in HIFT athletes, which means that this specific adaptation unlikely occurs in HIFT athletes. Other specific findings in healthy HIFT athletes compared to overall healthy athletes have not able to be detected because the HIFT athletes represented almost 90% of the athletes, while all non-HIFT athletes practiced both endurance and strength sports modalities, which makes these athletes possibly resemble HIFT athletes' responses.

In regard to the OTS particularities in HIFT, OTS-related HIFT apparently seem to be slightly milder than overall OTS, as HIFT-OTS tended to show earlier and less flattened hormonal responses to an ITT, milder muscle compromising, lower psychological and sleep dysfunctions, and less affected body metabolism and composition. The detection of these differences was performed despite the relative low number of participants, which shown that these peculiarities are highly likely present. Despite the attenuated changes evident in OTS HIFT athletes, more than 90% of the markers of OTS revealed by the EROS study were still present in OTS "HIFT athletes [25–28].

Besides the specific training patterns, peculiar diets, particularly those with low carbohydrate content, have been spread and become popular among HIFT athletes [29–31]. Low carbohydrate intake has become part of the current sports culture, whose alleged benefits have been supposedly reinforced by scientific and non-scientific literature, despite contradictory findings [32, 33].

Low carbohydrate intake was found to be part of both healthy and OTS-affected HIFT athletes' routine. However, as mentioned earlier, healthy HIFT athletes tended to have three times more unplanned high-carbohydrate meals than OTS-HIFT athletes, which raised the mean daily carbohydrate intake of healthy HIFT athletes when analyzed through a 7-day dietary record [27] that could be unnoticed if the record was performed for fewer days. The compensatory high-carbohydrate meals could be an unconscious protective process against starvation, which may have been critical to prevent these individuals from OTS.

Low calorie and low carbohydrate intake, as a sort of "philosophy of life," likely played a central role in development of OTS among HIFT athletes, which may currently be the most important trigger of OTS in HIFT. Compensatory moments of increased carbohydrate intake may be necessary to protect from burnout and OTS and to maintain continuous improvement of sports performance [31, 34].

HIFT regimens that aim to develop overall physical condition are perhaps the most comprehensive sport modality because of the different skills required [24, 35]. They may induce adaptive changes that are not caused by pure endurance or strength sports [24]. Therefore, the findings observed in healthy HIFT athletes may not be present in healthy athletes of other sports and require therefore cautious to extrapolate its findings.

Fig. 7.1 General and specific adaptations that occur in healthy high-intensity functional training (HIFT) athletes

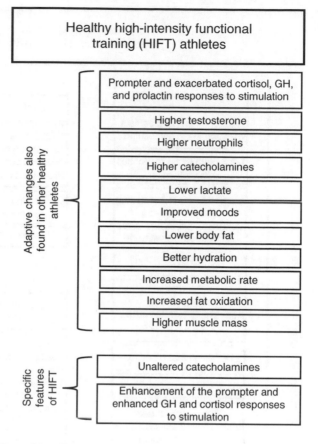

Conversely, OTS in HIFT seems to present the major findings of OTS, although less compromised than overall OTS [25–28]. The summary of the general and specific findings and in healthy and OTS CF athletes, and are shown in Figs. 7.1 and 7.2, respectively.

7.2.5   Recommendations for High-Intensity Training Functioning (HIFT) Athletes to Prevent Overtraining Syndrome

Specific recommendations for HIFT athletes can be provided from the particular findings in this specific population of athletes. However, it is noteworthy the same recommendations are likely applicable to other sports.

1. A long-term low-carbohydrate diet (with carbohydrate intake <5.0 g/kg/day) for more than 8 weeks, without compensatory high-carbohydrate meals, may be a

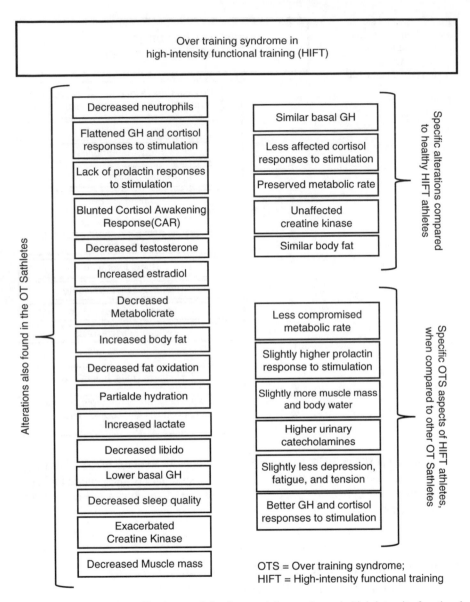

Fig. 7.2 General and specific characteristic of overtraining syndrome in high-intensity functional training (HIFT)

major cause of OTS, or even states of fatigue or any unexplained loss of performance.

2. For low-carbohydrate diet enthusiasts, allowing free meals (i.e., uncontrolled for the amount of each macronutrient) may have a protective role against OTS and may boost performance and beneficial adaptive changes.

3. HIFT athletes may not notice early signs of OTS, including fatigue and mood changes, as they are highly motivated by this sports – as observed in our findings;

thus, a more active search for early signs of OTS may be helpful to prevent this condition.

4. Once following a training periodization plan, excessive training may not be a major concern for OTS

5. Atypical signs of imminent OTS, including paradoxical muscle loss or fat gain, despite a restricted diet, or the need for longer resting periods (due to prolonged muscle recovery) may be important tools to avoid OTS.

6. Cognitive activities (studying or working) for more than 8 h a day may be an outsider trigger of OTS, as recovery process during the resting period may be impaired in case of excessive concomitant cognitive effort.

7. Sleep quality, rather than sleep duration, plays an important role for the prevention of OTS; hence, sleep hygiene, including lights, smart phones, social medias and TVs turned off, and relaxation exercises are useful tools for high-performance HIFT athletes.

8. For an OTS preventive approach including a calorie intake >35 kcal/kg/day, carbohydrate intake >5.0 g/kg/day, protein intake >1.6 g/kg/day, working and/or studying <8 h/day following the training plan, and having good sleep quality will likely prevent almost all cases of OTS in HIFT.

7.3 Particularities of the Diagnosis of Overtraining Syndrome in Female Athletes

There are additional challenges for the research, investigation, and diagnosis of OTS in female athletes, since hormonal responsiveness, metabolic aspects including fat oxidation, metabolic rate and glucose metabolism, and several other parameters have wide variations throughout the menstrual cycle [36].

Although the use of oral contraceptives tends to maintain more stable hormonal levels, as hormonal responsiveness become acyclic, their hormonal and metabolic profile are distinct from those females not using these hormones, which precludes proper comparisons between women taking and not taking oral contraceptives.

In addition, polycystic ovary syndrome (PCOS), a highly prevalent condition in females (between 15% and 20% of clinically present PCOS among women in menacme) [37, 38], has substantial influences on hormonal and metabolic evaluation, since increased androgen and insulin levels inherently found in PCOS have large interferences on other hormonal responsiveness, fat and glucose metabolism and oxidation, and body composition. Particularly in sports, the prevalence of PCOS seems to be even higher, once female athletes with PCOS tend to have better performance in most sports due to the typical hyperandrogenism [39]. Since the basal and physiological hormonal profile of females with PCOS is different from those without PCOS, it is expected that its biochemical presentation in OTS is also different between PCOS and non-PCOS.

Besides, while OTS has been majorly studied in male athletes [1], the few studies in females did not yield specific biomarkers, mechanisms, or understanding of OTS in women. As the biological differences of hormonal production, action, and secondary signaling and metabolic profile between males and females are great, studies on OTS should not evaluate males and females altogether, and alterations in OTS are likely sex-specific and therefore cannot be extrapolated for the understanding of OTS in the other sex.

In summary, phases of menstrual cycle, use of oral contraceptives, presence of PCOS, and shortage of studies in females hamper the evaluation of OTS in this population. Instead, women have been extensively studied for the female athlete triad, also termed as relative energy deficiency of the sport (RED-S), as discussed in another chapter (Chap. 9).

7.4   The Sports on the Extremes

Challenges are part of human nature, who is always seeking for more difficult tasks to overcome. A growing number of athletes are undergoing exhausting sports demands and extreme situations, both leading to potential long-term harms. However, little is known regarding specific endocrine and metabolic effects under extreme circumstances.

Standardized parameters and tests should be employed in observational studies of a variety of situations of exposure to extremes. The use of standardized methods will allow appropriate comparisons of the hormonal and metabolic behaviors between usual and extreme circumstances.

Data should be collected throughout the process that will eventually lead to the extreme situation, including its beginning, during the process, during the extreme moment per se, immediately after, and in the long-run, as dysfunctions during the recovery may be as or more severe than the dysfunctions during the event and may take longer to recover. Testing for the proposed parameters during all these moments would provide sufficient information for the understanding of the biochemical and metabolic behaviors along the processes for different extreme scenarios.

Some of the extreme circumstances that deserve specific research, because of the large number of practitioners, or due to the possible interesting and insightful learnings from these extremes include:

1. Fighters undergo severe dehydration process for rapid weight loss immediately before fights, aiming to participate in lower weight classes [40–42].
2. Bodybuilders undergo an intensification of the "cutting" process aiming to compete on the stage with the thinnest subcutaneous they can achieve [43–46].
3. Ultramarathons and ultra-triathlons, which overpass all the organism capacity to repair oxidative damage and trigger multiple "alternative" metabolic processes in order to maintain functioning despite the loss of all energy sources and mechanisms of repair [47–51].
4. Mountain climbers, in which the shortage of oxygen induces multiples adaptations, which, together with the exhaustive physical effort under the almost inhos-

pitable condition, becomes among the most challenging circumstances [52–54] that athletes undergo.

Fighters and bodybuilders undergo similar dehydration process, with severe salt and water restrictions for weeks and days prior to the competition, respectively. Ultramarathons, IronMan, and some other extreme competitions surpass any human regular ability to deal with lack of energy and mechanisms of repair and excessive oxidative stress.

Besides the data collected sequentially throughout the processes, an estimation of the area under the curve (AUC) of each of the hormones and other parameters will help understand the overall level of exposure to each parameter, while the angles of curve along the process should help understand the hormonal and metabolic behavioral patterns.

In summary, observational studies of extreme situations may tell how metabolism behaves during and after the events, short- and long-term consequences, and the recovery process.

With regard to the correlation between sports on the extremes and OTS, one should assume that any extreme situation will trigger at least a functioning overreaching. All athletes that have undergone at least one of the abovementioned situations should promptly receive support to overcome the hostile circumstances that will inherently lead to a short-term depletion of energy and/or mechanisms of repair. Actual OTS is likely underreported but prevailing in ultra-performance athletes, in both endurance and resistance sports.

7.5 Conclusion

Overtraining syndrome in resistance training has been poorly described, and the few studies that assessed this population of athletes showed conflicting findings, although the continuous training to muscular failure has been consistently associated with more severe *maladaptations* that occur in OTS.

High-intensity functional training (HIFT) athletes disclosed unique findings of multiple hormonal, biochemical, and metabolic adaptations to sports, possibly resulted from the unique patterns of this sport modality, that requires multiple abilities. Conversely, HIFT athletes affected by OTS presented impaired cortisol, GH, and prolactin responses to an ITT, decreased testosterone and increased estradiol, increased catecholamines, impaired psychological moods, decreased fat oxidation, and decreased muscle mass and hydration when compared to healthy HIFT athletes, while similar when compared to sedentary. Compared to overall OTS, HIFT-OTS disclosed milder dysfunctions. Finally, OTS in HIFT was remarkably triggered by a long-term low-carbohydrate, relative low-protein, and low-calorie diet, without compensatory high-carbohydrate meals.

Particularities in the research on females affected by OTS include the phases of menstrual cycle, use of oral contraceptives, and presence of PCOS. Shortage of studies in females hamper the evaluation of OTS in this population, although females have been much less assessed than males in OTS research.

Further observational studies may tell how metabolism behaves during and after sports events that force athletes to undergo extreme situations, short- and long-term consequences, and the recovery process.

References

1. Grandou C, Wallace L, Impellizzeri FM, Allen NG, Coutts AJ. Overtraining in resistance exercise: an exploratory systematic review and methodological appraisal of the literature. Sports Med. 2020;50(4):815–28.
2. Grandou C, Wallace L, Coutts A, Bell L, Impellizzeri FM. Symptoms of overtraining in resistance exercise: international cross-sectional survey. In review.
3. Fry AC, Kraemer WJ, Ramsey LT. Pituitary-adrenal-gonadal responses to high-intensity resistance exercise overtraining. J Appl Physiol. 1998;85(6):2352–9.
4. Fry AC, Kraemer WJ, Van Borselen F, et al. Catecholamine responses to short-term high-intensity resistance exercise over- training. J Appl Physiol. 1994;77(2):941–6.
5. Fry AC, Kraemer WJ, Van Borselen F, et al. Performance decrements with high-intensity resistance exercise overtraining. Med Sci Sports Exerc. 1994;26(9):1165–73.
6. Sterczala A, Fry A, Chiu L, et al. β2-adrenergic receptor maladaptations to high power resistance exercise overreaching. Hum Physiol. 2017;43(4):446–54.
7. Margonis K, Fatouros IG, Jamurtas AZ, et al. Oxidative stress biomarkers responses to physical overtraining: implications for diagnosis. Free Radic Biol Med. 2007;43(6):901–10.
8. Nicoll JX, Fry AC, Galpin AJ, et al. Changes in resting mitogen-activated protein kinases following resistance exercise overreaching and overtraining. Eur J Appl Physiol. 2016;116(11–12):2401–13.
9. Warren BJ, Stone MH, Kearney JT, et al. Performance measures, blood lactate, and plasma ammonia as indicators of overwork in elite junior weightlifters. Int J Sports Med. 1992;13(5):372–6.
10. Hecksteden A, Skorski S, Schwindling S, et al. Blood-borne markers of fatigue in competitive athletes—results from simulated training camps. PLoS One. 2016;11(2):e0148810.
11. Haddock CK, Poston WS, Heinrich KM. The benefits of high-intensity functional training fitness programs for military personnel. Mil Med. 2016;181(11):e1508–14.
12. Internet data. https://www.crossfit.com/what-is-crossfit. Last access: 18 Apr 2020.
13. Tafuri S, Notarnicola A, Monno A, Ferretti F, Moretti B. CrossFit athletes exhibit high symmetry of fundamental movement patterns. A cross-sectional study. Muscles Ligaments Tendons J. 2016;6(1):157–60.
14. Internet data. https://en.wikipedia.org/wiki/CrossFit. Last access: 18 Apr 2020.
15. Fisher J, Sales A, Carlson L. A comparison of the motivational factors between CrossFit participants and other resistance exercise modalities: a pilot study. J Sports Med Phys Fitness. 2017;57(9):1227–34.
16. Partridge JA, Knapp BA, Massengale BD. An investigation of motivational variables in CrossFit facilities. J Strength Cond Res. 2014;28(6):1714–21.
17. Nieuwoudt S, Fealy CE, Foucher JA. Functional high-intensity training improves pancreatic β-cell function in adults with type 2 diabetes. Am J Phys Endocrinol Metab. 2017;313(3):E314–20.
18. Perciavalle V, Marchetta NS, Giustiniani S, Borbone C, Perciavalle V, Petralia MC, et al. Attentive processes, blood lactate and CrossFit®. Phys Sportsmed. 2016;44(4):403–6.

19. Cadegiani FA, Kater CE, Gazola M. Clinical and biochemical characteristics of high-intensity functional training (HIFT) and overtraining syndrome: findings from the EROS study (the EROS-HIFT). J Sports Sci. 2019;20:1–12. https://doi.org/10.1080/02640414.2018.1555912.
20. Knapik JJ. Extreme conditioning programs: potential benefits and potential risks. J Spec Oper Med. 2015;15(3):108–13.
21. Meyer J, Morrison J, Zuniga J. The benefits and risks of CrossFit: a systematic review. Workplace Health Saf. 2017;65(12):612–8.
22. Murawska-Cialowicz E, Wojna J, Zuwala-Jagiello J. Crossfit training changes brain-derived neurotrophic factor and irisin levels at rest, after wingate and progressive tests, and improves aerobic capacity and body composition of young physically active men and women. J Physiol Pharmacol. 2015;66(6):811–21.
23. Choi EJ, So WY, Jeong TT. Effects of the crossfit exercise data analysis on body composition and blood profiles. Iran J Public Health. 2017;46(9):1292–4.
24. Kliszczewicz BM, Esco MR, Quindry JC, Blessing DL, Oliver GD, Taylor KJ, Price BM. Autonomic responses to an acute bout of high-intensity body weight resistance exercise vs. treadmill running. J Strength Cond Res. 2016;30(4):1050–8.
25. Cadegiani FA, Kater CE. Hypothalamic-pituitary-adrenal (HPA) axis functioning in overtraining syndrome: findings from Endocrine and Metabolic Responses on Overtraining Syndrome (EROS) – EROS-HPA axis. Sports Med Open. 2017;3(1):45.
26. Cadegiani FA, Kater CE. Growth hormone (GH) and prolactin responses to a non-exercise stress test in athletes with overtraining syndrome: results from the Endocrine and metabolic Responses on Overtraining Syndrome (EROS) – EROS-STRESS. J Sci Med Sport. 2018;21(7):648–53.
27. Cadegiani FA, Kater CE. Body composition, metabolism, sleep, psychological and eating patterns of overtraining syndrome: results of the EROS study (EROS-PROFILE). J Sports Sci. 2018;36(16):1902–10.
28. Cadegiani FA, Kater CE. Basal hormones and biochemical markers as predictors of overtraining syndrome in male athletes: the EROS-BASAL study. J Athl Train. 2019;54:906. https://doi.org/10.4085/1062-6050-148-18.
29. Burke LM, Ross ML, Garvican-Lewis LA, Welvaert M, Heikura IA, Forbes SG, Mirtschin JG, Cato LE, Strobel N, Sharma AP, Hawley JA. Low carbohydrate, high fat diet impairs exercise economy and negates the performance benefit from intensified training in elite race walkers. J Physiol. 2017;595(9):2785–807.
30. Escobar KA, Morales J, Vandusseldorp TA. The effect of a moderately low and high carbohydrate intake on crossfit performance. Int J Exerc Sci. 2016;9(3):460–70.
31. Outlaw JJ, Wilborn CD, Smith-Ryan AE, Hayward SE, Urbina SL, Taylor LW, Foster CA. Effects of a pre-and post-workout protein-carbohydrate supplement in trained crossfit individuals. Springerplus. 2014;21(3):369.
32. Chang CK, Borer K, Lin PJ. Low-carbohydrate-high-fat diet: can it help exercise performance? J Hum Kinet. 2017;12(56):81–92.
33. Heatherly AJ, Killen LG, Smith AF. Effects of ad libitum low carbohydrate high-fat dieting in middle-age male runners. Med Sci Sports Exerc. 2018;50(3):570–9.
34. Webster CC, Swart J, Noakes TD. A carbohydrate ingestion intervention in an elite athlete who follows a LCHF diet. Int J Sports Physiol Perform. 2017;18:1–12.
35. Sprey JW, Ferreira T, de Lima MV, Duarte A, Jorge PB, Santili C. An epidemiological profile of CrossFit athletes in Brazil. Orthop J Sports Med. 2016;4(8):2325967116663706.
36. Melin AK, Ritz C, Faber J, et al. Impact of menstrual function on hormonal response to repeated bouts of intense exercise. Front Physiol. 2019;10:942.
37. Lizneva D, Suturina L, Walker W, Brakta S, Gavrilova-Jordan L, Azziz R. Criteria, prevalence, and phenotypes of polycystic ovary syndrome. Fertil Steril. 2016;106(1):6–15.
38. Lauritsen MP, Bentzen JG, Pinborg A, et al. The prevalence of polycystic ovary syndrome in a normal population according to the Rotterdam criteria versus revised criteria including anti-Mullerian hormone. Hum Reprod. 2014;29(4):791–801.

39. Hagmar M, Berglund B, Brismar K, Hirschberg AL. Hyperandrogenism may explain reproductive dysfunction in olympic athletes. Med Sci Sports Exerc. 2009;41(6):1241–8.
40. Jetton AM, Lawrence MM, Meucci M, et al. Dehydration and acute weight gain in mixed martial arts fighters before competition. J Strength Cond Res. 2013;27(5):1322–6.
41. Matthews JJ, Nicholas C. Extreme rapid weight loss and rapid weight gain observed in UK mixed martial arts athletes preparing for competition. Int J Sport Nutr Exerc Metab. 2017;27(2):122–9.
42. Pettersson S, Berg CM. Hydration status in elite wrestlers, judokas, boxers, and taekwondo athletes on competition day. Int J Sport Nutr Exerc Metab. 2014;24(3):267–75.
43. Steele IH, Pope HG Jr, Kanayama G. Competitive bodybuilding: fitness, pathology, or both? Harv Rev Psychiatry. 2019;27(4):233–40.
44. Spendlove J, Mitchell L, Gifford J, et al. Dietary intake of competitive bodybuilders. Sports Med. 2015;45(7):1041–63.
45. Mitchell L, Hackett D, Gifford J, Estermann F, O'Connor H. Do bodybuilders use evidence-based nutrition strategies to manipulate physique? Sports (Basel). 2017;5(4):76.
46. Chappell AJ, Simper TN. Nutritional peak week and competition day strategies of competitive natural bodybuilders. Sports (Basel). 2018;6(4):126.
47. Knechtle B, Nikolaidis PT. Physiology and pathophysiology in ultra-marathon running. Front Physiol. 2018;9:634.
48. Knechtle B, Chlíbková D, Papadopoulou S, Mantzorou M, Rosemann T, Nikolaidis PT. Exercise-associated hyponatremia in endurance and ultra-endurance performance-aspects of sex, race location, ambient temperature, sports discipline, and length of performance: a narrative review. Medicina (Kaunas). 2019;55(9):537.
49. Knechtle B. Ultramarathon runners: nature or nurture? Int J Sports Physiol Perform. 2012;7(4):310–2.
50. Danielsson T, Carlsson J, Schreyer H, et al. Blood biomarkers in male and female participants after an Ironman-distance triathlon. PLoS One. 2017;12(6):e0179324. Published 13 Jun 2017. https://doi.org/10.1371/journal.pone.0179324.
51. Sousa CV, Aguiar SDS, Olher RDR, et al. Hydration status after an ironman triathlon: a meta-analysis. J Hum Kinet. 2019;70:93–102. Published 30 Nov 2019. https://doi.org/10.2478/hukin-2018-0096.
52. Heggie V. Experimental physiology, Everest and oxygen: from the ghastly kitchens to the gasping lung. Br J Hist Sci. 2013;46(1):123–47. https://doi.org/10.1017/S0007087412000775.
53. Watts PB, Daggett M, Gallagher P, Wilkins B. Metabolic response during sport rock climbing and the effects of active versus passive recovery. Int J Sports Med. 2000;21(3):185–90.
54. MacDonald MJ, Green HJ, Naylor HL, Otto C, Hughson RL. Reduced oxygen uptake during steady state exercise after 21-day mountain climbing expedition to 6,194 m. Can J Appl Physiol. 2001;26(2):143–56.

Chapter 8
Recovery from Overtraining Syndrome: Learnings from the EROS-Longitudinal Study

Outline

Recovery from overtraining syndrome – learnings from the EROS-longitudinal study: Recovery from overtraining syndrome is challenging and not always complete. In addition, very few longitudinal researches have been conducted with actual OTS athletes, which precludes the identification of biomarkers that could indicate the stage of the healing process of OTS. In this chapter, the author brings the clinical, biochemical, and metabolic biomarkers that occur earlier and later during the OTS recovery process, as well as practical recommendations for sports coaches and related health practitioners that may face this hazardous situation.

8.1 Introduction

Since overtraining syndrome is a complex and multifactorial disorder [1–9], with a wide range and individuality of its manifestations, its mitigation is challenging. Indeed, once triggered, OTS rarely improves completely, which may affect many athlete careers indefinitely.

To be fully recovered from OTS, all clinical and biochemical manifestations of OTS are needed to be overcome, which goes beyond paradoxical decrease in sports performance unexplained by overt dysfunctions, since it is associated with non-recoverable fatigue, changes in immunity, pathologically easiness to develop injuries, and psychological disturbances, including low energy levels, blurry mind and confusion state, anxiety, and symptoms of depression. Similarly, biochemical manifestations of OTS encompass a range of complex markers and alterations, including blunted hormonal responses to stimulations [10–14], decreased testosterone (in male athletes), impaired and prolonged muscle recovery, decreased metabolic rate and fat oxidation, muscle loss, body fat gain, and relative dehydration [13–19]. The large number of dysfunctions, among which some are troublesome and harmful in the short- and long-term for sports performance, precludes the successfulness of the attempts to remit from OTS.

Oppositely to biomarkers of OTS and tools for its diagnosis, the follow-up and natural history of the recovery, as well as the clinical, biochemical, and hormonal behaviors have barely been evaluated in previous researches. Expectedly, evaluation of the responses to specific interventions on OTS is yet to be reported in the literature.

In a similar manner of the natural course of OTS, in which some dysfunctions occur earlier stages whereas others only appear later stages, when OTS is deeply installed and becomes severe, the recovery process also presents a sequence of improvements, in which some biomarkers quickly improve right after the recovery process begin, while others become normal only after complete recovery.

Unlike the characterization of OTS, the recovery process from OTS have been poorly assessed, including the necessary time needed to recover clinically, biomarkers of improvement that appears earlier and later during the process, and biochemical behaviors throughout the recovery period. In addition, the evaluation of the effectiveness of proposed approaches to OTS have not been reported.

The EROS study followed up athletes affected by OTS for 3 months after the implementation of some interventions, aiming to describe the restoration process biochemical and hormonal markers that recovered and those that failed to improve, as well as the improvement of sports performance, during the follow-up [20].

8.2 Proposed Protocols for Recovery from OTS

The protocol proposed for the improvement of athletes with OTS include increased overall caloric, carbohydrate, and protein intake and full interruption of the training sessions, followed by beginning of the trainings with slow progression in volume and intensity of training, non-pharmacological recommendations for improvement of sleep quality and hygiene, and individual management of personal stressors, when needed, as comprehensive support for the improvement of OTS.

The multi-intervention comprehensive support protocol for the improvement of OTS was based on the fact that a combination of stressors, instead of a sole trigger, was the critical environment for the development of OTS. These factors influenced and acted synergistically to trigger OTS, and for its mitigation, all factors needed to be addressed. Counteracting all these factors was an intuitive approach to mitigate the underlying mechanisms that lead to OTS.

However, it is unfeasible to isolate and to correctly quantify the level of contribution of each of the interventions. Whether the improvement observed was solely due to the increase of caloric, carbohydrate, and protein intake, improvement of sleep hygiene, or cognitive stress management is uncertain. From an ethical perspective, not addressing one of the factors that may have contributed for the development of OTS, aiming to isolate and identify which specific factor had more important roles in the pathogenesis of OTS, is questionable.

Overall, all risk factors that may have influenced should be addressed. Whenever interventions are unfeasible to be implemented altogether, the management of which changes are more likely to be priorities and a scheduled sequence of changes should be employed.

An initial combination of managements of the training, eating, social, and sleeping patterns should be employed for 12 weeks and include, respectively:

1. Training management:

   - First 4 weeks: full resting, no exercise or any physical activity allowed
   - Weeks 5–8: training at half of the previous training volume and intensity
   - Weeks 9–12: full training load, similarly to training patterns before the intervention

2. Instead of a specific diet, nutritional interventions in relation to the previous dietary patterns tend to be better adopted and include:

   - Increase of overall caloric intake by 50%
   - Increase of carbohydrate intake by 100%
   - Increase of protein intake by 30%
   - Optionally, two ad libitum meals per week

3. Sleep interventions

   - Improvement of sleep hygiene
   - Strict self-control on the regularity of the time to sleep
   - Sleeping duration

4. Mental health:

   - Stress management
   - Avoidance of excessive cognitive demands
   - Psychotherapy when appropriates

Athletes should be evaluated after 12 weeks for the following aspects:

- Self-reported subjective sense of effort for a same training load
- Self- or coach-supervised evaluation of sports-specific performance
- Persistence or remission of signs and symptoms that were present at the moment of the diagnosis of OTS

Among the interventions, an interesting recommendation is to encourage athletes to have two ad libitum meals every week, particularly under lower carbohydrate and caloric intake, since it has been observed that several athletes under restricted diets unaffected by OTS had an average of two compensatory high carbohydrate loads during the week. Although these free meals were unplanned, they seem to have protected against the development of OTS. While unplanned carbohydrate uploads may have protected against OTS, the specific role of these ad libitum meals during the recovery process has not been assessed.

8.3   Early Versus Late Markers of Recovery

Different levels of recovery from OTS have been observed during a 12-week intervention protocol, which are summarized in Fig. 8.1. Levels of recovery were based
on the differences between levels in OTS-affected and healthy athletes, in which 0%
of recovery reflects the same levels of those at the diagnosis of OTS, whereas 100%
of recovery represents levels that equaled to those in healthy athletes. This graph
made explicit that markers could be clustered into either early or late/non-recovers.
It has been proposed that early markers were those that demonstrated at least 80%
recovery compared to baseline levels, while late or non-recovers had a recovery rate
of less than 40%. Whether the late or non-recovers would disclose further improvements on the long run is uncertain. The lack of full recovery of all markers after
3 months of follow-up reinforces the challenging recovery as a hallmark of OTS.

Specific parameters of cortisol, ACTH, GH, and prolactin response to an insulin
tolerance test (ITT), basal hormones, basal non-hormonal biochemical parameters,
and body composition and metabolism in the recovery process from overtraining
syndrome (OTS) are described in Tables 8.1, 8.2, 8.3, and 8.4, respectively.

Among the parameters evaluated, early cortisol and prolactin responses and
overall GH responses to stimulations, total testosterone, estradiol, testosterone/
estradiol ratio, salivary cortisol 30 minutes after awakening and cortisol awakening
response (CAR), and CK levels improved significantly and became similar to levels
in healthy athletes. In addition, IGF-1, freeT3, and usCRP levels improved in an
exacerbated manner and outranged healthy athletes.

Conversely, markers of body composition and metabolism and long-term hormonal responses failed to demonstrate improvements during the intervention period.

8.4   Hormonal and Other Biochemical Behaviors
in the Recovery Process

Early hormonal responses to demands seem to be the first hormonal behavior to
recover, while the late responses tend to persist blunted for longer, except for
GH. This is in full accordance with the clinically reported observation from athletes
and coaches that the explosive response recovers faster compared to the ability to
persist on an adequate pace. This means that athletes under the recovery process
from OTS may be able to achieve their previous pace in up to 12 weeks, while the
time-to-fatigue persists shorter than that in healthy athletes. The inability to recover
the previous time-to-fatigue can be explained by the inability to sustain hormonal
responses to stimulation for longer periods, i.e., the fact that the early responses
normalized justify the ability to recover the pace, whereas the inability to maintain
the responses justifies the lack of recovery from the shortened time-to-fatigue that
these athletes experiment.

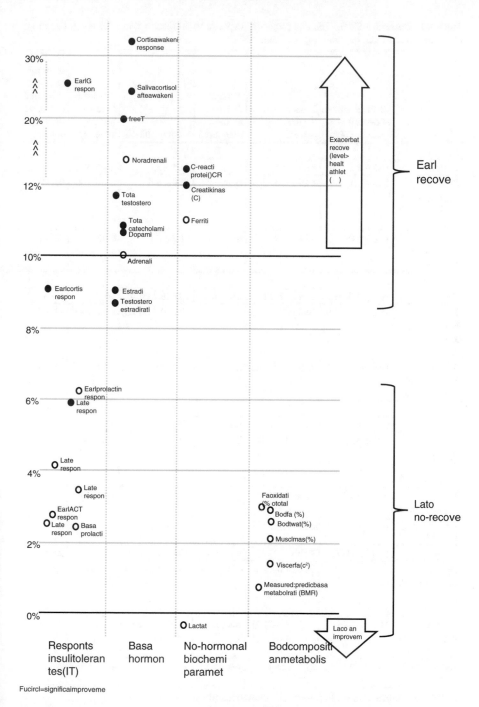

Fig. 8.1 Level of recovery from overtraining syndrome after a 12-week intervention (0 = levels at the diagnosis of OTS; 100 = levels in healthy athletes)

Table 8.1 Cortisol, ACTH, GH, and prolactin response to an insulin tolerance test (ITT) in the recovery process from overtraining syndrome (OTS)

	On the diagnosis of overtraining syndrome (OTS)	3 months after the recovery process	P-value (3 mo vs at the diagnosis)	Healthy athletes	P-value (3 mo vs healthy athletes)	Level of recovery (%, compared to levels of healthy athletes)
Cortisol						
Basal	11.3 ± 2.5	11.3 ± 2.2	n/s	12.5 ± 3.1	n/s (0.92)	n/a
During hypoglycemia (early response)	11.8 ± 3.1	15.6 ± 4.0	0.026	15.9 ± 5.6	n/s (0.99)	92.7%
30′ after hypoglycemia (late response)	17.6 ± 3.0	18.7 ± 3.8	n/s	21.7 ± 3.2	0.02	26.8%
ACTH						
Basal	19.2 (11.3–31.9)	20.4 (8.7–32.3)	n/s	18.1 (6.4–38.4)	n/s	n/a
During hypoglycemia (early response)	25.6 (7.9–172.9)	34.6 (11.9131.2)	n/s	57.8 (9.2–219.0)	n/s (0.44)	27.9%
30′ after hypoglycemia (late response)	30.3 (9.3–95.1)	42.9 (13.0–68.2)	n/s	60.6 (22.0196.3)	0.042	41.6%
GH						
Basal	0.1 (0–0.93)	0.2 (0–3.3)	n/s	0.3 (0.1–1.26)	n/s (0.43)	50%
During hypoglycemia (early response)	0.4 (0–4.9)	7.8 (0.1–31.5)	0.004	2.5 (0.2–34.5)	n/s (0.41)	252.3%
30′ after hypoglycemia (late response)	3.3 (0–14.1)	8.9 (1.8–36.5)	0.037	12.7 (2.5–37.6)	n/s (0.40)	59.6%
Prolactin						
Basal	8.0 (5.2–20.8)	8.7 (5.1–27.6)	n/s	10.8 (7.2–23.0)	n/s	25%
During hypoglycemia (early response)	8.4 (4.6–52.9)	14.6 (5.9–40.5)	n/s (0.069)	18.3 (10.0–63.4)	n/s (0.183)	62.6%
30′ after hypoglycemia (late response)	12.1 (4.5–26.0)	16.2 (7.5–51.7)	n/s (0.18)	23.3 (10.5–67.5)	0.043	36.6%
Hypoglycemic peak	19.7 ± 10.7	21.2 ± 11.3	n/s	18.0 ± 8.4	n/s	n/a

ACTH adrenocorticotrophic hormone, *GH* growth hormone

Table 8.2 Basal hormones in the recovery process from overtraining syndrome (OTS)

	On the diagnosis of overtraining syndrome (OTS)	3 months after the recovery process	P-value (3 mo vs at the diagnosis)	Healthy athletes	P-value (3 mo vs healthy athletes)	Level of recovery (%, compared to levels of healthy athletes)
Estradiol	41.0 ± 11.4	30.9 ± 11.4	0.0002	29.8 ± 14.0	n/s (0.57)	90.2%
Total testosterone	440.5 ± 181.5	558.1 ± 160.5	0.014	540.3 ± 174.6	n/s (0.65)	117.8%
Testosterone/estradiol (T:E) ratio	11.1 ± 4.0	19.6 ± 7.5	0.0005	20.9 ± 10.0	n/s (0.97)	86.7%
TSH	2.26 ± 1.00	2.10 ± 1.14	n/s (0.42)	1.82 ± 0.81	n/s (0.60)	n/a
freeT3	0.30 ± 0.06	0.34 ± 0.04	0.043	0.32 ± 0.05	n/s (0.82)	200%
IGF-1	182.1 ± 43.4	232.3 ± 42.6	0.003	176.6 ± 49.9	0.004	n/a; exacerbated increased, not needed to recover
Total catecholamines	266.6 ± 176.4	165.8 ± 93.9 (1 missing data)	0.043	174.6 ± 70.8	n/s (0.51)	109.6%
Noradrenaline	27 ± 10.1	20.1 ± 8.1	0.09 (n/s)	22.3 ± 13.0	n/s (0.83)	146.8%
Adrenaline	3 (1.5–6.3)	2 (1.0–11.4)	0.26 (n/s)	2 (1.0–10.1)	n/s (0.56)	100%
Dopamine	236 ± 169.5	141.9 ± 89.2	0.049	149.3 ± 60.9	n/s (0.38)	108.5%
Salivary cortisol at awakening	0.39 ± 0.11	0.32 ± 0.24	0.26 (n/s)	0.34 ± 0.13	n/s (0.37)	n/a
Salivary cortisol 30' after awakening	0.32 ± 0.11	0.75 ± 0.33	0.0003	0.50 ± 0.17	n/s (0.098)	238.9%
Cortisol awakening response (CAR)	28.3 (−36.3 − +166.3)	81.9 (31.2–164.4)	0.001	41.1 (1.21–170.4)	n/s (0.19)	318.7%
Salivary cortisol at 4 PM	0.16 ± 0.13	0.16 ± 0.09	n/s	0.14 ± 0.08	n/s	n/a
Salivary cortisol at 11 PM	0 (0–0.18)	0.07 (0–0.24)	n/s	0.08 (0.08–0.18)	n/s	n/a

IGF-1 insulin-like growth factor-1, *freeT3* free thyronine

Table 8.3 Basal non-hormonal biochemical parameters in the recovery process from overtraining syndrome (OTS)

	On the diagnosis of overtraining syndrome (OTS)	3 months after the recovery process	P-value (3 mo vs at the diagnosis)	Healthy athletes	P-value (3 mo vs healthy athletes)	Level of recovery (%, compared to levels of healthy athletes)
usCRP	0.10 (0.04–2.58)	0.04 (0.01–0.5)	0.019	0.06 (2.0–12.6)	n/s (0.16)	133%
ESR	2.0 (1.6–12)	2.0 (1.6–8.7)	n/s (0.32)	2.0 (2.0–12.6)	n/s (0.65)	n/a
CK	751 (124–3045)	258 (93–825)	0.045	340 (92–783)	n/s (0.33)	120%
Ferritin	201 (60–418)	162 (33–442)	n/s (0.42)	166 (61–378)	n/s (0.93)	111.4%
Vitamin B12	541.5 ± 241.7	503.4 ± 157.1	n/s	553.2 ± 191.4	n/s	n/a
Lactate	1.26 ± 0.5	1.38 ± 0.4 (3 missing data)	n/s	0.88 ± 0.28	0.001	−31.6%
Creatinine	1.11 ± 0.16	1.16 ± 0.20	n/s	1.14 ± 0.17	n/s	n/a

usCRP ultrasensitive C-reactive protein, *ESR* erythrocyte sedimentation rate, *CK* creatine kinase

Table 8.4 Body composition and metabolism in the recovery process from overtraining syndrome (OTS)

	On the diagnosis of overtraining syndrome	3 months after recovery process	P-value (3 mo vs at the diagnosis)	Healthy athletes	P-value (3 mo vs healthy athletes)	Level of recovery (%, compared to levels of healthy athletes)
Measured predicted basal metabolic ratio (BMR) (%)	102.6 ± 8.3 (2012 ± 246)	103.2 ± 8.3 (2025 ± 207)	n/s	109.7 ± 9.3 (1980 ± 217)	0.0005	8.5%
Fat oxidation (% of total BMR)	33.5 ± 21.0	41.7 ± 35.7	n/s	58.7 ± 18.7	n/s	32.5%
Body fat (%)	17.0 ± 6.0	15.0 ± 5.2	n/s	10.8 ± 4.2	n/s (0.084)	32.3%
Muscle mass (%)	47.2 ± 3.8	47.9 ± 3.0	n/s	50.5 ± 2.3	n/s (0.058)	21.2%
Body water (%)	59.5 ± 3.9	61.0 ± 3.7	n/s	64.7 ± 2.7	0.0082	28.8%
Visceral fat (cm²)	67.5 ± 36.5	62.8 ± 20.8	n/s	35.7 ± 20.6	0.0076	14.8%

Unlike other hormonal responses, GH disclosed full improvement within 12 weeks, being the earliest stress-sensitive hormone to respond, and could be therefore considered therefore a marker of early improvement for further clinical use.

Alongside with the improvement of the GH releasing patterns, an increased sensitivity to GH could explain the exacerbation of IGF-1, which became higher than levels in healthy athletes after 12 weeks of recovery. The increased sensitivity could be resulted from an upregulation of the density of the GH receptors (GHR) in response to chronic low GH levels and peaks, as an attempt to increase the efficiency of its actions, despite the lower ability of GHR to change its density compared to steroid receptors, since GHR is a G-coupled-receptor attached to the cell membrane, while steroid receptors are located intracitoplasmatically.

Similarly to IGF-1, freeT3, initially similar between OTS and ATL, increased after the 12-week intervention and also became higher than ATL. The process that fT3 undergoes is apparently similar to the recovery period of the euthyroid sick syndrome (ESS), when circulating T3 acutely increases when the underlying condition that led to the syndrome starts to heal [21]. Indeed, both OTS and ESS are conditions in which a sort of an "energy saving mode" is a remarkable mechanism to focus the energy expenditure to the most vital functions. OTS is fundamentally a disease caused by dysfunctional adaptations (*maladaptations*) forced to occur in order to maintain functioning and survival under a chronic and severe deprivation of energy and mechanisms of repair. Analogically similar mechanisms occur in ESS, leading to a sort of central hypothyroidism, aiming to prevent non-essential energy demands.

The fact that IGF-1 and freeT3 are both bioactive hormones with multiple metabolic roles may justify why these hormones demonstrated similar behaviors during the recovery process.

The CAR, which unlike freeT3 and IGF-1, was markedly reduced in OTS; showed an exacerbated recovery, compared to healthy athletes; and became almost three times higher compared to the levels at the moment of diagnosis of OTS. Although the CAR behavior in OTS was different from IGF-1 and freeT3, these markers have corresponding bioactive actions in the, which may help understand why these three parameters behaved similarly during the recovery process.

The concurrent decrease of testosterone and increase of estradiol levels, with a consequent important reduction of the T:E ratio, was a remarkable parameter shown to be specific of OTS, since neither healthy athletes nor sedentary showed alterations in the T:E ratio. As mentioned in a previous chapter, the concurrent changes in both testosterone and estradiol indicate that this alteration likely occurred as a result of a pathological exacerbation of aromatase activity, with enhanced conversion from testosterone into estradiol, possibly as an anti-anabolic mechanism to avoid additional energy expenditure and decrease the highly metabolically active muscle tissue.

The complete normalization of testosterone, estradiol, and T:E ratio was found to occur early in the recovery process, possibly as a mechanism of "reconstruction" of the damaged muscle that occurs in OTS, and to enhance the metabolism recovery during the healing process.

A full normalization of the catecholamine levels occurred in the OTS athletes after the 12-week intervention and has become similar to levels in healthy athletes. Since the chronic lack of energy and mechanisms of repair was corrected by increase of caloric and carbohydrate intake and rebalance of the training-resting-nutrition-sleep balance, respectively, the attempt to keep functioning detected with the exacerbation of catecholamines is no longer necessary, which allow these hormones to return to baseline.

Although not initially high, usCRP reduced substantially during the recovery, which may indicate an anti-inflammatory pattern typical of the recovery period, oppositely to the reperfusion-like behavior, in which multiple paradoxical harms may occur during recovery process from deprivation states. Hence, the reduction of usCRP may correspond to a sort of loss of the inflammation that is present in OTS, although the inflammation present in OTS is only detectable through the measurement of non-classic inflammatory markers.

The overactive CK found in OTS can be completely normalized to the expected levels after 12 weeks of interventions, in case a 4-week period of complete rest is employed followed by an additional period of four fours training in average half of the previous training load, before returning to the previous training program.

The marked reduction of circulating CK represents the remission of the impaired, prolonged, dysfunctional overreacting muscle response to exercise found in OTS.

Unexpectedly and unlike CK, lactate remained relatively altered after 12 weeks, when compared to levels of healthy athletes. Due to the complex roles of lactate that goes beyond those attributed to muscle metabolism, including a wide range of pathways of which its end-product is lactate, this parameter may take longer to normalize, since OTS is only partially remitted aster 12 weeks, at least under the proposed interventions. Lactate could be employed as a marker of full remission of OTS, although further confirmation with longer periods of follow-up is needed.

Unlike hormonal and biochemical parameters, both body metabolism and composition failed to fully improve after 12 weeks, showing that changes in the metabolic patterns may require longer periods or recovery.

8.5   Practical Applications

The novel markers of recovery identified in the present study will allow coaches and other sports-related health providers to have a more precise follow-up of athletes affected by OTS. In summary, the early markers of recovery that can be employed in clinical practice include IGF-1, fT3, cortisol awakening response, normalization of testosterone, estradiol, and testosterone/estradiol ratio, and CK and catecholamines, reduction of usCRP, and improvement of early cortisol and prolactin responses and overall GH response to stimulations. These early markers of recovery identified could be used as markers for further athletes with OTS, during the recovery process, to predict the successfulness of their recovery, and plan on the future of their careers.

In addition, the present study also encourages athletes to have two ad libitum meals every week when under lower carbohydrate and caloric intake, since they seem to have provided protection for these athletes against the development of OTS.

8.6 Conclusion

Athletes affected by actual OTS that undergo a recovery protocol including nutritional, exercise, sleeping, and social interventions during 12 weeks showed substantial improvements of multiple hormonal and other biochemical parameters, while body metabolism and composition recovered partially. Remarkable early markers of recovery include exacerbated increase of IGF-1 free T3 and CAR, exacerbated reduction of usCRP, normalization of testosterone, estradiol, testosterone/estradiol ratio, CK and NUC, improvement of early cortisol and prolactin responses, and overall GH response to stimulations. FreeT3, usCRP, and IGF-1 seem to be the sentinel markers of recovery from OTS. These markers can be further employed for athletes recovering from OTS, at least until further prospective studies are conducted.

References

1. Meeusen R, Duclos M, Foster C, European College of Sport Science, American College of Sports Medicine, et al. Prevention, diagnosis, and treatment of the overtraining syndrome: Joint Consensus Statement of the European College of Sport Science and the American College of Sports Medicine. Med Sci Sports Exerc. 2013;45(1):186–205.
2. Lehmann M, Foster C, Keul J. Overtraining in endurance athletes: a brief review. Med Sci Sports Exerc. 1993;25(7):854–62.
3. Rietjens GJ, Kuipers H, Adam JJ, et al. Physiological, biochemical and psychological markers of strenuous training-induced fatigue. Int J Sports Med. 2005;26(1):16–26.
4. Cadegiani FA, Kater CE. Hormonal aspects of overtraining syndrome: a systematic review. BMC Sports Sci Med Rehabil. 2017;9:14.
5. Kreider R, Fry AC, O'Toole M. Overtraining in sport: terms, definitions, and prevalence. In: Kreider R, Fry AC, O'Toole M, editors. Overtraining in sport. Champaign: Human Kinetics; 1998. p. VII–X.
6. Kreher JB, Schwartz JB. Overtraining syndrome: a practical guide. Sports Health. 2012;4(2):128–38.
7. Fry RW, Morton AR, Keast D. Overtraining in athletes. An update. Sports Med. 1991;12(1):32–65.
8. Budgett R. Fatigue and underperformance in athletes: the overtraining syndrome. Br J Sports Med. 1998;32:107–10.
9. Budgett R, Newsholme E, Lehmann M, et al. Redefining the overtraining syndrome as the unexplained underperformance syndrome. Br J Sports Med. 2000;34:67–8.
10. Meeusen R, Nederhof E, Buyse L, Roelands B, De Schutter G, Piacentini MF. Diagnosing overtraining in athletes using the two-bout exercise protocol. Br J Sports Med. 2010;44(9):642–8.

11. Meeusen R, Piacentini MF, Busschaert B, Buyse L, De Schutter G, StrayGundersen J. Hormonal responses in athletes: the use of a two bout exercise protocol to detect subtle differences in (over)training status. Eur J Appl Physiol. 2013;91(2–3):140–6.
12. Urhausen A, Gabriel HH, Kindermann W. Impaired pituitary hormonal response to exhaustive exercise in overtrained endurance athletes. Med Sci Sports Exerc. 1998;30(3):407–14.
13. Cadegiani FA, Kater CE. Hypothalamic-pituitary-adrenal (HPA) axis functioning in overtraining syndrome: findings from Endocrine and Metabolic Responses on Overtraining Syndrome (EROS) – EROS-HPA axis. Sports Med Open. 2017;3(1):45.
14. Cadegiani FA, Kater CE. Growth Hormone (GH) and prolactin responses to a non-exercise stress test in athletes with overtraining syndrome: results from the Endocrine and metabolic Responses on Overtraining Syndrome (EROS) – EROS-STRESS. J Sci Med Sport. 2018;21(7):648–53.
15. Cadegiani FA, Kater CE. Novel causes and consequences of overtraining syndrome: the EROS-DISRUPTORS study. BMC Sports Sci Med Rehabil. 2019;11:21.
16. Cadegiani FA, Kater CE. Eating, Sleep, and Social Patterns as Independent Predictors of Clinical, Metabolic, and Biochemical Behaviors Among Elite Male Athletes: The EROS-PREDICTORS Study. Front Endocrinol (Lausanne). 2020;11:414. Published 2020 Jun 26. https://doi.org/10.3389/fendo.2020.00414.
17. Cadegiani FA, Kater CE. Inter-correlations among clinical, metabolic, and biochemical parameters and their predictive value in healthy and overtrained male athletes: the EROS-CORRELATIONS study. Front Endocrinol (Lausanne). 2019;10:858.
18. Cadegiani FA, Kater CE. Body composition, metabolism, sleep, psychological and eating patterns of overtraining syndrome: results of the EROS study (EROS-PROFILE). J Sports Sci. 2018;36(16):1902–10.
19. Cadegiani FA, Kater CE. Basal hormones and biochemical markers as predictors of overtraining syndrome in male athletes: the EROS-BASAL study. J Athl Train. 2019;54:906. https://doi.org/10.4085/1062-6050-148-18.
20. Cadegiani FA, Kater CE. Novel markers of recovery from overtraining syndrome: the EROS-LONGITUDINAL study. Int J Sports Physiol Perform. In review.
21. Lee S, Farwell AP. Euthyroid sick syndrome. Compr Physiol. 2016;6(2):1071–80.

Chapter 9
Female Athlete Triad (TRIAD), Relative Energy Deficiency of the Sport (RED-S), and Burnout Syndrome of the Athlete (BSA): Concepts, Similarities, and Differences from Overtraining Syndrome (OTS)

Outline
TRIAD, RED-S, BSA: concepts, similarities, and differences from over-training syndrome (OTS), in which the author briefly describes each of these syndromes, from a perspective that allows the further comparative analysis between them. Surprisingly, TRIAD, RED-S, BSA, and OTS/PDS share multiple similarities, and the hypothesis that they actually belong to a wider, more comprehensive condition may be feasible, as will be shown in the future by forthcoming researches.

9.1 Introduction

While OTS was majorly studied in males, women have been extensively evaluated for the presence of the female athlete triad (TRIAD) [1, 2], which has been further proposed to be termed as relative energy deficiency of the sport (RED-S) [3]. The fact that women tend to have a sentinel early sign of amenorrhea when under extensive training and deprived eating [1–3] may have motivated researchers to consider the existence of another condition than OTS, since OTS had been unrelated to any eating pattern.

Conversely, burnout syndrome, characterized by an atypical presentation of features of depression, anxiety, and loss of self-identification, has been proposed to be also present in athletes, with some adaptations of its characteristics, including a possible loss of performance and unwillingness to believe and continue in the athletes' career, and which has been termed as burnout syndrome of the athlete (BSA).

Both TRIAD/RED-S and BSA share some common characteristics with OTS, as all these conditions derive from some sorts of deprivation, including resting, eating, and/or sleeping, due to either self-inability to allow a better balance, resulted from

F. Cadegiani, *Overtraining Syndrome in Athletes*,
https://doi.org/10.1007/978-3-030-52628-3_9

an "excessive motivation" with implicates in lower or avoidance of self-perception of fatigue and other signs, or due to a toxic sports environment. Since PDS is a more comprehensive understanding of OTS [4], and includes insufficient caloric intake, excessive concurrent cognitive demands and unrefreshing sleep as important risk factors, PDS allows better comparisons and share more similarities with these states.

From a broader perspective, OTS/PDS, RED-S/TRIAD, and BSA have remarkable similarities, which allow the hypothesis that all these disorders are actually different faces of a *spectrum* of a broader complex condition. However, researches on these conditions have been performed independently from each other, with distinct focuses, and to date no articles have compared their characteristics, except for RED-S, as being a more comprehensive understanding of TRIAD, although this proposal is still controversial [5–8].

9.2   The Female TRIAD (TRIAD)

The TRIAD is a largely researched syndrome for over 30 years and consists of three interrelated conditions: menstrual dysfunction, decreased mineral bone density, and low energy availability (LEA) [9]. However, whenever one of these conditions is present, unless in specific cases, the other two should also be present, as these three key characteristics play inter-synergistically [1–3].

Disordered eating is usually the key factor that leads to the TRIAD. Additional signs include mental and physical exhaustion, severe weight loss, hair loss, cold and dry hands and feet, increased time for recovery from injuries of any sort, increased risk of bone fracture, hypogonadism hypogonadotrophic (functional hypothalamic amenorrhea), anemia, orthostatic hypotension, electrolyte irregularities, hypoestrogenism, vaginal atrophy, and bradycardia [1–3, 10–12]. Although not highlighted, performance may be severely affected, resembling OTS/PDS in this specific aspect.

However, compared to TRIAD, in which psychological issues are among its triggers, in OTS/PDS mood disturbances tend to be consequences, not causes, as a major risk factor of OTS/PDS is a sort of "excessive" motivation, which avoids the self-perception of mild and early signs of OTS/PDS [13–15]. Indeed, once disordered eating is the primary cause for the TRIAD, underlying strong psychological components of eating misbehaviors, including low self-esteem and depression, are the actual root causes for this syndrome.

LEA, which plays a major role for the occurrence of TRIAD, is also the underlying mechanism that leads to functional hypothalamic amenorrhea (FHA) in TRIAD [1–3, 10–12], since FHA is one of the multiple adaptive mechanisms to preserve energy for vital processes under severe energy deprivation [9]. Depending on the age of beginning, female athletes can experience early and severe hypoestrogenism state, preventing them from a proper bone mass peak, which approximately occurs by the age of 20 years old.

TRIAD typically occurs in athletes of aesthetic sports, rather than ball games or sports purely focused on speed, intensity, or volume performance, since in aesthetical sports a lower body weight and leaner shape are highly desired, which motivates progressively strict diets [3, 5–7]. However, in some cases, the LEA experienced in TRIAD may be due to increased training load without a concomitant compensatory increase of calorie intake.

9.3   Relative Energy Deficiency of the Sport (RED-S)

RED-S was proposed to describe a condition that encompasses the TRIAD and expands its concepts to male athletes and to a wider range of dysfunctions, not originally comprised in TRIAD [6, 8].

Similar to TRIAD, the key trigger and hallmark of RED-S is LEA, due to insufficient caloric intake and/or excessive energy expenditure, but not necessarily under negative energy balance. In addition to the alterations described in the TRIAD, RED-S leads to overall impaired physiological functioning, with consequences on metabolic rate, HPG axis in both males and females, immune system, bone health (other characteristics than bone density), and abnormalities on other systems.

However, since RED-S is a proposal of a new understanding of the TRIAD, studies were predominantly conducted with females, while some claim that the existing data still does not allow to extend the understanding of TRIAD to male athletes. Indeed, despite the consensus from the International Olympic Committee (IOC) on RED-S, specific data on male, non-white, disabled, and clinical athletes is insufficient to include these populations as definite part of RED-S [5–8].

A key reason why TRIAD was recognized in females before RED-S, and why LEA-induced conditions have been primarily researched in females the disruption of regular menstrual cycles is a very early sign of any abnormality in the HPG axis, once an intact cyclicity of the hormones is crucial for regular cycles. Disrupted menstrual cycles are therefore a sentinel signal for forthcoming dysfunctions [1–3, 5–12], which has no parallels in males.

Despite the controversies, when analyzed through a wider perspective, TRIAD can be considered the female specific expression of RED-S, while RED-S needs further clarification to characterize its existence in male athletes.

Many of the more recently described findings on OTS/PDS are also present in RED-S, including decreased muscle strength and recovery, depleted glycogen and energy availability, psychological disturbances including depression and irritability, and loss of sports performance (which is the hallmark of OTS/PDS), including decreased coordination, blunted prompt response to stimulations, and decreased endurance performance. Hormonal responses were also found to be blunted at a similar level in both disorders, particularly those of the HPG and GH axes. Although increased risk of muscle and joint injuries is alleged to be present in both RED-S and OTS/PDS, there is insufficient evidence demonstrating a causal relationship between injuries and these states.

Also, insufficient caloric intake of 30 kcal/kg FFM/day and 31.5 kcal/kg FFM/day (from 35 kcal/kg/day) were identified as major risk factors for RED-S and OTS/

PDS, respectively [5–8, 13–16]. The cutoff for both conditions was unexpectedly similar, reinforcing their pathogenic similarities.

Conversely, there are important differences between RED-S and OTS/PDS including hormonal responses of the thyrotrophic and adrenocorticotrophic axes, which are unaltered and decreased in OTS/PDS, while abnormally low and high in RED-S, respectively. Upper respiratory tract infections (URTIs), while strongly correlated with LEA, affecting up to 100% of athletes with RED-S, lack conclusive correlations with OTS/PDS.

In addition, self-perception of body image is more correlated with RED-S, while in OTS/PDS the focus is mainly on performance. However, more recently OTS-affected athletes moved toward a dual and contradictory goal on performance and body shape [13–16], which is inherently incompatible, as dietary recommendations for improvement of performance tend to be distinct from recommendations for body fat loss.

An additional risk factor for RED-S is the low body mass index (BMI), which, although not fully accurate, can be a helpful marker to detect those at imminent risk of a forthcoming RED-S, and includes BMI ≤ 17.5 kg/m^2, <85% and expected body weight for adolescents or $\geq 10\%$ weight loss for females [1–3, 5–8]. Cutoffs have not yet been established for males. Conversely, as athletes affected by OTS/PDS do not seem to be as affected by dysfunctional eating, the majority of this population does not present low BMI.

Although data from the latest guidelines on RED-S on the LEA effects on male athletes was scarce, the physiological and clinical consequences of LEA in male athletes were addressed in the EROS study [4, 13–19], in which depleted energy was shown to be the most important independent predictor of OTS/PDS [13, 14]. Also, although authors who contested the 2014 IOS guideline on RED-S claimed that effects of LEA on male reproductive system are unlikely [8], there is growing and strong evidence showing several harms of oxidative stress and energy deprivation on male sperm [20–22], while the reproductive axis also seems to be impaired in OTS/PDS [4, 17, 23, 24].

Thus, after the improvements of the concepts and understanding of both RED-S and OTS/PDS, the level of similarity between these conditions now allows us to hypothesize that these may be different clusters of manifestations of a same condition. It means that current data sufficiently supports the fact that the male representative disease of the RED-S/TRIAD is OTS/PDS, OTS/PDS as a type of RED-S, or RED-S as a type of OTS/PDS. However, as these dysfunctions were studied in distinct populations evaluating distinct aspects, further studies evaluating the previously non-assessed types of parameter, i.e., eating patterns in OTS/PDS and training patterns in RED-S/TRIAD, are crucial to better compare these conditions.

9.4   Burnout Syndrome of the Athlete (BSA)

Burnout syndrome, another state related to athletes that leads to fatigue and loss of performance, has been first described in the early 1970s, as a behavioral pattern suffered by volunteers and workers of unwieldy workplaces. These people were reported as presenting a progressive loss of energy, demotivation, anhedonia, and eventually exhaustion [25].

After innumerable researches, burnout was defined as a syndrome characterized by three dimensions: emotional exhaustion, feelings of cynicism or depersonalization, and reduced personal accomplishment [26]. Burnout may also result in a wide variety of physiologic symptoms, including headaches, insomnia, fatigue, decreased appetite, and higher susceptibility to infections; psychological manifestations, such as increased negative self-talk; difficulty in interpersonal relationships; behavioral symptoms, including decreased performance in all aspects of life, withdrawal from activity due to loss of interest, increased rigidness behavior, and neuromotor symptoms, including reduced efforts in the movements; and delayed reflexes [26–28].

Unlike other sport-related syndromes, burnout has been strictly linked to workplace, affecting approximately 6% of workers on its most severe stage. However, despite some estimations, the precise prevalence of burnout is unknown, due to a lack of predefined diagnostic criteria [26, 28].

Burnout is seen as a process along time: increased coping efforts with external demands lead to emotional exhaustion, which is a trigger for depersonalization, which in turn leads to diminished personal accomplishment, adding further emotional exhaustion, causing a vicious cycle. Negative social experiences promote additional risk of burnout, irrespective of other factors [25, 26].

Although robustly researched in workplaces, studies on burnout syndrome lacked control and evaluation for external factors, as well as their influences, including the practice of physical activity. Many of those people who experience burnout related to workplace are also physically active, which can be positive depending on the level of stress, but also harmful, when absolute resting is essential. Despite the assumption that exercise seems to be a successful approach to burnout, specific studies failed to demonstrate its benefits [29].

Athletes, who were presumably protected from burnout due to the practice of exercise, are oppositely at increased risk for burnout syndrome [30, 31]. Despite the few original studies published in sport-related burnout, it is clear that this specific population suffers from a cognitive-affective syndrome, termed as burnout syndrome of the athlete (BSA). BSA has some particularities, including many negative outcomes such as decreased motivation, emotional and physical exhaustion, reduced sense of accomplishment and sport devaluation leading to further reduced performance, and ultimately sport dropout [30, 32–34].

BSA occurs as a consequence of a dysfunctional social and professional context. Some athletes experience high demands such as high training loads, excessive performance expectations, insufficient resources for the level of demands, low social support, low autonomy, scarce reward, or, in the opposite direction, low demands that take to the boredom [30–34]. This environment precludes these athletes from the development of multi-faceted identities, redirecting them to a harmful unidimensional athlete identity, which in turn leads to burnout syndrome, when associated with the sense of loss of control over their sport participation. Through a perspective of an Integrated Model of Athlete Burnout, these athletes are poorly self-determined, with perceptions of lack of control and self-endorsement of their activities, false perception of poor competence, and weak connection with others, leading to worse sense of well-being and lower functioning [35–38].

Particular characteristics of some athletes, including perfectionistic concerns and "excessive" commitment – when athletes believe they "have to" instead of "want to" – have been proposed as important and independent risk factor for burnout [30, 34, 35]. Oppositely, hope, perception of self-control, optimism, and coping skills are inversely associated with burnout [34, 35].

9.5   Similarities Between RED-S/TRIAD, BSA, and OTS/PDS

The joint comparative analysis and particularities of OTS/PDS, RED-S/TRIAD, and athlete-related burnout are detailed in Fig. 9.1. When analyzed together, these conditions disclose similar hallmark characteristics, including underlying complex and multi-factorial pathophysiology, which occur as *continuum* and progressive processes, with the existence of preclinical, subclinical, and occult signs. For all these syndromes, both statistical and diagnostic criteria have been transformed from a continuous scale to a dichotomy that discriminates between "cases" and "non-cases," for practical and clinical reasons. However, this categorization does not

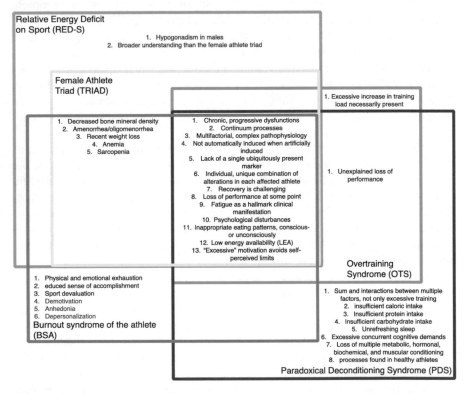

Fig. 9.1 Differences and similarities between overtraining syndrome (OTS), paradoxical deconditioning syndrome (PDS), relative energy deficiency in sport (RED-S), female athlete triad (TRIAD), and burnout state of the athlete (BSA)

reflect the natural history of these dysfunctions, since they all occur and progress in a continuous manner and should be understood as *spectrum* diseases.

All these conditions lack a ubiquitous single marker, once RED-S/TRIAD, BSA and OTS/PDS are characterized by an individual and single combination of dysfunctions, without a sine qua non abnormality. Indeed, besides the assessment methods, another challenge of researches on OTS/PDS, RED-S/TRIAD, and BSS is the large within- and between-participant variability. Hence, more and better quality of research is needed, with standardized tests, with direct comparisons between the conditions described, particularly in men.

Moreover, in common all these conditions tend to be diagnosed late, when the most severe stages are installed, which precludes a complete recovery.

For all states, including OTS/PDS, RED-S/TRIAD, and BSA, sports, social, and familiar environments play critical roles for the development of these states, since relative low caloric, protein, or carbohydrate intake, inadequate food availability, including food insecurity from cultural practices or lack of financial resources contribute for all these conditions, particularly in places affected by shortage of food availability, which is highly influenced by religious-related fasting or starvation practices.

The presence of signs of restrictive eating, which may not fully meet clinical criteria for any specific eating disorder, is present in all these conditions, since in OTS/PDS and RED-S/TRIAD, abnormal eating behaviors occur deliberately, whereas in BSA a natural loss of appetite is a consequence of its occurrence. However, while LEA is an important trigger for OTS/PDS and RED-S/TRIAD, it is yet to be identified as a cause for BSA.

Although the assessment of energy availability (EA) is seemingly a powerful diagnostic tool for these conditions, a more comprehensive approach including assessment of sleeping, professional, and social patterns may be more effective.

Indeed, important underlying psychological aspects are almost always present in similar patterns in OTS/PDS, RED-S/TRIAD, and BSA. The concurrent desires to be leaner and to enhance performance, the perfectionism, competitiveness, and pain tolerance seem to predict most of the fatigue- and sport-related conditions, including OTS/PDS, RED-S/TRIAD, and BSA, while disordered eating and unsupportive and conflicting coach-athlete relationship are major risk factors of RED-S/TRIAD and BSA, respectively. The coexistence of an anxiety state and mental exhaustion, typically observed in BSA, is also largely present in both OTS/PDS and RED-S/TRIAD.

In common, OTS/PDS, RED-S/TRIAD, and BSA may affect not only professional but also recreational athletes. In addition, an unfavorable workplace or social and familial environment, although only highlighted in BSA, may also direct or indirectly aggravate OTS/PDS and RED-S/TRIAD states.

In addition, RED-S/TRIAD/LEA and OTS/PDS occur due to similar mechanisms, including impaired and prolonged muscle recovery and reduced function, impaired protein synthesis, chronic depletion of glycogen stores, reduced time-to-fatigue, and psychological and mental disturbances.

Despite the major similarities, groups of researches classically analyzed each condition apart from each other. Although reduced performance is barely described in RED-S/TRIAD and BSA, this physical characteristic has not been assessed in athletes affected by these two states, which precludes a conclusive answer. Conversely, although dietary characteristics have only been described in RED-S/TRIAD, studies on OTS and BSA rarely controlled for these variables. If the assessment methods were crossed between these conditions, it is possible that many of the disturbances from a variety of aspects may equally affect OTS/PDS, RED-S/TRIAD, and BSA.

In summary, OTS/PDS, RED-S/TRIAD, and BSA related to the athlete share sufficient similar characteristics to hypothesize them as different poles of a multidimensional spectrum of a broader condition. However, comparative research should be performed with populations with similar characteristics and tested for the same parameters, using respective and basal control groups.

9.6   Conclusion

While overtraining syndrome (OTS) has been typically described in males, female athletes have been extensively reported for the diagnosis of the female athlete triad (TRIAD) that encompasses the triad of menstrual dysfunction and decreased mineral bone density and also include low energy availability (LEA), which in turn leads functional hypothalamic amenorrhea (FHA). RED-S is a proposal that encompasses the TRIAD and expands its concepts to male athletes and a wider range of dysfunctions, although the male version of TRIAD is still controversial. Although BSA has been described as an independent condition of the athlete, while it is focused on a hostile sports and social environment, the description of BSA is not different from those of OTS and RED-S.

OTS/PDS, RED-S/TRIAD, and BSA have multiple similar characteristics that allow the hypothesis that these are actually different manifestations of a broader condition.

References

1. Nattiv A, Loucks AB, Manore MM, et al. American College of Sports Medicine position stand. The female athlete triad. Med Sci Sports Exerc. 2007;39(10):1867–82.
2. Matzkin E, Curry EJ, Whitlock K. Female athlete triad: past, present, and future. J Am Acad Orthop Surg. 2015;23(7):424–32.
3. Mountjoy M, Sundgot-Borgen J, Burke L, et al. The IOC consensus statement: beyond the Female Athlete Triad – relative energy deficiency in sport (RED-S). Br J Sports Med. 2014;48:491–8.
4. Cadegiani FA, Kater CE. Novel insights of overtraining syndrome discovered from the EROS study. BMJ Open Sport Exerc Med. 2019;5(1):e000542. https://doi.org/10.1136/bmjsem-2019-000542.
5. De Souza MJ, Williams NI, Nattiv A, et al. Misunderstanding the female athlete triad: refuting the IOC consensus statement on Relative Energy Deficiency in Sport (RED-S). Br J Sports Med. 2014;48:1461–5.

6. Mountjoy M, Sundgot-Borgen J, Burke L, et al. Authors' 2015 additions to the IOC consensus statement: Relative Energy Deficiency in Sport (RED-S). Br J Sports Med. 2015;49:417–20.
7. Mountjoy M, Sundgot-Borgen JK, Burke LM, et al. IOC consensus statement on relative energy deficiency in sport (RED-S): 2018 update. Br J Sports Med. 2018;52:687–97.
8. Tenforde AS, Barrack MT, Nattiv A, et al. Parallels with the female athlete triad in male athletes. Sports Med. 2016;46:171–82.
9. Logue D, Madigan SM, Delahunt E, et al. Low energy availability in athletes: a review of prevalence, dietary patterns, physiological health, and sports performance. Sports Med. 2018;48:73–96.
10. Williams NI, Leidy HJ, Hill BR, et al. Magnitude of daily energy deficit predicts frequency but not severity of menstrual disturbances associated with exercise and caloric restriction. Am J Physiol Endocrinol Metab. 2015;308:E29–39.
11. Loucks AB, Thuma JR. Luteinizing hormone pulsatility is disrupted at a threshold of energy availability in regularly menstruating women. J Clin Endocrinol Metab. 2003;88:297–311.
12. Waters DL, Qualls CR, Dorin R, et al. Increased pulsatility, process irregularity, and nocturnal trough concentrations of growth hormone in amenorrheic compared to eumenorrheic athletes. J Clin Endocrinol Metab. 2001;86:1013–9.
13. Cadegiani FA, Kater CE. Novel causes and consequences of overtraining syndrome: the EROS-DISRUPTORS study. BMC Sports Sci Med Rehabil. 2019;11:21.
14. Cadegiani FA, Kater CE. Eating, Sleep, and Social Patterns as Independent Predictors of Clinical, Metabolic, and Biochemical Behaviors Among Elite Male Athletes: The EROS-PREDICTORS Study. Front Endocrinol (Lausanne). 2020;11:414. Published 2020 Jun 26. https://doi.org/10.3389/fendo.2020.00414.
15. Cadegiani FA, Kater CE. Inter-correlations among clinical, metabolic, and biochemical parameters and their predictive value in healthy and overtrained male athletes: the EROS-CORRELATIONS study. Front Endocrinol (Lausanne). 2019;10:858.
16. Cadegiani FA, Kater CE. Body composition, metabolism, sleep, psychological and eating patterns of overtraining syndrome: results of the EROS study (EROS-PROFILE). J Sports Sci. 2018;36(16):1902–10.
17. Cadegiani FA, Kater CE. Basal hormones and biochemical markers as predictors of overtraining syndrome in male athletes: the EROS-BASAL study. J Athl Train. 2019;54:906. https://doi.org/10.4085/1062-6050-148-18.
18. Cadegiani FA, Kater CE. Hypothalamic-pituitary-adrenal (HPA) axis functioning in overtraining syndrome: findings from Endocrine and Metabolic Responses on Overtraining Syndrome (EROS) – EROS-HPA axis. Sports Med Open. 2017;3(1):45.
19. Cadegiani FA, Kater CE. Growth Hormone (GH) and prolactin responses to a non-exercise stress test in athletes with overtraining syndrome: results from the Endocrine and metabolic Responses on Overtraining Syndrome (EROS) – EROS-STRESS. J Sci Med Sport. 2018;21(7):648–53.
20. Wright C, Milne S, Leeson H, et al. Sperm DNA damage caused by oxidative stress: modifiable clinical, lifestyle and nutritional factors in male infertility. Reprod Biomed Online. 2014;28(6):684–703.
21. Jayasena CN, Radia UK, Figueiredo M, et al. Reduced testicular steroidogenesis and increased semen oxidative stress in male partners as novel markers of recurrent miscarriage. Clin Chem. 2019;65(1):161–9.
22. Rosety I, Elosegui S, Pery MT, et al. Association between abdominal obesity and seminal oxidative damage in adults with metabolic syndrome. Rev Med Chil. 2014;142(6):732–7.
23. Cadegiani FA, Kater CE. Hormonal aspects of overtraining syndrome: a systematic review. BMC Sports Sci Med Rehabil. 2017;9:14.
24. Meeusen R, Duclos M, Foster C, European College of Sport Science, American College of Sports Medicine, et al. Prevention, Diagnosis, and Treatment of the Overtraining Syndrome: Joint Consensus Statement of the European College of Sport Science and the American College of Sports Medicine. Med Sci Sports Exerc. 2013;45(1):186–205.
25. Freudenberger HJ. Staff burn-out. J Soc Issues. 1974;30:159–65.

26. Koutsimani P, Montgomery A, Georganta K. The relationship between burnout, depression, and anxiety: a systematic review and meta-analysis. Front Psychol. 2019;10:284.
27. Aronsson G, Theorell T, Grape T, et al. A systematic review including meta-analysis of work environment and burnout symptoms. BMC Public Health. 2017;17(1):264.
28. Fernandez-Montero A, García-Ros D, Sánchez-Tainta A, Mourille AR, Vela A, Kales SN. Burnout syndrome and increased insulin resistance. J Occup Environ Med. 2019;61:729.
29. Maslach C, Jackson SE. The measurement of experienced burnout. J Org Behav. 1981;2:99–113.
30. Ochentel O, Humphrey C, Pfeifer K. Efficacy of exercise therapy in persons with burnout. A systematic review and meta-analysis. J Sports Sci Med. 2018;17(3):475–84.
31. Smith RE. Toward a cognitive-affective model of athletic burnout. J Sport Psychol. 1986;8:36–50.
32. Raedeke TD. Is athlete burnout more than just stress? A sport commitment perspective. J Sport Exercise Psychol. 1997;19:396–41.
33. Goodger K, Gorely T, Lavallee D, Harwood C. Burnout in sport: a systematic review. Sport Psychol. 2007;21:127–51.
34. Gustafsson H, DeFreese JD, Madigan DJ. Athlete burnout: review and recommendations. Curr Opin Psychol. 2017;16:109–13.
35. Francisco C, Arce C, Vilchez MP, Vales A. Antecedents and consequences of burnout in athletes: perceived stress and depression. Int J Clin Health Psychol. 2016;16(3):239–46.
36. Smith AL, Gustafsson H, Hassmén P. Peer motivational climate and burnout perceptions of adolescent athletes. Psychol Sport Exerc. 2010;11:360–453.
37. Cresswell SL, Eklund RC. The convergent and discriminant validity of burnout measures in sport: a multi-trait/multi-method analysis. J Sports Sci. 2006;24:209–20.
38. Manuela M, Pfeffer MM, Paletta A, Suchar G. New perspectives on burnout: a controlled study on movement analysis of burnout patients. Front Psychol. 2018;9:1150.

Chapter 10
Special Topics on Overtraining Syndrome (OTS)/Paradoxical Deconditioning Syndrome (PDS)

Outline

The high prevalence of "pseudo-OTS": more than 90% of the diagnosis of OTS are imprecise, particularly because the majority of the athletes diagnosed with "OTS" actually does not present the sine qua non alterations, which precludes the diagnosis. Hence, the book brings the characteristics of "pseudo-OTS," since this is more prevalent than the actual OTS.

"Deconditioning" and "hypometabolism" as two neologisms that describe the novel processes identified in OTS: the author proposed these two neologisms, which have been both already published in different papers, as they precisely summarize and explain the phenomena these words represent, respectively.

Deprivations, excess of commitments, and perfectionism: the harms of human attempts to become robot machines – it is likely that burnout syndrome, not depression, is the actual epidemic of the twenty-first century. Current overtraining syndrome could perhaps be understood as the burnout syndrome when manifested in athletes. Attempts to have high performance at multiple aspects simultaneously have been progressively encouraged in the highly competitive environment we are living nowadays. However, these over-human attempts naturally lead to exhaustion, eventually leading to burnout-related conditions, including OTS. The author depicts the potential mechanisms that may have led to the current burnout and OTS epidemics.

Low energy availability (LEA) as the underlying link between RED-S, OTS, BSA, burnout syndrome, and other burnout-related conditions: in this topic, the author provides the substantiation for the fact that LEA is the fundamental underlying pathogenic environment that leads to multiple interrelated conditions.

© The Editor(s) (if applicable) and The Author(s), under exclusive license to Springer Nature Switzerland AG 2020
F. Cadegiani, *Overtraining Syndrome in Athletes*,
https://doi.org/10.1007/978-3-030-52628-3_10

> **The (energy-deprived derived) paradoxical deconditioning syndrome (PDS):** OTS goes beyond excessive training. However, in common, all cases presented LEA as a triggering factors and eventually lead to PDS, irrespective of overtraining. The author discuss on the reasons why we should rethink about the terms we have been using to name the burnout-related dysfunctions.

10.1   The High Prevalence of "Pseudo-OTS"

The vast majority of sports-related health providers are unaware of the correct OTS definition, characteristics, and diagnosis. A decreased performance that cannot be justified by underlying conditions as the sine qua non criteria for OTS is unfamiliar to most, while the occurrence of URTI, prolonged or exacerbated muscle soreness, or use of non-validated markers, including cortisol-to-testosterone ratio or CK, is deemed to be pathognomonic signals of OTS. Detection of OTS in the general sports practice gets highly compromised, as both under- and overdiagnoses may be more prevalent than a proper diagnosis of OTS [1, 2].

In fact, almost 90% of the athletes suspected for OTS are excluded after the assessment of the manifestations that were believed to be caused by OTS, the possible underlying reasons for the loss of performance, and the exclusions of confounding factors [3]. Addressing the underlying reasons that may lead to underperformance other than OTS is crucial, once when these abnormalities are corrected, performance tends to fully recover.

Despite the misleading concepts such as sudden muscle distension, bone fracture, decreased humor or libido, repeated URTIs, or new-onset sleep disturbances as identifiers of OTS, even in the absence of verified reduction in sports-performance, what mostly allows athletes and coaches of all types of sports to consider the diagnosis of OTS is the lack of a clear trigger and the disproportional severity of any of the abovementioned dysfunctions.

Since athletes with decreased performance without the exclusion of confounding factors and those with any "unexplained" event, regardless of the reasons, are misdiagnosed with OTS, for practical purposes these misleading cases can be termed as "pseudo-OTS" or "OTS-like" conditions. The massive popularity of the term "overtraining syndrome" to label all these cases, among most of which are not adequately "diagnosed", compromises the correct identification of those with actual OTS [1–3].

Hence, grouping "pseudo-OTS" or "OTS-like" conditions into a sort of "impaired athlete syndrome" (IAS), when together with OTS, paradoxical deconditioning syndrome (PDS), overreaching states, relative energy deficiency of the sport (RED-S), and burnout syndrome of the athlete (BSA), once IAS encompasses all sorts of events that prevent athlete from training at full potential, should be the first step to standardize the assessment of these athletes in order to improve the level of investigation of the etiology for these dysfunctions. The approach to the "pseudo-OTS" athlete is therefore a way to help athletes that present actual complains or dysfunctions that are not with OTS.

Figure 10.1 depicts the manifestations commonly attributed to OTS and how to differentiate between OTS and OTS-related states, which altogether have been proposed to be termed as the impaired athlete syndrome (IAS).

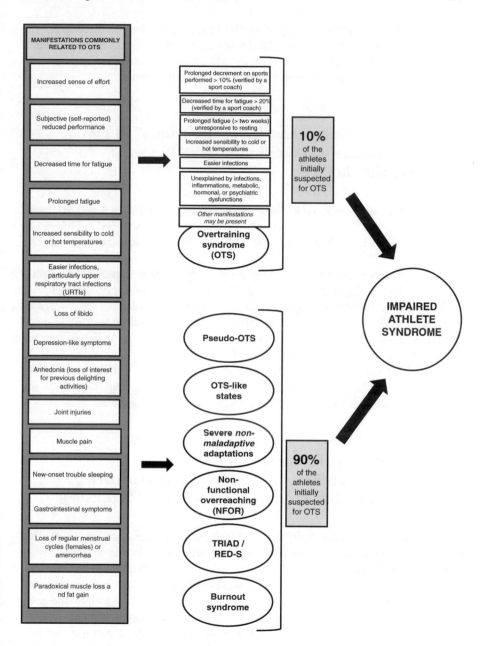

RED-S = relative energy deficiency of the sport; TRIAD = female athlete triad

Fig. 10.1 Manifestations commonly attributed to OTS, differentiation between OTS and OTS-related states, and impaired athlete syndrome (IAS)

Conversely, the education of sports-related health providers on the characteristics of OTS as well as how to correctly identify or suspect OTS among their athletes is one of the hardest challenges in OTS, as this condition is not usually part of the education processes for those who will further deal with athletes.

10.2   "Deconditioning" and "Hypometabolism" as Two Neologisms that Describe the Novel Processes Identified in OTS

The use of *deconditioning and hypometabolism* as neologisms was found to be suitable for the descriptions of two complex processes recently identified, which can be better illustrated and understood with the use of these intuitive new expressions [3, 4].

The use of "deconditioning" as a single-word expression is important to simplify the new understanding that athletes under OTS lose their previous conditioning.

The use of the neologism "hypometabolism" and its related expression – "hypometabolic" – describes the result of different processes that occurs diffusely of forced mitigation of the metabolism, as an additional strategy to maintain survival under severely and chronically depletion of energy.

"Maladaptation" is a neologism previously reported that describes the occurrence of forced adaptations to maintain survival and functioning, but which are highly dysfunctional, and lead to further and more complex diseases.

10.3   Deprivations, Excess of Commitments, and Perfectionism: The Harms of Humans Attempts to Become Robot Machines (From EROS Predictors)

Collectively, the subjective analysis of the findings of the present study shows that concurrent strict lifestyle in the long run may bring more harms than previously thought. Despite the benefits of adequate caloric and carbohydrate intake, food deprivation and carbohydrate phobia ("carbphobia") are present in some athletes, especially those in sports in which categories are based on body weight and body shape is culturally acclaimed, such as high-intensive functional training (HIFT), e.g., CrossFit®, which attempts to simultaneously lower body fat and improve performance [5–8]. These behaviors can lead to fatigue and temporary underperformance, consistent with our finding that lower caloric intake reduces alertness in the morning and impairs muscle recovery, while lower carbohydrate intake may lead to a paradoxical decrease in pace and strength; together these findings are termed overreaching [9]. If overreaching is not addressed by an increase in caloric and carbohydrate intake and compensatory rest, athletes can progress to a state of prolonged and hard to recover from decrease in performance, chronic fatigue, and mood distur-bances, which characterize classic OTS. In one of the EROS studies, a relatively low caloric (not hypocaloric) and low carbohydrate intake were the two major OTS triggers [8].

Excessive work or studying might lead to multiple harmful effects in athletes, including worsening of hormonal levels, libido, sleep quality, and performance [5, 6, 8, 10, 11]. Sleep quality impairs performance, libido, and all psychological func-tions. We recommend, therefore, against concurrent intense levels of physical and cognitive activity during championships or intensified training. Athletes should decrease the intensity and duration of studying and/or working, and when more intense studying or working is needed, the volume of training should be decreased. During intensification of training, a maximum work or study duration of 7 hours is recommended, following the findings of the EROS study [3].

Multiple modifiable patterns were found to modulate clinical and biochemical behaviors, and we learned answers are unlikely to be found if studies evaluate each aspect separately. The level of importance of each modifiable factor varies by the type of sport. For instance, carbohydrate intake plays an important role in explosive, stop-and-go, and short and intense sports, in which prompter and enhanced hor-monal responses and prompter energy availability are the two major factors influ-encing performance. An overall balance between training, eating, and resting is the most important factor for endurance sports, when prolonged optimization of hor-monal responses are desired for a longer time-to-fatigue and a maximum mainte-nance of pace throughout the training session.

10.4   Low Energy Availability (LEA) as the Underlying Link Between RED-S, OTS, BSA, Burnout Syndrome, and Other Burnout-Related Conditions

Increased commitment to simultaneous improvements in body shape, performance, and overall life activities, encouraged by a progressively competitive society, may become excessive and toxic, leading to multiple harms. Natural consequences include disturbance in eating behaviors, orthorexia, excessively strict eating, and alterations in moods and personality. In consequence, burnout and burnout-related syndromes are emerging, despite the increasing level of consciousness on the harm-ful consequences of living in extremes.

Low energy availability (LEA) seems to be the key underlying mechanism of all deprivation-triggered syndromes, including non-functional overreaching (NFOR), overtraining syndrome (OTS), relative energy deficiency of the sport (RED-S) and female athlete triad (TRIAD), and burnout syndrome of the athlete (BSA). Probably, LEA is the hallmark of burnout and burnout-related conditions [12–20].

In general, LEA is triggered by dysfunctional eating, including insufficient caloric intake, insufficient carbohydrate intake, insufficient protein intake, wrong timing of eating, insufficient hydration, disturbed eating behaviors, inaccuracy of self-report caloric and macronutrient intake, and indirectly triggered by body image dissatisfaction, sleep disturbances, alteration of mood states, personality disorders, excessive training, overcompensating physical activity, and attempts to have better performances in all major aspects of life (Fig. 10.2).

Analogically, LEA can be understood by the consequences of the shortage of gas for a driving car that needs to keep working (surviving and functioning) despite the absolute lack of gas (energy), leading to the disastrous beginning of utilization of the car own pieces and parts (muscle tissue) to be used as the source of gas (energy). Besides, adaptations to consume less gas (reduce metabolism) are forced to occur in this car, but these adaptations lead to a dysfunctional car (multiple dysfunctions), with lower power and speed (reduced performance and increased sense of effort), and break easier (overt infirmities occur more easily).

However, insufficient food is distinct from hypocaloric diets. Usually, LEA occurs under non-hypocaloric diets, but inadequate quality of eating. In this case, high-performance and elite athletes, who have additional caloric needs, are at higher risk of LEA. Most likely affected athletes include those of endurance sports, particularly sprinters, although any sport may cause LEA, as LEA depends on nutritional patterns in more extension than training characteristics. In addition, hypocaloric diets and LEA lead to distinct sources of energy for utilization, metabolic effects, and health consequences, although hypocaloric diets may induce to LEA in the long run.

Methodological issues on the quality of the assessment of energy availability (EA) precludes more conclusive results [13, 14, 16]. Despite conflicting findings, LEA has been independently associated with increased risk of menstrual irregularity, worse bone health, functional hypothalamic amenorrhea, hematological abnormalities, metabolic dysfunctions, psychological disturbances, and gastrointestinal misalignments, which are also the major features of RED-S. In addition, loss of performance including decrease of training response, reflex, coordination, concentration, and endurance performance, while irritability and depression are exacerbated, which are also typical characteristics of OTS.

Common to both RED-S/TRIAND and OTS/PDS, LEA impairs recovery, muscle mass and neuromuscular function, and increases the risk of joint and muscle pathological injuries.

In the long run, LEA induces metabolic and physiological adaptations to prevent further weight loss, and athletes may therefore be weight stable yet have impaired physiological function secondary to LEA. Hence, weight loss in not necessarily present in LEA, since it may depend on the duration of the LEA, as well as caloric balance. The maintenance of lower body mass or fat levels through long-term LEA may result in impaired health and reduced performance, as proposed for both RED-S and PDS models [19].

Ultimately, because LEA indistinctly causes manifestations of both RED-S/TRIAD and OTS/PDS (Fig. 10.3), LEA may be the missing link that demonstrates that these conditions are actually part of the same condition.

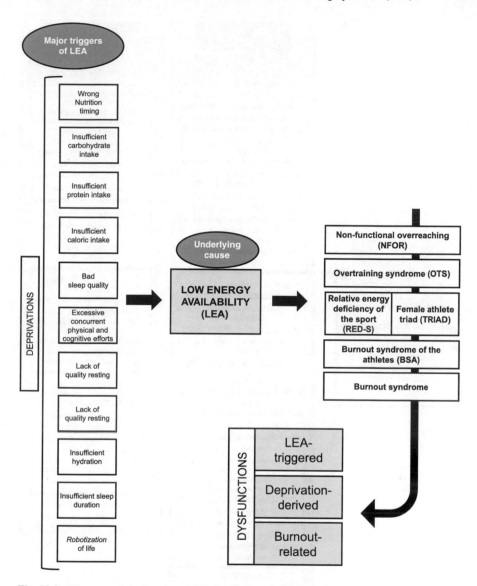

Fig. 10.2 Triggers and dysfunctions related to low energy availability (LEA)

10.5   The (Energy-Deprived Derived) Paradoxical Deconditioning Syndrome (PDS)

Paradoxical deconditioning syndrome (PDS) is a concept developed from the unification of all novel concepts on OTS together with classical theories on the syndrome and can be described as the result of combination of the multiple alterations detected in OTS that elucidated some previously unclear pathogenic pathways. In

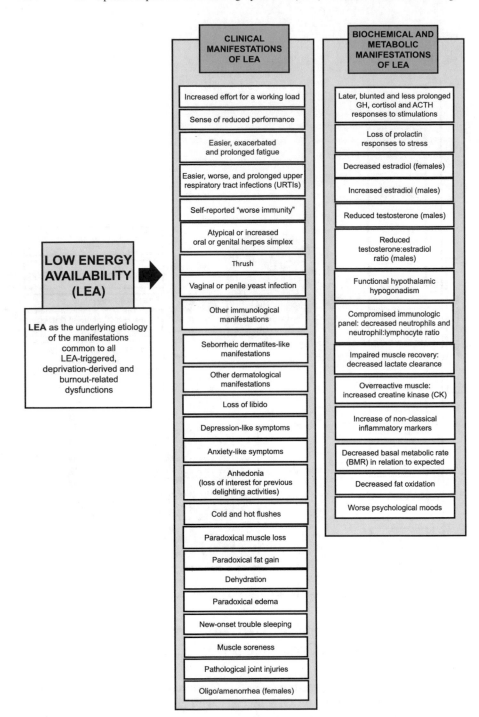

Fig. 10.3 Clinical, biochemical, and metabolic manifestations of low energy availability (LEA)

summary, PDS is the result of chronic physical and mental fatigue, loss of performance in terms of impulse, and explosion and in terms of endurance (stability) and a state of a sort of *hypometabolism*. These dysfunctions are consequences of the sum of the number, intensity, and interaction between sympathetic overcompensation and global reduction of metabolic pathways while preserving fat storage; anti-anabolic and procatabolic states; flattened responses to *fight-or-flight* reaction; overreactive, prolonged, and impaired muscle recovery; and exacerbated proteolysis. These are some of the multiple chronic *maladaptations* aiming to function toward energy saving under chronic glycogen depletion, excessive oxidative stress, cytokines released, and immunological abnormalities. These disturbances are induced by caloric-, protein-, and carbohydrate-insufficient intake, excessive training or too-fast training intensity or volume increase, unrefreshing sleep, and excessive overall concerns and cognitive demands.

Since PDS is ultimately derived from chronic energy deprivation, it is logical to suppose that LEA is the key circumstance for its occurrence.

The fact that overtraining syndrome was no longer able to justify the majority of the athletes affected, since excessive training has become a minor trigger of OTS after the learning of periodization of trainings, allowed the suggestion for more suitable terms for this syndrome. The first suggested term was "unexplained underperformance syndrome," which indeed better reflects the hallmark of OTS, although has not been extensively adopted. The expression proposed herein reinforces the characteristic of the opposite performance than the expected as the main finding on OTS.

In conclusion, PDS is an expression to intuitively improve the understanding of OTS hallmarks and may encompass other related conditions such as BSA.

10.6 Conclusion

Despite the massive popularity of the term "overtraining syndrome" (OTS), few are familiar with its definition, characteristics, and diagnosis. While the sine qua non criteria is underemployed, random have been inappropriately although highly used to diagnose OTS. The vast majority of the athletes claimed to present OTS actually present "pseudo-OTS" or "OTS-like" conditions. Altogether, actual OTS, paradoxical deconditioning syndrome (PDS), overreaching states, relative energy deficiency of the sport (RED-S), burnout syndrome of the athlete (BSA), "pseudo-OTS," or "OTS-like" have substantial similiarities in their pathophysiology and underlying mechanisms, in an extension that allow to cluster them into a sole condition, as a sort of "impaired athlete syndrome" (IAS) that encompasses all sorts of events that prevent athlete from training at full potential.

The expressions *deconditioning and hypometabolism* are neologisms designated to describe a loss of the conditioning processes that athletes undergo and the result of different processes resulted in the mitigation of the metabolism, respectively.

Society is increasingly encouraged to perform well in physical, social, athletic, financial, intellectual, and sexual aspects which simultaneously led to the current epidemic of burnout syndrome. When in athletes, burnout is better represented by OTS. In all cases, atypical abnormalities in multiple systems occur similarly.

Low energy availability (LEA) seems to be the underlying mechanism of dysfunction that leads to multiple burnout-related dysfunctions, including elative energy deficiency of the sport (RED-S), OTS, burnout syndrome of the athlete (BSA), burnout syndrome, and other burnout-related conditions.

Paradoxical deconditioning syndrome (PDS) better describes OTS, once PDS encompasses beyond those with OTS triggered by excessive training and describes within one expression the key pathogenic path of OTS/PDS.

References

1. Grandou C, Wallace L, Impellizzeri FM, Allen NG, Coutts AJ. Overtraining in resistance exercise: an exploratory systematic review and methodological appraisal of the literature. Sports Med. 2020;50(4):815–28.
2. Cadegiani FA, Kater CE. Hormonal aspects of overtraining syndrome: a systematic review. BMC Sports Sci Med Rehabil. 2017;9:14.
3. Cadegiani FA, Kater CE. Novel insights of overtraining syndrome discovered from the EROS study. BMJ Open Sport Exerc Med. 2019;5(1):e000542. https://doi.org/10.1136/bmjsem-2019-000542.
4. Cadegiani FA, Kater CE. Inter-correlations among clinical, metabolic, and biochemical parameters and their predictive value in healthy and overtrained male athletes: the EROS-CORRELATIONS study. Front Endocrinol (Lausanne). 2019;10:858.
5. Crewther B, Keogh J, Cronin J, Cook C. Possible stimuli for strength and power adaptation: acute hormonal responses. Sports Med. 2006;36(3):215–38.
6. Durand RJ, Castracane VD, Hollander DB, et al. Hormonal responses from concentric and eccentric muscle contractions. Med Sci Sports Exerc. 2003;35(6):937–43.
7. Cadegiani FA, Kater CE, Gazola M. Clinical and biochemical characteristics of high-intensity functional training (HIFT) and overtraining syndrome: findings from the EROS study (The EROS-HIFT). J Sports Sci. 2019 Feb;20:1–12.
8. Cadegiani FA, Kater CE. Body composition, metabolism, sleep, psychological and eating patterns of overtraining syndrome: results of the EROS study (EROS-PROFILE). J Sports Sci. 2018;36(16):1902–10.
9. Meeusen R, Duclos M, Foster C, European College of Sport Science, American College of Sports Medicine, et al. Prevention, Diagnosis, and Treatment of the Overtraining Syndrome: Joint Consensus Statement of the European College of Sport Science and the American College of Sports Medicine. Med Sci Sports Exerc. 2013;45(1):186–205.
10. Hayes LD, Grace FM, Baker JS, Sculthorpe N. Resting steroid hormone concentrations in lifetime exercisers and lifetime sedentary males. Aging Male. 2015;18(1):22–6.
11. Shaner AA, Vingren JL, Hatfield DL, Budnar RG Jr, Duplanty AA, Hill DW. The acute hormonal response to free weight and machine weight resistance exercise. J Strength Cond Res. 2014;28(4):1032–40.
12. Loucks AB, Verdun M, Heath EM. Low energy availability, not stress of exercise, alters LH pulsatility in exercising women. J Appl Physiol (1985). 1998;84(1):37–46.
13. Mata F, Valenzuela PL, Gimenez J, et al. Carbohydrate availability and physical performance: physiological overview and practical recommendations. Nutrients. 2019;11(5):1084.
14. Logue D, Madigan SM, Delahunt E, Heinen M, Mc Donnell SJ, Corish CA. Low energy availability in athletes: a review of prevalence, dietary patterns, physiological health, and sports performance. Sports Med. 2018;48(1):73–96.

15. Melin AK, Heikura IA, Tenforde A, Mountjoy M. Energy availability in athletics: health, performance, and physique. Int J Sport Nutr Exerc Metab. 2019;29(2):152–64.
16. Sygo J, Coates AM, Sesbreno E, Mountjoy ML, Burr JF. Prevalence of indicators of low energy availability in elite female sprinters. Int J Sport Nutr Exerc Metab. 2018;28(5):490–6.
17. Heikura IA, Uusitalo ALT, Stellingwerff T, Bergland D, Mero AA, Burke LM. Low energy availability is difficult to assess but outcomes have large impact on bone injury rates in elite distance athletes. Int J Sport Nutr Exerc Metab. 2018;28(4):403–11.
18. Brook EM, Tenforde AS, Broad EM, et al. Low energy availability, menstrual dysfunction, and impaired bone health: a survey of elite Para athletes. Scand J Med Sci Sports. 2019;29(5):678–85.
19. Ackerman KE, Holtzman B, Cooper KM, et al. Low energy availability surrogates correlate with health and performance consequences of relative energy deficiency in sport. Br J Sports Med. 2019;53(10):628–33.
20. Black K, Slater J, Brown RC, Cooke R. Low energy availability, plasma lipids, and hormonal profiles of recreational athletes. J Strength Cond Res. 2018;32(10):2816–24.

Chapter 11
Practical Approach to the Athlete Suspected or at High Risk for OTS

Outline

New diagnostic tools for OTS: once multiple novel markers and mechanisms have been unveiled in OTS, it is natural that novel diagnostic tools would be proposed, as consequences of these novel findings. The author explains each of these tools, and how they can be greatly helpful for the everyday practice.

Preliminary score for the diagnosis of OTS: following the specific scientific publication on this, the author proposes practical scales, including those easy to use for clinical practice, even in places where some tools for the diagnosis are not available.

Tools for the identification of athletes at high risk for OTS: once OTS is hard to be fully remitted, working on a prevention basis is much more effective. And the paramount method is to effectively identify those at high or imminent risk of OTS, in order to offer specific preventive approaches. The book brings accurate tools for the detection of these athletes. The ultimate objective is to reduce and ideally extinct the occurrence of OTS.

Advances in the diagnosis of OTS: direct comparisons with the researches of the past highlight the advances that science has had in the field of OTS and help readers to better understand the main messages of the book.

Effective prevention of OTS: regardless of the risk of developing OTS, a long-term, solid prevention approach should be the most effective modality to substantially decrease the development of OTS. The author enumerates the tools for its prevention, in a quite practical and clinical manner, so that the increasing number of health professionals that deal with athletes may help this population in the long run.

> *Recommendations to recover from OTS*: while research on OTS has been primarily focused on the identification of its biomarkers, protocols and evaluation of the course of OTS once installed lacks. In this section, the author provides a scientific-based protocol to optimize the recovery process from OTS, as well as what to expect from biochemical biomarkers.

11.1   New Diagnostic Tools for OTS

With the currently available tools, OTS can be diagnosed only in later and more clinically meaningful staged, when treatment and recovery are less successful [1]. This should be reframed toward a prevention, identification of those at risk, and early identification basis. Earlier diagnosis is crucial to avoid irreversible alterations of OTS, which usually happen in more severe stages of OTS, including important impaired sports performance and burnout-like manifestations. In addition, ideally OTS should be eradicated, since its consequences include long-term post-traumatic stress disorder, end of the athletes' career, and unrecoverable energy levels. For this objective, we should fully understand all risk factors for OTS and propose a preventive approach.

Based on the OTS key feature as an underperformance syndrome unexplained by alterations or dysfunctions that could lead to decrease in performance, the latest guideline on OTS proposed a diagnostic flowchart recommending that some biochemical alterations and confounding conditions should be excluded prior to the diagnosis of OTS [1]. However, this flowchart does specify which parameters and diseases should be assessed except for few markers that do not cover the majority of the cases. Also, in the clinical practical diagnostic flowchart, the main tool offered for the diagnosis of OTS in the guideline, cutoffs for potential markers to diagnose or exclude OTS, questionnaires, or scores has not been suggested.

Given the limitations of the current diagnostic tools for OTS, proposals for new proposals for the diagnosis of OTS have been shown to be crucial, preferably if these tools help detect at earlier stages. Meanwhile, with the novel findings of which parameters are abnormal and more severely and commonly affected in OTS [2–8], the development of new diagnostic tools has become feasible, which should help spread the scientifically driven diagnosis of OTS even in places without access to biochemical or body composition exams. Two sorts of diagnostic tools for earlier detection and effective prevention of OTS have been proposed, two-step diagnostics tool and diagnostic scores, and are divided into two populations: those suspected of OTS and those at high risk of OTS.

11.2   Tools for the Diagnosis of Athletes Suspected of OTS

For athletes suspected of OTS, its diagnosis used to be based on a multiple-step diagnostic flowchart proposed by the latest guideline on OTS [1]. The guideline based the recommendation of the diagnostic flowchart on correct assumptions of unexplained underperformance as the *sine qua non* criteria for the diagnosis of OTS. However, this diagnostic tool does not specify how performance should be assessed and does not provide a list of disorders and biochemical abnormalities, which precludes from a final confirmation of the diagnosis of OTS from a perspective of full checked list of confounding diagnoses and abnormalities to exclude.

Oppositely, the present proposal of a two-step OTS diagnosis provides a list with the specific aspects of performance that should be assessed and the full list of disorders and biochemical tests that should be employed prior to the diagnosis of OTS.

The second proposal of scores for the diagnosis of OTS has been based on parameters that together were able to distinct healthy from OTS athletes in 100% of the cases, providing a 100% accuracy, at least in a population of approximately 40 athletes.

11.2.1   Two-Step OTS Diagnosis

11.2.1.1   Step #1 – Definition of the Clinical Manifestations of OTS

Manifestations in Training Patterns

The reduction of at least 10% in sports performance, verified by a sports coach or sports health-related professional, of at least one of ten major dimensions of sports performance, should be present. The reduction of one or more of the aspects below should be verified in order to formally diagnose as loss of performance:

(a) Training volume (reduction >10%)
(b) Training intensity (reduction >10%)
(c) Explosion (reduction of >10% of the maximum speed, pace or weight lifted)
(d) Reduction of the time-to-fatigue (of at least 20%)
(e) Sense of increased effort for a same training volume and intensity (>10% of the increased self-perceived effort).

Although the reduction of at least one of the domains of sports performance is classically *sine qua non* for the diagnosis of OTS, earlier stages may not present alterations in performance. Hence, the presence of any reduction helps the diagnosis; from a more preventive perspective, this should not be necessarily present.

Clinical Manifestations

Since reduced performance usually appears later in the *continuum* process of OTS, the diagnosis of OTS without changes in performance is feasible and possibly desirable, as stages in which performance is not compromised tend to be incipient, earlier and easier to recover. In these cases, some early signs may be present and helpful for the detection of OTS and are listed below:

(a) Fatigue
(b) Loss of libido
(c) Upper tract respiratory infections (in atypical seasons or with uncommon symptoms)
(d) Psychological disturbances with no previous history of psychiatric or psychological issues
(e) Muscle soreness
(f) An apparent contradictory combination of anxiety and fatigue, typical of burnout states
(g) Non-typical signs, such as unexplained fever or flu-like symptoms

Predisposition

Athletes suspected of OTS should have a minimum risk predisposition for OTS, including recent changes in any of these aspects: training overload, restricted eating, sleeping disturbances, and/or social or professional increased demand. At least one of the characteristics below should be present:

(a) Excessive intensification of training patterns (beyond the usual increase that typically occurs right before beginning of season periods)
(b) Carbohydrate intake (which should be above 5.0 g/kg/day)
(c) Protein intake (which should be above 1.6 g/kg/day)
(d) Overall caloric intake (which should be above 35 kcal/kg/day)
(e) Self-reported sleep quality and level of refreshing sleep lower than 6, in a scale from 0 to 10
(f) Extra-exercise demands: number of working or studying hours, social or familial issues

Although this specific *criterium* is not mandatory, it is unlikely that an athlete without any of the characteristics above develop OTS. At least one characteristic of two of the three aspects above should be present for the diagnosis of OTS.

The presence of any of the signs above is indicative of OTS, unless other diagnoses are present, as depicted below.

11.2.1.2   Step #2 Exclusion of Alterations that Could Lead to Decreased Performance

As already mentioned, OTS can only be diagnosed when the reduction of performance cannot be explained by any pre-existing, uncontrolled, or previously non-diagnosed dysfunction. Hence, there are multiple clinical and biochemical criteria and conditions that can also cause reduction in performance, which must be excluded before the diagnosis of OTS.

In the case of pre-existing disorders, these should be under control or remission for a diagnosis of OTS, once currently active or uncontrolled pre-existing conditions tend to influence athletes' physical capacity.

After the correction of the alterations and dysfunctions, performance should be reassessed, and in case reduction persists, further steps must be employed for the diagnosis of OTS. However, sports performance tends to be restored after ceasing the underlying alteration, so persistence of reduced capacity is unlike.

The assessment of clinical alterations and diagnoses detailed below should precede biochemical ones in order to prevent unnecessary exams, as OTS may be excluded in the clinical step. Eighty of the possible confounders of OTS that must be considered are listed in Table 11.1. The clinical criteria include:

(a) Current or recent use (<6 months before the investigation) of androgenic anabolic steroid (AAS), other hormones, glucocorticoids, oral contraceptives, centrally acting, muscle relaxing, and anti-inflammatory drugs, and peptides
(b) Use of supplements, which should be investigated, particularly those that are not pure substances, bought from non-verified sources, and that induced unexpected effects.
(c) Overuse of caffeine or caffeine-containing foods, like green tea (equivalent to >400–600 mg/day of caffeine)
(d) Presence of previously diagnosed conditions, including psychiatric diseases, like depression and anxiety disorders, or diseases of any sort
(e) Symptoms of upper respiratory tract infection (URTI), chronic sinusitis, or infections of any nature, osteoarticular
(f) Physical examination, which should include analysis of the skin, signs of liver, adrenal, or kidney failure, or autoimmune disorders, screening of heart rate (HR) for HR variability abnormalities, seated and orthostatic blood pressure (BP) for orthostatic hypotension, heart auscultation for valvular and rhythm alterations, lung auscultation for respiratory obstructions, neurologic examination for motor, sensitive, and reflex abnormalities, and testicle size and aspect (in case of males) for hypogonadism, genetic hormone alterations, and steroid use or abuse, and oxygen saturation, for oxygen exchange abnormalities (usually related to pulmonary dysfunctions)

Since clinical exclusion criteria are not present, further biochemical investigation should also be employed for the detection of atypical presentation of any metabolic dysfunction that could reflect in sports performance. Biochemical, functional, and imaging investigation should therefore include:

Table 11.1 List of clinical and biochemical confounders, dysfunctions and diseases that should be excluded prior to the diagnosis of OTS

Overweight, obesity, and excessive body fat	New-onset type 1 diabetes mellitus	Major depression	Guillain-Barré syndrome
Hypothyroidism	Uncontrolled type 1 or 2 diabetes mellitus	Generalized anxiety disease	Ankylosing spondylitis
Hashimoto thyroiditis	Symptomatic undiagnosed type 2 diabetes mellitus	Osteoarticular	Polymyositis
Hypogonadism	Chronic kidney disease	Unnoticed muscle lesions	Cytomegalovirus infection
Hyperprolactinemia	Liver failure	Rheumatoid arthritis	HIV or AIDS
Overt or relative adrenal insufficiency	Congestive heart failure	Multiple sclerosis	Chronic type B autoimmune
Asthma	Chronic obstructive pulmonary disease (COPD)	Lung fibrosis, either idiopathic or chemically induced	Chronic type C autoimmune
Growth hormone (GH) deficiency	Lyme's disease	Neurodegenerative diseases	Chronic diffuse allergies
Current or recent use of androgen anabolic steroids (AAS)	Epstein-Barr chronic infection	Muscle degenerative diseases	Variety of paraneoplastic syndrome
Current or recent use of other hormones	Tuberculosis	Sleep disorders, including sleep obstructive apnea	Different types of lymphoma
Current or recent use of centrally acting drugs	Atypical pneumonia	Neuromuscular autoimmune disorders – Myasthenia gravis	Different types of leukemia
Current or recent use of statins	Chronic sinusitis	Uncontrolled attention-deficit/ hyperactive disorder	Incipient neoplasia
Current or recent use of muscle relaxers	Chronic myocarditis	Abuse of non-steroidal anti-inflammatory	Trauma-driven long term panhypopituitarism
Inflammatory bowel diseases	Irritable bowel syndrome	Unexplained hypoglycemia in non-diabetics	Vitamin B12 deficiency
Chronic fatigue syndrome	Fibromyalgia	Inborn errors of metabolism	Iron deficiency and ferropenic anemia
Post-traumatic stress syndrome	Repeated hypoglycemic episodes in T1DM	Severe hyperglycemia	Zinc deficiency
Vasovagal syndrome and syncope	Orthostatic hypotension	Uncontrolled hypertension	Heavy metal intoxication

Table 11.1 (continued)

Overweight, obesity, and excessive body fat	New-onset type 1 diabetes mellitus	Major depression	Guillain-Barré syndrome
Burnout syndrome	Sarcopenia	Post-bariatric nutritional deficiency	Paracoccidioidomicosis
Current or recent use of glucocorticoids	Current or recent use of not open-label supplements	Current or recent use of "peptides"	Exacerbation of hereditary anemias

In the absence of specific symptoms or clinical conditions that require specific investigation, screening tests for the following diseases should be performed: hypothyroidism (TSH > 5.0 mU/L), Hashimoto's thyroiditis (thyroid peroxidase antibody (TPOAb) > 35 IU/mL or thyroglobulin antibody (TgAb) > 20 IU/mL), hypogonadism (total testosterone <200 ng/dL), adrenal insufficiency (basal cortisol <5.0 mcg/dL, excluded for oral, nasal, spray, or topic use of glucocorticoids), iron deficiency (ferritin <20 ng/dL and/or serum iron <50 μg/dL) or anemia (CBC with hematocrit <39–40% for males and <36% for females), vitamin B12 deficiency (vit B12 < 200 pg/mL), hyperglycemia (random serum glucose >200 mg/dL), electrolytic abnormalities (particularly hypo- and hypercalcemia – total calcium <8.5 and >10.5 mg/dL, respectively, hyponatremia – sodium <135 mEq/L, hypo- or hyperkalemia – potassium <3.5 and >5.5, respectively), liver abnormalities (ALT >50 U/L, gamma-GT >75 U/L), chronic renal disease (CRD) (creatinine >1.4 mg/dL for males or >1.1 mg/dL for females, and/or calculated eGFR <60 ml/min), subacute or chronic infections (Epstein-Barr, cytomegalovirus, HIV, and hepatitis types B and C serologies), and classic inflammatory markers (ultrasensitive C-reactive protein (CRP) >3.0 mg/L or erythrocyte sedimentation rate (ESR) >20 mm/h).

Optionally, whenever available, additional screening tests can be performed, in case of specific dysfunctions are suspected, or further clarification is needed: autoimmune disorders (antinuclear antibody, ANA), chronic or subacute non-controlled or non-diagnosed less common infections (secondary syphilis – VDRL and FTA-ABS, Lyme's disease and chronic manifestations of Zika or Chikungunya viruses – specific serologies), heart failure (transthoracic echocardiogram), asthma and chronic obstructive pulmonary disease (COPD) (spirometry), pulmonary fibrosis (thorax CT), spontaneous pneumothorax (chest X-ray), and varicose veins and related diseases (arterial and venous blood flow studies of the legs, segmental Doppler pressures).

Whenever a condition is diagnosed, some specific further investigation for the elucidation of its etiology should be performed: sex hormone-binding globulin (SHBG), LH, and FSH for hypogonadism, freeT4 (fT4) for hypothyroidism or thyroiditis, and DHEA-S and ACTH for adrenal insufficiency, as some examples.

Once at least two of the three criteria for step #1 are met and all confounding factors described in step #2 are excluded, the diagnosis of OTS can be concluded.

11.2.2   Scores for the Diagnosis of OTS

Although more than 40 novel markers have been identified in OTS, none of them accurately distinguished between OTS-affected athletes and healthy athletes when analyzed individually, given the highly individual characteristic of manifestations and dysfunctions in OTS [5–8].

Despite the lack of a single marker that could diagnose OTS in all affected athletes, it has been hypothesized that different combinations of OTS biomarkers would yield accurate results for a diagnosis of OTS. Indeed, when certain combinations of parameters were tested, OTS could be accurately identified in all affected athletes while excluding OTS in all healthy athletes, without the need to exclude confounding disorders.

The new biomarkers identified for OTS, among which many showed very few overlapping results between OTS-affected and healthy athletes, allowed the proposal of preliminary diagnostic scores for OTS [9]. These scores preclude from undergoing the full diagnostic flowchart proposed by the latest guideline on OTS, which has not been validated and which has been shown to be unfeasible for clinical utilization in most sports centers [1].

Although preliminary, in the context of the absence of any other diagnostic tool, the proposed scores are useful for clinical practice, at least until further more consistent studies are conducted to improve the current options for OTS diagnosis, since these scores:

1. Have been shown to be 100% accurate to distinct all affected from healthy athletes that participated in the study.
2. Were derived from a study that evaluated a larger population of athletes, compared to previous studies.
3. Do not disclose any overlapping result and presented a quite substantial gap of the results of all OTS-affected athletes from those of all healthy athletes. This gap allowed the results to be clustered according to each group, without the presence of any outlier.
4. Are able to detect OTS in earlier stages, even before the presence of clinical symptoms or reduced performance, when recovery is more likely.
5. Can be helpful for clinical use, as they are easy-to-perform, not time-consuming, have no or very low cost.

For the selection of the parameters to be included in the scores, parameters that alone disclosed the highest ability for the detection of OTS and the fewest overlapping results between affected and healthy athletes were proposed as screening tests with specific cutoffs for higher specificity, sensitivity, or overall accuracy. The sequence of steps for the identification of the most appropriate markers for the EROS scores is shown in Fig. 11.1.

The idea of prevention of OTS or its detection in the early beginning and milder states is key for its eradication, as overt and clinical OTS is harsh to recover in a fully manner, with long-term consequences, including the end of the athletes career. From the initial 117 markers evaluated for OTS, selection processes of clinical and biochemical parameters for the development of the diagnostic tools are shown in Tables 11.2 and 11.3, respectively, and are depicted in Fig. 11.2.

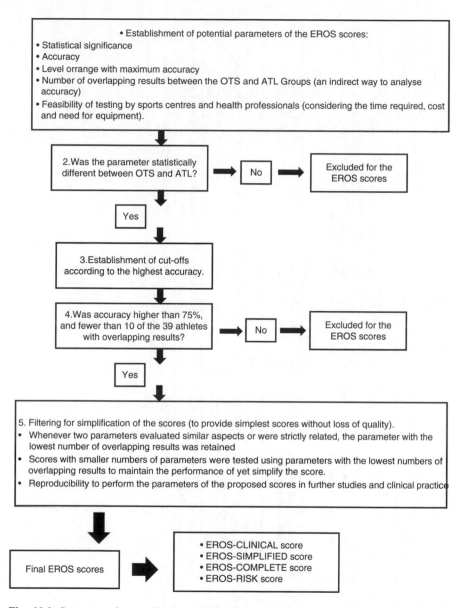

Fig. 11.1 Sequence of steps for the identification of the most appropriate markers for the EROS scores

Nine biochemical and 9 clinical parameters for the proposed scores have been selected, comprising a total of 18 parameters. Selected parameters included four subscales (anger, fatigue, tension, and vigor) of the Profile of Mood States (POMS) questionnaire, three eating patterns (daily calorie intake [kcal/kg/day], protein intake [g/kg/day], and carbohydrate intake [kcal/kg/day]), and two characteristics of body composition (body fat [%] and muscle mass [%], comprising nine clinical parameters. Also included were four basal hormones (GH, prolactin, total testosterone, and the testosterone-to-estradiol ratio) and five hormonal responses to an

Table 11.2 Clinical parameters evaluated for the development of overtraining syndrome diagnostic tools

Study/tests	Markers	p-score	Highest NPV or PPV (%) and accuracy	Ranges of the maximum accuracy and maximum true positive and true negative values	Useful for any of the questionnaires (No/potentially/yes)
EROS-PROFILE					
Eating patterns	Calorie intake (kcal/kg/day)	<0.001	Cutoff: 32–40 kcal/kg/day; PPV: 100%; NPV: 96.1%; accuracy: 97.4%	OTS: 13/14 if <32 kcal/kg/day; ATL: 25/25 if >40 kcal/kg/day or 22/25 if >47 kcal/kg/day; accuracy: 38/39 if 32–40 kcal/kg/day	Yes
	Protein intake (g/kg/day)	<0.001	Cutoff: 1.7 g/kg/day; PPV: 92.9%; NPV: 85.7%; accuracy: 84.6%	OTS: 14/14 if <2.5 g/kg/day, 13/14 if <2.2 g/kg/day or 10/14 if <1.6 g/kg/day; ATL: 24/25 if >1.6 g/kg/day or 19/25 if >2.5 g/kg/day; accuracy: 33/39 if 1.7 g/kg/day	Yes
	Carbohydrate intake (g/kg/day)	0.003	Cutoff: 5–5.4 g/kcal/day; PPV: 70%; NPV: 100%; accuracy: 84.6%	OTS: 14/14 if <5 g/kg/day or 10/14 if <3.2 g/kg/day; ATL: 23/25 if <3 g/kg/day or 18/25 if >6.5 g/kg/day; accuracy: 33/39 if 5–5.4 g/kg/day	Yes
	Fat intake (g/kg/day)	n/s	n/a	n/a	No
Social patterns	Self-reported sleep quality (0–10)	0.004	Cutoff: ≤5; PPV: 85.7%; NPV: 70.4%; accuracy: 82.1%	OTS: 12/14 if >8; ATL: 18/25 if <7; accuracy: 32/39 if ≤5	Potentially
	Duration of night sleep (h)	n/s	n/a	n/a	No
	Number of hours of activities (h/day)	<0.001	Cutoff: ≥9 h; PPV: 83.3%; NPV: 72.7%; accuracy: 74.4%	OTS: 10/14 if >7 h/day or 14/14 if >5 h/day; ATL: 25/25 if ≤10 h/day or 24/25 if ≤8 h/day; accuracy: 29/39 if ≥9 h/day	Potentially
	Self-reported libido (0–10)	0.024	Cutoff: ≤5; PPV: 66.7%; NPV: 73.3%; accuracy: 71.8%	OTS: 9/14 if <7 or 11/14 if <8; ATL: 22/25 if >5 or 20/25 if >6; accuracy: 28/39 if ≤5	Potentially

Psychological patterns	Total POMS questionnaire score (−32–120)	<0.001	Cutoff: 24–30; PPV: 100%; NPV: 89.3%; accuracy: 92.3%	OTS: 14/14 if >6, 13/14 if >18 or 11/14 if >24–30; ATL: 25/25 if <24 or 20/25 if <2; accuracy: 36/39 if 24–30	Yes
	POMS anger subscale (0–48)	0.003	Cutoff: ≥11; PPV: 81.8%; NPV: 82.2%; accuracy: 82.1%	OTS: 13/14 if >19 or 9/14 if >11; ATL: 23/25 if <11 or 19/25 if <8; accuracy: 32/39 if ≥11	Yes
	POMS confusion subscale (0–28)	0.001	Cutoff: ≥6; PPV: 85.7%; NPV: 75%; accuracy: 76.9%	OTS: 10/14 if >4 or 6/14 if >6; ATL: 24/25 if <6; accuracy: 30/39 if >6	Yes
	POMS depression subscale (0–60)	0.008	Cutoff: ≥9; PPV: 85.7%; NPV: 73.3%; accuracy: 76.9%	OTS: 6/14 if >9 or 8/14 if >6; ATL: 22/25 if <9 or 21/25 if <6; accuracy: 30/39 if <9	Yes
	POMS fatigue subscale (0–28)	<0.001	Cutoff: 5–7; PPV: 100%; NPV: 100%; accuracy: 100%	OTS: 14/14 if >8 or 12/14 if >13; ATL: 25/25 if <5; accuracy: 39/39 if 5–8	Yes
	POMS tension subscale (0–36)	<0.001	Cutoff: ≥13; PPV: 84.6%; NPV: 88.5%; accuracy: 87.2%	OTS: 14/14 if >5 or 11/14 if >13; ATL: 23/25 if <13; accuracy: 34/39 if 12 or 13	Yes
	POMS vigor subscale (0–32)	<0.001	Cutoff: ≤18; PPV: 100%; NPV: 96.2%; accuracy: 97.4%	OTS: 13/14 if <18; ATL: 25/25 if >18, 25/25 if >20 or 23/25 if >23; accuracy: 38/39 if >18–20	Yes
Body metabolism	Measured/predicted BMR (%)	0.013	Cutoff: <102%; PPV: 62.5%; NPV: 82.6%; accuracy: 74.4%	OTS: 13/14 if <108% or 10/14 if <102%; ATL: 19/25 if >102%; accuracy: 29/39 if 102%	Potentially
	Percentage of fat burned compared with total BMR (%)	<0.001	Cutoff: <38%; PPV: 75%; NPV: 74.2%; accuracy: 74.4%; Good for PPV and NPV; not good for accuracy	OTS: 14/14 if <58% or 6/14 if <40% and <30%; ATL: 23/25 if >20% or >38%; accuracy: 29/39 if <30–38%	Potentially
Body composition	Body fat (%)	<0.001	Cutoff: >17%; PPV: 80%; NPV: 70.6%; accuracy: 76.9%	OTS: 14/14 if >10%, 7/14 if >15% or 4/14 if >17%; ATL: 24/25 if <17% or 22/25 if <15%; accuracy: 30/39 if >17%	Yes
	Muscle mass (%)	0.008	Cutoff: <47%; PPV: 100%; NPV: 75.8%; accuracy: 79.5%	OTS: 12/14 if <50% or 6/14 if <47%; ATL: 25/25 if >47%; accuracy: 31/39 if 47%	Yes
	Body water (%)	<0.001	Cutoff: <60%; PPV: 100%; NPV: 75.8%; accuracy: 79.5%	OTS: 14/14 if <65%, 13/14 if <63.5% or 6/14 if <60%; ATL: 25/25 if >60%; accuracy: 31/39 if 60%	Yes
	Extracellular water (%)	n/s	n/a	n/a	No
	Visceral fat (cm²)	0.01	Cutoff: >68 cm²; PPV: 87.5%; NPV: 77.4%; accuracy: 79.5%	OTS: 13/14 if >38 cm² or 7/14 if >68 cm²; ATL: 24/25 if <68 cm², 22/25 if <56 cm², or 14/25 if <38 cm²	Potentially
	Chest-waist circumference ratio	n/s	n/a	n/a	No

Abbreviations: *ATL* healthy athletes, *BMR* basal metabolic rate, *EROS* Endocrine and Metabolic Responses on Overtraining Syndrome, *n/a* non-applicable, *NPV* negative predictive value, *n/s* non-significant, *OTS* overtraining syndrome-affected athletes, *POMS* Profile of Mood States, *PPV* positive predictive value

Table 11.3 Biochemical parameters evaluated for the development of overtraining syndrome diagnostic tools

Study/tests	Markers	p-score	Highest NPV or PPV (%) and accuracy (%)	Ranges of the maximum accuracy and maximum true positive and true negative values	Useful for any of the questionnaires (No/potentially/yes)
EROS-HPA axis					
Response to an ITT	Basal ACTH levels (pg/mL)	n/s	n/a	n/a	No
	ACTH levels during hypoglycemia (pg/mL)	n/s	n/a	n/a	No
	ACTH levels 30 min after hypoglycemia (pg/mL)	<0.001	Cutoff: <35 pg/mL; PPV: 75%; NPV: 81.5%; accuracy: 79.5%	OTS: 13/14 if <106 pg/mL or 9/14 if <35 pg/mL; ATL: 22/25 if >35 pg/mL; accuracy: 31/39 if >35 pg/mL	Yes
	ACTH increase during an ITT (pg/mL)[a]	<0.001	Cutoff: <20 pg/mL; PPV: 66.7%; NPV: 83.3%; accuracy: 76.9%	OTS: 14/14 if <75 pg/mL, 12/14 if <35 pg/mL or 10/14 if <20 pg/mL; ATL: 25/25 if >3 pg/mL or 20/25 if >20 pg/mL accuracy: 30/39 if 1–3 pg/mL or 20 pg/mL	Potentially
	Basal serum cortisol levels (µg/dL)	n/s	n/a	n/a	n/a
	Cortisol levels during hypoglycemia (µg/dL)	0.015	Cutoff: <13.5 µg/dL; PPV: 62.5%; accuracy: 74.4%	OTS: 12/14 if <16.9 µg/dL, 11/14 if <13.7 µg/dL or 10/14 if <13.2 µg/dL; ATL: 17/25 if >13.7 µg/dL or 19/25 if >13.5 µg/dL, accuracy: 29/39 if 13.2–13.5 µg/dL	No (10/39 overlapping results)
	Cortisol levels 30 min after hypoglycemia (µg/dL)	0.002	Cutoff: <19.1 µg/dL; PPV: 72.2%; NPV: 95.2%; accuracy: 84.6%	OTS: 13/14 if <19.1 µg/dL,;ATL: 25/25 if <17 µg/dL or 20/25 if <19.1 µg/dL; accuracy: 33/39 if <19.1 µg/dL	Yes
	Cortisol increase during an ITT (µg/dL)	0.008	Cutoff: <9 µg/dL; PPV: 57.1%; NPV: 89.9%; accuracy: 71.8%	OTS: 14/14 if <10 µg/dL, 12/14 if <9 µg/dL or 11/14 if <8 µg/dL; ATL: 16/25 if >9 µg/dL, 13/25 if >10 µg/dL or 10/25 if >11 µg/dL; accuracy: 28/39 if <9 µg/dL	No (11/39 overlapping results)
Response to a CST	Cortisol levels 30 min after cosyntropin (µg/dL)	n/s	n/a	n/a	No
	Cortisol levels 60 min after cosyntropin (µg/dL)	n/s	n/a	n/a	No
SCR	Waking salivary cortisol (ng/dL)	n/s	n/a	n/a	No
	Salivary cortisol 30 min after waking (ng/dL)	0.002	Cutoff: >390 ng/dL; PPV: 68.7%;	OTS: 14/14 if <520 ng/dL or 11/14 if <390 ng/dL; ATL: 20/25 if >390–400 ng/dL; accuracy: 31/39 if <390–400 ng/dL	Yes
	4 p.m. salivary cortisol (ng/dL)	n/s	NPV: 87%; accuracy: 79.5%	n/a	No
	11 p.m. salivary cortisol (ng/dL)	n/s	n/a	n/a	No
	CAR (%)	n/s	n/a	n/a	No

EROS-STRESS

		p	Statistics	Details	Useful?
Response to an ITT	Basal GH levels (µg/L)	0.009	Cutoff: <0.1 µg/L; PPV: 87.5%; NPV: 77.4%; accuracy: 79.5%	OTS: 11/14 if <0.2 µg/L or 7/14 < 0.07 µg/L; ATL: 24/25 if >0.1 µg/L or 18/25 if >0.2 µg/L; accuracy: 31/39 if <0.1 µg/L	Yes
	GH levels during hypoglycemia (µg/L)	0.018	Cutoff: <0.1 µg/L; PPV: 100%; NPV: 73.5%; accuracy: 76.9%	OTS: 12/14 if <1.8 µg/L, 8/14 if <0.7 µg/L, 6/14 if <0.4 µg/L or 5/14 if <0.25 µg/L; ATL: 25/25 if >0.1 µg/L, 24/25 if >0.17 µg/L, 22/25 if >0.3 µg/L, 18/25 if >0.4 µg/L or 16/25 if >0.5 µg/L; accuracy: 30/39 if <0.1 µg/L	Yes
	GH levels 30 min after hypoglycemia (µg/L)	0.001	Cutoff: <5–6 µg/L; PPV: 81.8%; NPV: 82.1%; accuracy: 82	OTS: 14/14 if <14.4 µg/L, 12/14 if <10.7 µg/L, 9/14 if <6 µg/L or 7/14 if <1.5 µg/L; ATL: 23/25 if >6 µg/L; accuracy: 32/39 if <1.5–560 µg/L	Potentially
	Basal prolactin levels (ng/mL)	0.014	Cutoff of <7.1 ng/mL; PPV: 85.7%; NPV: 75%; accuracy: 76.9%	OTS: 13/14 < 14.5 ng/mL or 6/14)if <7.1 ng/mL; ATL: 25/25 if >6.6 ng/mL, 24/25 if >7.1 ng/mL or 23/25 if >7.7 ng/mL; accuracy: 30/39 if <6.6–7.7 ng/mL	Yes
	Prolactin levels during hypoglycemia (ng/mL)	0.001	Cutoff: <11.5–12 ng/mL; PPV: 83.3%; NPV: 85.2%; accuracy: 84.6%	OTS: 12/14 if <17.5 ng/mL or 10/14 if >11.5–12 ng/mL; ATL: 23/25 if <12 ng/mL; accuracy: 33/39 if <11.5–12 ng/mL	Yes
	Prolactin levels 30 min after hypoglycemia (ng/mL)	0.001	Cutoff of <10 ng/mL; PPV: 100%; NPV: 78.1%; accuracy: 82.1%	OTS: 7/14)if <14 ng/mL or < 10 ng/mL; ATL: 25/25 if 10 ng/mL or 18/25 if >20 ng/mL; accuracy: 32/39 if <10 ng/mL	No (11/39 overlapping results)
	Prolactin increase during an ITT (ng/mL)	0.047	Cutoff: <0–2.5 ng/mL; PPV: 61.5%; NPV: 76.9%; accuracy: 71.8%	OTS: 14/14 if <14 ng/mL, 10/14 if <4 ng/mL or 8/14 if <0 ng/mL; ATL: 20/25 if >0–2.5 ng/mL, 8/14 if <0 ng/mL; ATL: 20/25 if >0–2.5 ng/mL; accuracy: 28/39 if <0–2.5 ng/mL	
Glucose behavior and related symptoms during an ITT	Basal serum glucose (mg/dL)	n/s	n/a	n/a	No
	Serum glucose during hypoglycemia (mg/dL)	n/s	n/a	n/a	No
	Time to hypoglycemia (min)	n/s	n/a	n/a	No
	Adrenergic symptoms (0–10)	0.034	Cutoff: ≤3: PPV: 56.2%; NPV: 78.3%; accuracy: 69.2%	OTS: 13/14 if ≤6 or 9/14 if ≤3; ATL: 20/25 if ≥3 or 18/25 if ≥4; accuracy: 27/39 if ≤3	No (12/39 overlapping results)
	Neuroglycopenic symptoms (0–10)	n/s	n/a	n/a	No

(continued)

Table 11.3 (continued)

Study/tests	Markers	p-score	Highest NPV or PPV (%) and accuracy (%)	Ranges of the maximum accuracy and maximum true positive and true negative values	Useful for any of the questionnaires (No/potentially/yes)
EROS-BASAL					
Hormonal markers	Total testosterone (ng/dL)	0.008	Cutoff: <400 pg/mL; PPV: 72.7%; NPV: 78.6%; accuracy: 76.9%	OTS: 10/14 if <440 pg/mL, 8/14 if <380 pg/mL; ATL: 24/25 if >350 ng/mL or 22/25 if >400 ng/mL; accuracy: 30/39 if <380–410 pg/mL	Potentially
	Estradiol (pg/mL)[a]	0.007	Cutoff: >35 pg/mL; PPV: 66.7%; NPV: 75%; accuracy: 75.7%	OTS: 13/14 if >28 pg/mL or 10/14 if <35.5 pg/mL; ATL: 18/23 if <34 pg/mL; accuracy: 28/37 if >34–35.5 pg/mL[a]	Potentially
	IGF-1 (pg/mL)	n/s	n/a	n/a	No
	TSH (µUI/mL)	n/s	n/a	n/a	No
	freeT3 (pg/mL)	n/s	n/a	n/a	No
	Total catecholamines (µg/12 h)	0.032	Cutoff: >220 µg/12 h; PPV: 58.3%; NPV: 74.1%; accuracy: 69.2%	OTS: 14/14 if >130 µg/12 h, 11/14 if >155 µg/12 h, 8/14 if >200 µg/12 h 7/14 if >220 µg/12 h or 5/14 if >270 µg/12 h; ATL: 20/25 if <220 µg/12 h, 15/25 if <200 µg/12 h, 11/25 if <150 µg/12 h or 10/25 if <130 µg/12 h; accuracy: 27/39 if >220 µg/12 h	No (12/39 overlapping results)
	Noradrenaline (µg/12 h)	n/s	n/a	n/a	No
	Dopamine (µg/12 h)	0.042	Cutoff: >200 µg/12 h; PPV: 58.3%; NPV: 74.1%; accuracy: 69.2%	OTS: 12/14 if >120 µg/12 h, 7/14 if >200 µg/12 h or 4/14 if >230 µg/12 h ATL: 20/25 if <200 µg/12 h, 13/25 if <160 µg/12 h or 11/25 if <120 µg/12 h accuracy: 27/39 if >200 µg/12 h	No (12/39 overlapping results)
	Epinephrine (µg/12 h)	n/s	n/a	n/a	No
	Total metanephrines (µg/12 h)	n/s	n/a	n/a	No
	Metanephrines (µg/12 h)	n/s	n/a	n/a	No
	Normetanephrines (µg/12 h)	n/s	n/a	n/a	No

Biochemical markers					
	ESR (mm/h)	n/s	*n/a*	n/a	No
	CRP (mg/dL)	n/s	*n/a*	n/a	No
	Vitamin B12 (pg/mL)	n/s	*n/a*	n/a	No
	Lactate (nMol/L)	0.007	Multiple Cutoffs: >0.9, 0.92–0.93, 0.95–0.98 or 1.06–1.1 nMol/L; PVV: 77.8% if >1.08 nMol/L; NPV: 100% if <0.75 nMol/L, accuracy: 74.4%	OTS: 14/14 if >0.75 nMol/L ATL: 21/25 if <1.05 nMol/L accuracy: 29/39 if 0.9, 0.92–0.93, 0.95–0.98 or 1.06–1.1 nMol/L	No (10/39 overlapping results)
	Ferritin (ng/mL)	n/s	*n/a*	n/a	No
	Neutrophils (/mm³)[b]	0.035	Cutoff: 2700/mm3; PPV: 61.5%; NPV: 76%; accuracy: 71,1%	OTS: 12/14 if <3500/mm³ or 8/14 if <2700/mm³; ATL: 19/24 if >2700/mm³ or 14/24 if >3500/mm³; accuracy: 27/38 if <2700/mm³, 26/38 if <2900/mm³ or 3500–3750/mm³	No (accuracy <75%)
	Lymphocytes (/mm³)	n/s	*n/a*	n/a	No
	Platelets (*1000/mm³)	n/s	*n/a*	n/a	No
	Hematocrit (%)[a]	n/s	*n/a*	n/a	No
	Eosinophils (/mm³)	n/s	*n/a*	n/a	No
	LDLc (mg/dL)	n/s	*n/a*	n/a	No
	HDLc (mg/dL)	n/s	*n/a*	n/a	No
	Triglycerides (mg/dL)	n/s	*n/a*	n/a	No
	CK (U/L)	0.043	Cutoff: 800–850 U/L; PPV: 85.7%; NPV: 75%; accuracy: 76.9%	OTS: 7/14 if >760 U/L, or 6/14 if >880 U/L; ATL: 24/25 if <780–910 U/L, or 22/25 if <580–730 U/L; accuracy: 30/39 if 730 U/L or between 800–850 U/L	Potentially
	Creatinine (mg/dL)	n/s	*n/a*	n/a	No

(continued)

Table 11.3 (continued)

Study/tests	Markers	p-score	Highest NPV or PPV (%) and accuracy (%)	Ranges of the maximum accuracy and maximum true positive and true negative values	Useful for any of the questionnaires (No/potentially/yes)
Ratios	Testosterone/estradiol ratio[a]	<0.001	Cutoff: 13.3; PPV: 83.3%; NPV: 84%; accuracy: 83.8%	OTS: 14/14 if <17 or 10/14 if <13.3; ATL: 21/23 if >13 or 16/23 if >16; accuracy: 31/37 if 13.3	Yes
	Testosterone/cortisol ratio	n/s	n/a	n/a	No
	Neutrophil/lymphocyte ratio[b]	0.017	Cutoff: <1.65; PPV: 54.2%; NPV: 92.3%; accuracy: 68.4%	OTS: 14/14 if <1.95 or 13/14 if <1.65; ATL: 13/24 if >1.65 or 11/24 if >1.95 accuracy: 26/38 if <1.65	No (accuracy <75%)
	Platelet/lymphocyte ratio[b]	n/s	n/a	n/a	No

Abbreviations: *ACTH* adrenocorticotropic hormone, *ATL* healthy athletes, *CAR* cortisol awakening response, *CK* creatine kinase, *CRP* C-reactive protein, *CST* cosyntropin stimulation test, *EROS* Endocrine and Metabolic Responses to Overtraining Syndrome, *ESR* erythrocyte sedimentation rate, *GH* growth hormone, *HDLc* high-density lipoprotein-cholesterol, *HPA* hypothalamic–pituitary–adrenal, *IGF*-1 insulin-like growth factor 1, *ITT* insulin tolerance test, *LDLc* low-density lipoprotein-cholesterol, *n/a* non-applicable, *NPV* negative predictive value, *n/s* non-significant, *OTS* overtraining syndrome-affected athletes, *PPV* positive predictive value, *SCR* salivary cortisol rhythm, *T3L* free T3, *TSH* thyroid-stimulating hormone

[a]Two missing data points

[b]One missing data point

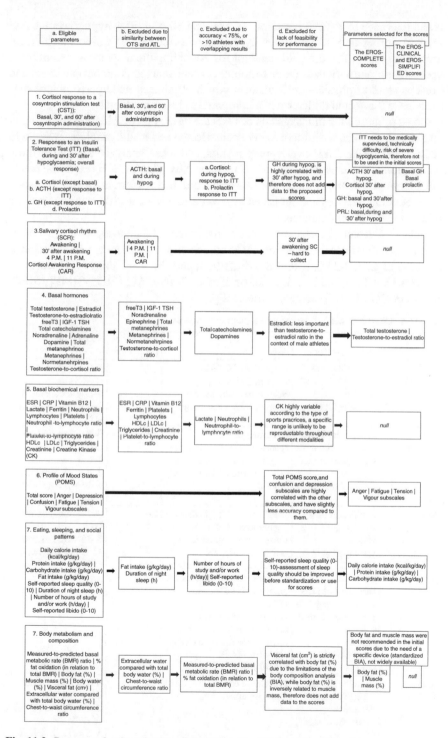

Fig. 11.2 Processes for the selection of the parameters for the proposed scores

insulin tolerance test (ITT) (cortisol, ACTH, GH 30 min after hypoglycemia, and prolactin during and 30 min after hypoglycemia), comprising nine biochemical parameters. The ITT is the gold standard for evaluating the responsiveness of three hypothalamic-pituitary axes (somatotropic, corticotropic, and lactotropic axes) and is independent of physical performance, which eliminates underperformance bias found in OTS-affected athletes when compared to healthy athletes.

Scores proposed for the diagnosis of OTS are based on combinations of markers that showed significant differences between affected and healthy athletes, with high accuracy (>75%) and few overlapping results, and when analyzed altogether these combinations were able to distinct OTS from healthy athletes in 100% of the cases, with a large gap between them. For this scale, reduced performance and exclusion of confounding diseases are not necessary, as the scoring for OTS was shown to be highly specific for its diagnosis.

Table 11.4 depicts the characteristics of the proposed diagnostic scores, whereas Fig. 11.3 demonstrates the selection of the combination of parameters to be included in the EROS-CLINICAL, EROS-SIMPLIFIED, and EROS-COMPLETE scores.

Tables 11.5, 11.6, and 11.7 describe the EROS-CLINICAL, EROS-SIMPLIFIED, and EROS-COMPLETE scores, respectively.

Table 11.4 Characteristics of the proposed diagnostic scores

Tool	Target athletes	Aim	Number of parameters	Score (points) and Criteria
EROS-CLINICAL	Suspected of OTS (possible signs of imminent or incipient OTS)	Diagnosis of OTS in suspected athletes, easy-to-perform and not time- or fund-consuming	9	0–2 = Excluded for OTS 3–5 = Inconclusive 6–9 = Diagnosis of OTS
EROS-SIMPLIFIED	Suspected of OTS when the diagnosis was not confirmed using the EROS-CLINICAL criteria	Diagnosis of OTS in suspected athletes when the diagnosis was not confirmed using the EROS-CLINICAL criteria	13	0–3 = Excluded for OTS 4–6 = Inconclusive 7–8 = Probable OTS 9–13 = Diagnosis of OTS
EROS-COMPLETE	Population-based screenings; athletes participating in research. Exception, suspected of OTS when the diagnosis was not confirmed using the EROS-SIMPLIFIED criteria	Diagnosis of OTS in large populations of athletes, irrespective of the risk, or probability of OTS. Identification of risk factors, biomarkers, and tools for the prevention and diagnosis of OTS. Exception, individual diagnosis of OTS when the diagnosis was not confirmed using the EROS-SIMPLIFIED criteria	20	0–4 = Excluded for OTS 5–10 = Inconclusive 11–20 = Diagnosis of OTS
EROS-RISK	At high risk for OTS.(absence of clinical or biochemical signs)	Prevention of OTS in high-risk athletes	11	0–1 = Low risk of OTS 2–4 = Moderate risk of OTS 5–6 = High risk of OTS 7–11 = Imminent risk of OTS

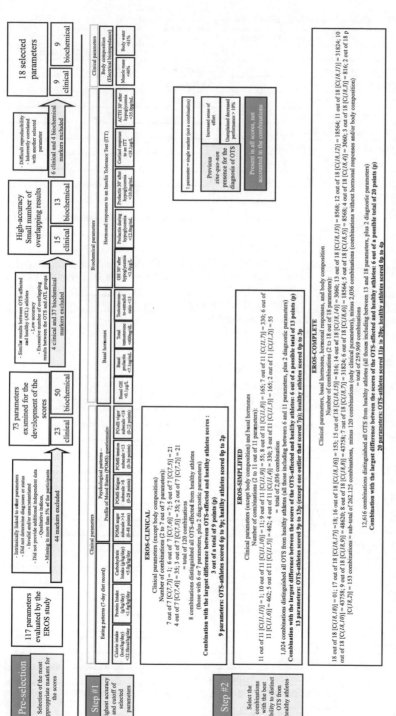

Fig. 11.3 Selection of the combination of parameters to be included in the EROS-CLINICAL, EROS-SIMPLIFIED, and EROS-COMPLETE scores

Table 11.5 The EROS-CLINICAL diagnostic score

(a) Scores with ranges and points to be entered

Parameter	Range	Score (to be entered)[a]
Calorie intake (kcal/kg/day)	<32.0 kcal/kg/day	
Protein intake (g/kg/day)	<1.6 g/kg/day	
Carbohydrate intake (g/kg/day)	<5.0 g/kg/day	
POMS anger subscale (0–48 points)	>14	
POMS fatigue subscale (0–28 points)	>8	
POMS tension subscale (0–36 points)	>13	
POMS vigor subscale (0–32 points)	<18	
Unexplained decreased performance >10%	Y/N	
Increased sense of effort	Y/N	
Total	0–9	

(b) Interpretation of the results

Score	Interpretation
0–2 points	Diagnosis excluded for overtraining syndrome
3–5 points	Inconclusive (at intermediate-to-high risk for OTS or an unusual presentation of OTS)
6–9 points	Diagnosis of overtraining syndrome

POMS Profile of Mood States, OTS overtraining syndrome
[a]Each parameter within the range is assigned 1 point.

Table 11.6 The SIMPLIFIED-OTS diagnostic score

(a) Scores with ranges and points to be entered

Parameter	Range	Points (to be entered)[a]
Calorie intake (kcal/kg/day)	<32.0 kcal/kg/day	
Protein intake (g/kg/day)	<1.6 g/kg/day	
Carbohydrate intake (g/kg/day)	<5.0 g/kg/day	
POMS anger subscale (0–48 points)	>14	
POMS fatigue subscale (0–28 points)	>8	
POMS tension subscale (0–36 points)	>13	
POMS vigor subscale (0–32 points)	<18	
Decreased performance >10%	Yes	
Increased sense of effort	Yes	
Basal GH (ug/L)	<0.1 ug/L	
Basal prolactin (ng/mL)	<7.1 ng/mL	
Total testosterone (ng/dL)	<400 ng/dL	
Testosterone to estradiol ratio	<13	
TOTAL (points)	0–13	

(b) Interpretation of the results

Score	Interpretation
0–3 points	Overtraining syndrome excluded
4–6 points	Inconclusive (imminent, incipient, or unusual presentation of overtraining syndrome)
7–13 points	Overtraining syndrome confirmed

[a]Each parameter within the range is assigned 1 point
POMS Profile of Mood States, GH Growth hormone, OTS Overtraining syndrome

Table 11.7 The EROS-COMPLETE diagnostic score

(a) Score with ranges and points to be entered

Risk factor	Range	Points (to be entered)[a]
Decreased performance >10%	Yes	
Increased sense of effort	Yes	
Calorie intake (kcal/kg/day)	<32.0 kcal/kg/day	
Protein intake (g/kg/day)	<1.6 g/kg/day	
Carbohydrate intake (g/kg/day)	<5.0 g/kg/day	
POMS anger subscale (0–48)	>14	
POMS fatigue subscale (0–28)	>8	
POMS tension subscale (0–36)	>13	
POMS vigor subscale (0–32)	<18	
Muscle mass (%)	<46%	
Body water (%)	<61%	
ACTH 30 min after hypoglycemia (pg/mL)	<35 pg/mL	
Cortisol response to ITT (μg/dL)	<19.1 μg/dL	
Basal GH (μg/L)	<0.1 ug/L	
GH 30 min after hypoglycemia (μg/L)	<1.0 ug/L	
Basal prolactin (ng/mL)	<7.1 ng/mL	
Prolactin during ITT (ng/mL)	<12 ng/mL	
Prolactin 30 min after hypoglycemia (ng/mL)	<10 ng/mL	
Total testosterone (ng/dL)	<400 ng/dL	
Testosterone to estradiol ratio	<13	
TOTAL	0–20	

(continued)

Table 11.7 (continued)

(b) Interpretation of the results

Score	Interpretation
0–4 points	Excluded diagnosis for overtraining syndrome
5–10 points	Inconclusive (at intermediate-to-high risk for OTS or an unusual presentation of OTS)
11–20 points	Diagnosis of overtraining syndrome

[a]Each parameter within the range is assigned 1 point.

POMS Profile of Mood States, *ITT* Insulin tolerance test, *ACTH* Adrenocorticotropic hormone, *GH* Growth hormone, *OTS* Overtraining syndrome

The use of the EROS-CLINICAL score as the initial diagnostic tool for the diagnosis of OTS in athletes suspected for OTS, without the need of biochemical markers, is recommended, since this is the simplest, the least expensive, and the easiest to employ, among diagnostic scores. Whenever the EROS-CLINICAL shows inconclusive results, the EROS-SIMPLIFIED score, with additional four basal hormone levels, should be employed. The EROS-COMPLETE score should be used for screening of OTS in population-based approaches or for research purposes, as it encompasses all major markers of OTS identified by the EROS study, and in larger population of athletes, more substantial discrepancies in the OTS presentations can be detected; exceptionally, when the EROS-SIMPLIFIED score is inconclusive, the EROS-COMPLETE score may be employed for an individual-based diagnostic tool.

11.3   Tools for the Identification of Athletes at High Risk for OTS

Preventive approaches to OTS are preferred over the detection after its occurrence, given the overwhelming consequences of OTS. Nonetheless, there are some challenging obstacles for effective actions in OTS prevention, including the relative inability of athletes to detect their own early signs of exhaustion, due to motivational reasons (high motivation may preclude from perception of signs that could be of concern in the sake of an objective), the inconsistency regarding which markers could be employed as alert signs, and the high heterogeneity and individuality of the symptoms, even prior to OTS, during its course.

Functional (FOR) and non-functional overreaching (NFOR) happily exist as important alerts that if properly addressed can provide overcompensatory improvements. The fundamental difference between FOR, NFOR, and OTS is that adaptations in the two first states are functional (even in NFOR), whereas adaptations become dysfunctional ("maladaptations") in OTS. In theory, FOR and NFOR prevent the progression to OTS. Indeed, the vast majority of athletes that develop any of these two conditions do not progress to OTS, since they adequately address the issues that led to FOR and NFOR.

However, subsequent episodes of FOR that are apparently recovered due to full improvement of clinical manifestations may actually play an accumulative role until OTS appears. First, many of the times that athletes report as "improved" are actually "partially improved but improved enough to go back to trainings," which is particularly harmful in the long run. Second, it is likely that clinical improvement occur earlier than molecular, biochemical, and metabolic recovery, which hinder from normalization of these parameters, leading to further consequences.

Furthermore, although FOR, NFOR and OTS can be understood as a *continuum*, some characteristics are only noticed in OTS. While physical exhaustion is the prevailing type of fatigue found in FOR and NFOR, mental fatigue is sometimes only reported in OTS. Perhaps, while mental fatigue tends to be more subjective and its presence can be therefore conscious or unconsciously avoided, physical fatigue cannot ben as ignored. While FOR and NFOR cannot be considered as burnout-related syndromes, once they recover after a period of resting, OTS fails to improve after specific interventions and can be therefore classified as part of the burnout spectrum of diseases.

Self-perceptions of the features of FOR and NFOR may be distinct from those in OTS. In FOR, acute changes are noticed, and improvement is clearly related to the level of resting. Contrariwise, the perceptions of the alterations in OTS, besides being distinct from FOR in terms of which and how moods are affected, tend to occur more insidiously, as a sort of accumulated unresolved issues during the course of OTS, and do not yield any improvement even after long periods of resting and appropriate nutrition and sleeping.

Hence, the assessment of psychological states plays an essential role to detect those athletes with imminent OTS, since moods may disclose early signs that can only be detected through questionnaires, but which would be denied if objectively questioned for these athletes. Besides, the now known non-training habits that disrupt physiology toward OTS may also be assessed in combination with the psychological states, as proposed below through two types of preventive tools: a two-step detection of forthcoming OTS and a specific score (EROS-RISK) for detection of those that are in the process to develop OTS.

11.3.1   Two-Step OTS Prevention

11.3.1.1   Step #1 – Identification of Athletes at High Risk

In addition to the proposed flowcharts for the diagnosis of OTS, once the prevention of OTS is highly desirable, other triggers have now been identified, and it has been observed that a minimum number and intensity of stressors is needed for the occurrence of OTS and a simple score for the identification of high and imminent risk has been proposed, based on the findings on the EROS study (Table 11.8) of the risk factors retrospectively identified in affected athletes that were absent in healthy ones. This is an easy-to-perform method that can be regularly employed as a screening test for different population of athletes. This score is also based on a predictive model to detect athletes that may be in the way for the development of OTS, as illustrated in Fig. 11.4.

11.3.1.2   Step #2 - Active Search for Symptoms

As the screening for athletes at high or imminent risk of OTS are initially performed in asymptomatic athletes, specific complains should not be expected at this point. However, an active search for clinical manifestations, particularly those atypically present in OTS, should be performed in all athletes identified as being at high or imminent risk for OTS. Specific active questions should be asked, including the presence of increased fatigue that is not recovered after a period of resting, symptoms of URTIs, decreased libido unrelated to a specific partner, and recent and unjustified mood changes.

Table 11.8 The OTS-RISK score for the assessment of risk level for the development of overtraining syndrome

(a) Score with ranges and points to be entered

Risk factor	Range	Points (to be entered)[b]
Calorie intake[a]	<32 kcal/kg/day	
Protein intake[a]	<1.6 g/kg/day	
Carbohydrate intake[a]	<5 g/kg/day	
Total POMS score (−32–120)	>19	
POMS anger subscale (0–48)	>14	
POMS confusion subscale (0–28)	>6	
POMS depression subscale (0–60)	>7	
POMS fatigue subscale (0–28)	>8	
POMS tension subscale (0–36) >13	>13	
POMS vigour subscale (0–32) <18	<18	
Self-reported sleep quality (0–10)	<6 (≤5)	
Total	0–11	

(b) Interpretation of the results

Score (points)	Interpretation
0–1	Low risk of overtraining syndrome
2–4	Moderate risk of overtraining syndrome
5–6	High risk of overtraining syndrome
7–11	Imminent risk of overtraining syndrome

POMS Profile of Mood States

[a]Using a 7-day dietary record

[b]Each parameter within the range is assigned 1 point

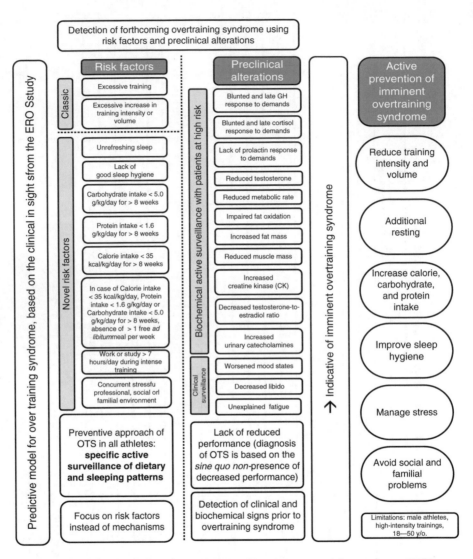

Fig. 11.4 Predictive model for the identification of imminent overtraining syndrome (OTS)

Whenever non-specific symptoms are identified, athletes may be experiencing overreaching or an initial progression to OTS, which must be promptly managed, by removing the stressor factors identified at step #1.

11.3.2 Score for the Prevention of Overtraining Syndrome (OTS) (EROS-RISK Score)

The use of combinations of different risk factors facilitated our identification of athletes at risk for OTS, thereby preventing its occurrence.

For the EROS-RISK score, self-reports of sleep quality (on a scale from 0 to 10, without the need for resource-intensive tests), total score and depression and confusion subscores of the Profile of the Mood States (POMS) questionnaire have been included.

The EROS-RISK score is likely the first tool to prevent OTS, based on for the identification of athletes at high or imminent risk of OTS. This proposal is based on the indisputable fact that preventive approaches to OTS have better outcomes compared to its treatment, once recovering from OTS is challenging and not always complete.

11.4 Discussion and Summary of the New Diagnostic Scores for Overtraining Syndrome (OTS)

Despite the lack of single markers to distinguish OTS-affected athletes from healthy athletes due to the high heterogeneity of OTS among the individual athletes and to the that each affected athlete disclose a unique combination of dysfunctions, the relative large number of true and naturally occurring OTS-affected athletes and the large number of parameters evaluated in the EROS study allowed the identification of more than 45 novel OTS biomarkers. This enabled the evaluation of a possible diagnosis of OTS using combinations of parameters, among which more than 240,000 were tested. Thus, OTS was precisely identified in all affected athletes while excluding OTS in healthy athletes, without the need to exclude confounding disorders in more than 4000 different combinations, as many of the identified markers yielded few overlapping results between the OTS and ATL groups [9].

In the absence of a sole marker that is able to distinct healthy from OTS-affected athletes and that each affected athlete discloses a unique combination of dysfunctions, scoring systems would presumably be more specific to diagnose OTS. Indeed, the new criteria for the diagnosis of OTS through proposed scores yielded 100% accuracy without the need to exclude confounding disorders. The proposed scores are easy-to-perform, not time-consuming, without the need of a complex structure for its employment, and can be used routinely.

Different levels of diagnostic scorings have been developed in order to avoid inconclusive results: in case a more simple score yields inconclusive results, further more complex scores may be used. Different levels of assessment for the diagnosis of OTS offer feasibility for sports centers of different structure levels. The choice of the most appropriate score to be employed for the diagnosis of OTS,should be based on a simple flowchart that has also been proposed in this chapter.

Although decreased performance and sense of increased effort for a same working load would have to be two additional criteria included in all the proposed tools, they may appear only in the more severe stages of OTS and progress unnoticed without proactive surveillance of athletes. At this stage, recovery is challenging and not always complete. Therefore, the identification of OTS at earlier stages is highly desirable, as recovery should be easier, faster, and less likely to compromise a training program. At this point, the presentation of OTS may be considered "unusual" because of the absence of a prolonged decline in performance, which is more likely to be underreported at this stage.

Indeed, the diagnostic tools proposed in the present study made the possibility of diagnosing OTS feasible, as the presence of an unexplained decline in performance and increased sense of effort were optional and would strengthen the diagnosis of OTS but were not required. However, some authors would consider the existence of OTS impossible without a decline in performance because fatigue and underperformance would be likely to appear in milder forms of OTS, including functional and non-functional overreaching, while a complete recovery would avoid the development of OTS. However, the false perception of full recovery from these earlier stages, which is common because of the high motivation to return to the training sessions, could hide an underlying process that could ultimately lead to OTS. In addition, the belief that overreaching happens in the very early stage of OTS might not be true, as minor symptoms in the early stage are usually ignored by athletes. Finally, although many authors consider overreaching and overtraining as different stages of the same process, others claim that the underlying mechanisms between these states differ, as OTS tends to result from a more chronic, deeper, and central process, while overreaching tends to be characterized as an acute physical process. Therefore, our proposed diagnostic tools could be helpful to identify athletes prior to clinical signs of OTS with true underperformance, regardless of previous episodes of overreaching.

The EROS-CLINICAL score is a diagnostic tool based on a combination of nine clinical parameters to yield a clinical score for athletes suspected of OTS, without the need to conduct biochemical tests. The exclusion of body fat and muscle mass has been demonstrated to disclose any loss of accuracy or reduction of the quality of the diagnostic assessment. The exclusion of muscle mass and body fat without loss of accuracy was important for the EROS-CLINICAL score because the main purpose of this specific diagnostic tool was to be an inclusive and not all facilities have devices to analyze body composition available.

When the EROS-CLINICAL is inconclusive, the EROS-SIMPLIFIED with its combination of four additional basal hormones may be successful to identify most of the affected athletes not diagnosed by the EROS-CLINICAL. Moreover, although the EROS-SIMPLIFIED score includes biochemical parameters, it is simple to use, once functional tests are unnecessary for this score.

The EROS-COMPLETE is a more comprehensive diagnostic tool encompassing the combination of all 18 parameters that yielded high accuracy and few overlapping results. This score is particularly valuable for:

1. Unusual presentations of OTS that might remain undiagnosed if the guidelines with the diagnostic flowchart are used, as it provides a broad variety of markers, covering the full spectrum of OTS presentations
2. Exceptions, for an individual-based diagnostic approach, when the EROS-CLINICAL and EROS-SIMPLIFIED fail to diagnose OTS
3. Population-based screenings, irrespective of the presence of risks or suspected features of OTS, as OTS presentations are highly diverse
4. Research purposes, as it includes current biomarkers of OTS and can be used for comparisons with potential markers or assessment methods to identify new tools or improve our understanding of OTS

5. Understanding the unique underlying pathophysiology of each OTS-affected athlete.

In the EROS study, other triggers of OTS than excessive training have been uncovered. Athletes at high risk for OTS are not only those who undergo excessive training without sufficient quality resting or excessive increase of training volume or intensity. Athletes who also:

1. Undergo long-term extreme diets, such as low carbohydrate, very low caloric intake, or a combination of intermittent-fasting with calorie-restricted diets, without a compensatory reduction of training load, which happens to be an insufficient (but not hypocaloric) caloric, protein, and carbohydrate intake
2. Have poor sleep hygiene and quality, usually related to the inability to disconnect from social media or TV; or
3. Work or study excessively in an unceasing stressful environment, leading to concurrent cognitive and physical efforts that are uncompensated are at high risk of developing OTS.

Similarly to the diagnosis of OTS, no single risk factor has been shown to be able to identify all athletes at high risk. However, a combination of some modifiable risk factors invariably led to OTS when analyzed collectively. Owing to the lack of other tools to detect athletes at high risk for OTS, the EROS-RISK can be a helpful tool to identify these athletes and prevent the occurrence of OTS. At the prevention level, OTS can be addressed more effectively than it can at the treatment level. The POMS subscales may predict future occurrences of OTS, and for this reason, all POMS subscales are present in the EROS-RISK tool.

In summary, these are the first diagnostic tools for OTS to achieve 100% accuracy in distinguishing OTS-affected athletes from healthy athletes without the need to exclude confounding disorders or the need to include the presence of decreased performance. Furthermore, the tools are easy to use and are not time-consuming. A flowchart for the selection of the most appropriate score to be employed is presented in Fig. 11.5.

11.5 Advances in the Diagnosis of OTS

A complete investigation for a precise diagnosis of OTS may be time-consuming, highly dependent on the availability of specialists, and financially unfeasible. For these reasons, a simplified list of conditions and exams to be performed is shown in Table 11.9, which encompasses the most important diseases and alterations to be excluded and/or characterized.

The proposed tools for the prevention and early diagnosis of OTS that do not require the evaluation for reduced performance or other complex criteria overcome the major barriers that lead to the lack of use of official diagnostic tools for OTS in clinical practice. Simple, easy-to-perform diagnostic scores for OTS should help to reduce the empirical and misdiagnosis of OTS, and spread the correct criteria for its detection.

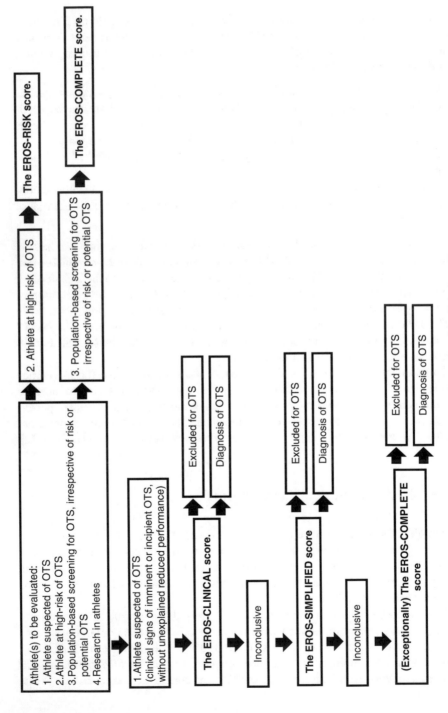

Fig. 11.5 Flowchart for the selection of the most suitable score to be employed in clinical practice

Table 11.9 Simplified list of abnormalities to be excluded prior to the diagnosis of overtraining syndrome

Aspects	Items to be assessed
Current or recent use of substances	Centrally-acting drugs Androgenic anabolic steroids (AAS) Glucocorticoids Other hormones Statins Muscle relaxing Non-steroidal anti-inflammatory agents Non-open label supplements "Peptides"
Pre-existent conditions	Existence, and if yes, current level of control
Signs and symptoms correlated	Full past and familial history Review of all systems, with active questions
Physical examination	Skin color and aspect Heart rate Respiratory rate Seated and orthostatic blood pressure Heart auscultation Body mass index Waist circumference Muscle strength tests
Overt or relative adrenal insufficiency	TSH >5.0 mU/L TPOAb >35 IU/mL or TgAb >20 IU/mL Total testosterone <200 ng/dL Basal cortisol <5.0 mcg/dL Ferritin <20 ng/dL and/or serum iron <50 μg/dL Hematocrit <39–40% for males and <36% for females) Vitamin B12 <200 pg/mL Random serum glucose >200 mg/dL Total calcium <8.5 and >10.5 mg/dL Sodium <135 mEq/L Potassium <3.5 and >5.5 mEq/L Alkaline phosphatase (ALT) >50 U/L Gamma glutamyl transferase (GGTP) >75 U/L Serum creatinine >1.4 mg/dL for males or >1.1 mg/dL for females, and/or calculated eGFR <60 ml/min Epstein-Barr, cytomegalovirus, HIV, and hepatitis types B and C serologies Ultrasensitive CRP >3.0 mg/L Erytrocyte sedmentation rate (ESR) >20 mm/h

Education on OTS should be spread among health practitioners aiming to reduce the misdiagnosis of OTS, since currently approximately 90% of the athletes alleged to be diagnosed with OTS actually present different sorts of "*pseudo*-OTS" or "OTS-like" syndromes. Conversely, athletes with actual OTS are likely underdiagnosed, since its hallmark of unexplained reduced performance tends to be less appreciated compared to commonly alleged signs of OTS, such as joint injuries, muscle soreness, and infections, without a context to support that these findings could be attributed to OTS.

11.6   Effective Prevention of Overtraining Syndrome

Results from the most recent studies and responses to different interventional therapeutic options showed that OTS is highly complex, unique, and in which responses to interventions are largely heterogeneous and many times unsuccessful, recovery is many times incomplete, with refractory psychiatric manifestations as a sort of post-traumatic stress disorder, and usually affects athletes career.

Given the new knowledge on OTS and the undesired consequences of this syndrome, it is clear that further researches should primarily focus on prevention and early identification, when alterations are reversible overuse injuries can be prevented, or ideally be eradicated, rather than treatments for severe and late OTS.

Indeed, novel tools to identify athletes at risk for developing OTS and for its prevention is more efficient than the management of the challenges associated with recovery from OTS.

Now that risk factors have been now more extensively described, effective preventive approaches have become feasible. The list of preventive recommendations is detailed in Table 11.10. If fully followed, the risk of OTS becomes apparently null. These preventive approaches should be followed by a systematic surveillance to become truly effective [10] and are illustrated in Fig. 11.6.

Table 11.10 List of recommendations for an effective prevention of overtraining syndrome

1. High intensity or volume of training should be accompanied with compensatory quality recovery resting
2. Avoid excessive or sudden increments in any of the training patterns
3. For elite athletes, diets should contain at least 5.0 g/kg/day of carbohydrate, 1.6 g/kg/day of protein and 35 kcal/kg/day of overall calorie intake
4. In case of more strict diets, particularly when the carbohydrate content is very low, athlete should have compensatory high carbohydrate meals, optionally ad libitum, at least once every 3 days
5. Reduce training during intensification of cognitive activity, or when facing social, familiar, or professional problems
6. During training overload, avoid concurrent work or study of >8 hours/day or activities that demand intense cognitive effort or important decisions to make
7. Constantly vary the training type and characteristics, aiming to avoid monotony
8. Employ a self-sleep hygiene, by turning off the lights, silencing the room, leave smart phones far, unreachable from arms, and turned off, in order to avoid blue light from these devices. Also avoid any sort of connectivity or interaction, including avoidance of televisions, movies, social medias, and computers, voiding mental activity and thoughts that may prevent sleep or specific and/or individual practices that may improve sleep quality
9. Maintain good hydration constantly. Avoid long periods without access to water, since even mild dehydration may take longer than expected to recover

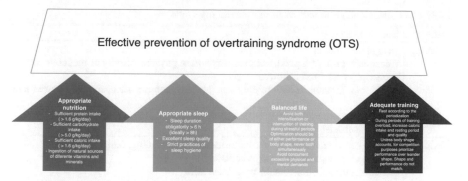

Fig. 11.6 Illustrative recommendations for the prevention of overtraining syndrome (OTS)

11.7   Recommendations to Recover from Overtraining Syndrome

Once OTS has now been learned to be a multifactorial disorder, interventions should go beyond interruption of training and must encompass nutritional, sleeping, social, and stress managements. For practical purposes, it is recommended that athletes are more aggressively approached in terms of interventions for its improvement, rather than a step-by-step sequence of attempts, since the synergistic interactions between the stressors likely play a more important role in the development of OTS than each triggers alone. Table 11.11 depicts the recommended change in the living patterns of athletes diagnosed with OTS.

After 12 weeks, athletes affected by OTS should be reassessed for the evaluation of improvement of clinical and biochemical manifestations, psychological aspects, and sports-specific performance. In the occasion, biochemical markers of early improvement can be assessed and are specified in Table 11.12.

In case early markers do not show improvement, training interventions should be extended, and nutritional interventions may be adapted. However, there is not an established number of markers among the early recovery parameters to be improved in order to consider the recovery process successful.

Table 11.11 Interventions to recover from overtraining syndrome

Patterns	Interventions
Training	3-step intervention: Weeks 1–4: Full resting. No exercise, any sort of physical activity or moderate to intense physical efforts allowed Weeks 5–8: Training at 50% of previous intensity, volume, pace, strength, power, and frequency From week 9 on: Full training as before the diagnosis of OTS, unless clinical symptoms persist
Eating	Increase of overall caloric intake by at least 50%, including, obligatorily: Increase of protein intake by at least 50% Increase of carbohydrate intake by at least 50% Allow two ad libitum (free) meals per week Good hydration: at least 2.5 liters (or 0.3 l/10 kg) of water per day (in addition to the water contained in food)
Sleeping	Strict control over the time to bed and number of hours sleep Force new habits and do not allow exceptions Improvement of sleep hygiene: Turn off all lights Turn off all digital electronic devices, avoiding any blue light Turn off TV and any sound system for a completely silent Sleeping environment with a comfortable air temperature Practice mind-silencing approaches (mindfulness, meditation) Have all problems written down for the following day and leave them there, instead of taking them together to bed Optionally, track sleep with technological devices
Social and stress	In the short term, avoid non-essential tasks Avoid toxic environment, people, and situations Yoga, meditation, and other relaxing practices are highly recommended Identify stressful aspects on social environment, and perform changes accordingly Psychotherapy may be desirable In the long run, avoid concurrent high physical and cognitive demands. During periods of more cognitive effort physical activity should not be too intense; conversely, during high physical demands, cognitive tasks should be reduced

Table 11.12 Markers of early recovery from OTS (compared to levels at the diagnosis)

1. Early cortisol response to stimulations (during the peak of the stimulation, stress, or during or immediately after exercise)
2. Early prolactin response to stimulations (during the peak of the stimulation, stress, or during or immediately after exercise)
3. Overall GH response to stimulations (during the peak of the stimulation, stress, or during or immediately after exercise, or until 30–60 min after the peak of the stimulation or the end of exercise)
4. Increased testosterone
5. Reduced estradiol
6. Reduced testosterone/estradiol ratio
7. Exacerbated increase of IGF-1 (regardless of basal levels being normal or not)
8. Increase of freeT3 (regardless of basal levels being normal or not)
9. Two to three times increase of the cortisol awakening response (CAR)
10. Increased salivary cortisol 30 min after awakening
11. Reduced CK levels (measured in the same interval since last training session with similar training patterns as before the diagnosis)
12. Reduced usCRP (regardless of basal levels being normal or not)

11.8 Conclusion

A range of new innovative tools for earlier diagnosis of OTS, without the need to exclude confounding disorders, have been proposed with 100% accuracy to distinct healthy from OTS-affected athletes. In addition, unprecedented tools to prevent OTS by the detection of athletes at high risk to develop OTS have also been proposed. In common, all tools, particularly the four different scores – EROS-CLINICAL, EROS-SIMPLIFIED and EROS-COMPLETE for OTS diagnosis, and EROS-RISK for OTS prevention – are easy to be employed, intuitive, and do not require highly specialized centers, which allow their use in clinical practice.

Specific recommendations for both prevention and recovery from OTS have been described, which should provide effective help to prevent and improve from this highly complex syndrome, respectively.

References

1. Meeusen R, Duclos M, Foster C, European College of Sport Science, American College of Sports Medicine, et al. Prevention, diagnosis, and treatment of the overtraining syndrome: joint consensus statement of the European College of Sport Science and the American College of Sports Medicine. Med Sci Sports Exerc. 2013;45(1):186–205.
2. Nicoll JX, Hatfield DL, Melanson KJ, Nasin CS. Thyroid hormones and commonly cited symptoms of overtraining in collegiate female endurance runners. Eur J Appl Physiol. 2018;118(1):65–73.
3. Joro R, Uusitalo A, DeRuisseau KC, Atalay M. Changes in cytokines, leptin, and IGF-1 levels in overtrained athletes during a prolonged recovery phase: a case-control study. J Sports Sci. 2017;35(23):2342–9.
4. Lewis NA, Redgrave A, Homer M, et al. Alterations in redox homeostasis during recovery from unexplained underperformance syndrome in an elite international rower. Int J Sports Physiol Perform. 2018;13(1):107–11.
5. Cadegiani FA, Kater CE. Novel causes and consequences of overtraining syndrome: the EROS-DISRUPTORS study. BMC Sports Sci Med Rehabil. 2019;11:21.
6. Cadegiani FA, Kater CE. Eating, sleep, and social patterns as independent predictors of clinical, metabolic, and biochemical behaviors among elite male athletes: the EROS-PREDICTORS study. Front Endocrinol (Lausanne). 2020;11:414.
7. Cadegiani FA, Kater CE. Inter-correlations among clinical, metabolic, and biochemical parameters and their predictive value in healthy and overtrained male athletes: the EROS-CORRELATIONS study. Front Endocrinol (Lausanne). 2019;10:858.
8. Cadegiani FA, Kater CE. Novel insights of overtraining syndrome discovered from the EROS study. BMJ Open Sport Exerc Med. 2019;5(1):e000542. https://doi.org/10.1136/bmjsem-2019-000542.
9. Cadegiani FA, da Silva PHL, Abrao TCP, Kater CE. Diagnosis of overtraining syndrome: results of the endocrine and metabolic responses on overtraining syndrome study: EROS-DIAGNOSIS. J Sports Med (Hindawi Publ Corp). 2020;3937819. Published 2020 Apr 22. https://doi.org/10.1155/2020/3937819.
10. Raedeke TD, Smith AL. Development and preliminary validation of an athlete burnout measure. J Sport Exerc Psychol. 2001;23:281–306.

Chapter 12
Research and Future Perspectives on Overtraining Syndrome

Outline

Limitations and challenges, design of the research, and future perspectives on overtraining syndrome: from the remaining questions and challenges on OTS, the author brings the current limitations of the proposed understanding of OTS, as well as the limitations for clinical practice with athletes, the issues that researches still have to face for appropriate methodological approaches aiming consistent findings and conclusions, what studies on OTS should characterize, and practical suggestions for future directions in the research on the field are described in this chapter.

12.1 Limitations and Challenges

Multiple limitations still surround OTS research. The majority of the studies evaluated males did not classify the type of sports practiced by the researched athletes, whereas almost none of studies on OTS controlled and adjusted for eating, sleeping, and social patterns, which has now been proved to be crucial for appropriate analysis of OTS.

In opposition to studies on relative energy deficiency of the sport (RED-S)/female athlete triad (TRIAD), most researches in OTS were conducted with male athletes, foreclosing the understanding of OTS on female athletes. It is important to highlight the limited extrapolation of the data obtained in males for females. Instead, criteria from burnout or RED-S/TRIAD can be adjusted instead.

The great majority of the athletes suspected for OTS tend to be excluded from studies that employ the appropriate criteria for the diagnosis of OTS, due to lack of decreased performance, or the presence of confounding disorders, when

F. Cadegiani, *Overtraining Syndrome in Athletes*, https://doi.org/10.1007/978-3-030-52628-3_12

properly screened. Hence, in high-quality studies, the number of athletes affected by actual OTS tends to be low, affecting the power of analysis and the effect size. Thus, for statistical purposes, OTS should be considered as a rare condition, exceptionally allowing slight extrapolations, since these are highlighted when data is disclosed.

Limitations on the research on OTS go beyond design and methodology. The inherent characteristics of OTS as being highly individual, whereas each diagnosed athletes present a combination of manifestations and dysfunctions that are unlikely to be found in other affected athletes and to encompass multiple systems simultaneously for its occurrence impose researches to be mandatorily thorough and specific.

Despite the advances in the methodology, the EROS study only evaluated male athletes that practiced both endurance and strength exercises, either together (as in high-intensive functional training or CrossFit) or separately (e.g., when athletes practice both weight lifting and middle distance running) [1]. Whether these findings are applicable for pure endurance, pure resistance, and pure explosive (e.g., ball games) and for females is unknown, in particular for basal and stimulated hormonal and metabolic levels, as these parameters are highly sex specific, and possibly sport specific. In the absence of more specific findings, the clinical applications of the present findings can be for clinical practice, as many of the adaptive changes and behaviors found in this study likely occur in other populations of athletes.

12.2   Design of the Research on Overtraining Syndrome

Researches with larger samples of athletes and adequately designed and controlled for variables are crucial to confirm the current findings, whereas longitudinal studies are key to draw conclusions regarding the sequence of events in response to interventions in modifiable patterns, including training, eating, and social aspects.

Future studies in OTS/PDS should search for clarification of its pathophysiology and effective approaches for its prevention and treatment and eventually eradicate this syndrome. OTS/DPS should be reframed from the understanding as an excessive training centered to a multifactorial disorder. Possibly, moving from the name "overtraining syndrome" (OTS) to "paradoxical deconditioning syndrome" (PDS) would help practitioners and researchers to reconsider their views on OTS/PDS.

The learnings from the methodology employed in the EROS study should be reproduced for specific evaluations of each type of sport: endurance, strength, explosive (stop-and-go), and mixed and for each of the fatigue-related sports diseases (RED-S/TRIAD, OTS/PDS, and BSA). Tests should be standardized and endorsed by guidelines and medical societies and must be similar among them. Similar and standardized tests between different types of sport and different conditions allow appropriate comparisons, help identify specific adaptive changes to each type of sport, and will be key to unveil particularities, similarities, and differences between OTS/PDS, RED-S/STRESS, and BSA.

Future researches and discussions should also focus on the overwhelming similarities between OTS/PDS, RED-S/TRIAD, and BSA and reconsider whether these are distinct conditions, or at least clarify the overlaps between them, since in the majority of the cases athletes could be potentially classified into more than one between these conditions. In the near future, these conditions will not be able to be analyzed separately.

In regard to the identification of novel markers, the unexpected findings of the EROS study provided plausibility to broaden researches on OTS. One example is the unexpected correlation between testosterone and estradiol levels, which pointed for the need of more precise markers of the HPG axis, including sex hormone binding globulin (SHBG), luteinizing hormone (LH), and follicle stimulating hormone (FSH). The elucidation of the other major markers of the HPG axis should elucidate the role of this axis in the pathophysiology of OTS.

Additional parameters should be considered for further studies: follicle-stimulating hormone (FSH), luteinizing hormone (LH), sex hormone binding globulin (SHBG), IGF binding globulin-3 (IGFBP-3), tumor necrosis factor-alpha (TNF-alpha), interleukin-1 beta (IL-1beta), IL-6, cluster of differentiation (CD)3, CD4, CD8, CD8:CD4 ratio, lactate dehydrogenase (LDH), free thyroxin (fT4), intra-tissue cortisone:cortisol ratio, and cortisol binding globulin (CBG). Comparisons between exercise-dependent and independent stimulations should also be performed. Compared to liquid chromatography mass spectrometry/tandem mass (LC/MS- MS/MS), electrochemiluminescence (CLIA) has sufficient relative precision for in-between (pairwise) group comparisons [2–7].

Non-classic inflammatory markers, particularly cytokines, such as IL-1beta, TNF-alpha, and IL-6, would improve the quality of assessment of studies on OTS, once the classical inflammatory markers, ESR and CRP, were found to be unaltered. Muscular IL-6, irisin, and other muscle markers could also provide additional data for the clarification of the underlying mechanisms of OTS, particularly for the detection of altered muscular metabolism and functioning.

The recommendations for the evaluation of other markers should be extended for the other sports-related conditions – RED-S/TRIAD and BSA – which can provide additional elucidation of the similarities between these conditions.

The employment of exercise-independent stimulation tests, such as the ITT, could bring an understanding of the roles of the hormonal responsiveness in all the fatigue-related conditions in athletes.

In summary, with the data now available regarding OTS pathophysiology, it is assumed that further researches must disclose and specify baseline and training characteristics, as depicted in Fig. 12.1. Characteristics described in Fig. 12.1 should be adequately reported and adjusted for variables, and tests should be standardized and adapted accordingly.

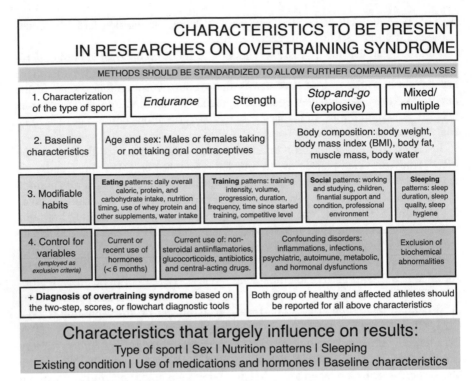

Fig. 12.1 Characteristics that must be present and specify in researches on overtraining syndrome

12.3 Future Perspectives

Much remains to be clarified in the OTS pathophysiology. The feasibility of novel technologies and markers may allow elucidate many of the unanswered questions, what is already happening in several disorders without full clarification.

Due to the progressive growing of the amateur athletes, who trains similarly to professional athletes, and the increasing number of regular active subjects, future researches should also consider to evaluate those who practice sports on a regular basis but as intensively as elite athletes, once OTS has now been identified in non-elite athletes since its occurrence no longer requires high training load, as recently elucidated.

After the elucidation of the basic characteristics of OTS/PDS, RED-S/TRIAD, and BSA, these prevailing conditions should also be assessed for metabolomics, as the full intermediate metabolites and end products better reflect the combination of biological phenotype and the momentaneous state of the organism, resulted from the interactions between genome, transcriptome, proteome, and environment. Both untargeted and targeted metabolomics should be employed, in particular steroid metabolome and protein, lipid, and carbohydrate profiles. The biological systems resulted from the collective analysis of all *omics* should uncover the pathological paths of OTS.

Changes in metabolome in response to specific interventions in training patterns or nutrition may provide information regarding early responses, whether these responses are beneficial or not, and then predict how athletes organisms will respond. Also, the variability between metabolomic responses to a specific intervention may contribute to the understanding of the level of individuality of the recommendations to be further proposed.

In addition, hormones produced by adipocytes and myocytes, as well as adipokines, myokines, and hepatokines, have important roles in the athlete health and are likely altered in OTS and RED-S.

Future researches with larger population of athletes with long-term longitudinal follow-up aiming to detect those athletes who will further develop OTS are desirable to understand the course of OTS and other sports-related dysfunctions, and what to physiologically expect from these athletes. Multicentric studies, which are still rare in the field of sports medicine, including different populations of athletes stratified according to the type of sport, may also answer to the question of reproducibility and sport specificity of the OTS clinical and biochemical presentation. A unified system for data collection worldwide and structured for the abovementioned characteristics would provide massive data and not only identify whether and which patterns are present in OTS but also highlight differences according to expected but also unexpected aspects.

Since OTS, RED-S/TRIAD, BSA, and similar disorders are complex and present overlapping characteristics, further researches should reduce the over-labeling of these burnout-associated diseases and reframe the eyes on research from a more comprehensive perspective, by avoiding presuming that OTS, RED-S, and BSA are separated and inter-excluding disorders. Also, whether these disorders, as well as FOR, NFOR, and even OTS-like syndromes, are actually part of a spectrum of conditions related to burnout could be further clarified.

Specific questions that may be better elucidated with specific researches include:

- Are dysfunctions in OTS sport specific?
- Are dysfunctions in OTS sex specific?
- How does OTS behave in female athletes?
- Is TRIAD/RED-S the female version of OTS?
- Is OTS the athlete version of burnout syndrome?
- Are dysfunctions specific to each type of deprivation, i.e., do carbohydrate, protein, overall caloric, sleeping, and resting deprivations, each of them, lead to specific dysfunctions?
- Are there genetic predispositions for the development of OTS?
- Does each type of deprivation play different level of influences according to genetic and baseline characteristics?
- Does OTS have long-term consequences, besides the psychological ones?

A personalized approach aiming to prevent OTS and related dysfunctions, based on the genome and metabolomic profile of each athlete and how this athlete responds to specific approaches, should become the gold standard follow-up of these patients, that should occur gradually, with the progressive reduction of the costs and the dissemination of the technologies.

12.4   Conclusion

Multiple limitations including the lack of categorization of sports, inappropriate diagnosis, and lack of adjustment for eating, sleeping, and social patterns, as well as the highly heterogeneity of OTS manifestations preclude from more consistent understanding of OTS, whereas whether current findings are applicable for other types of sport and for females is unknown. Characterization of type of sport, sex, age, height, weight, eating, training, social, and sleeping patterns, the use of hormones or medications, and the presence of known or new-onset dysfunctions must be specified in further researches on OTS. The assessment for genome, transcriptome, proteome, lipidome, untargeted and targeted metabolome, hormones produced by adipocytes and myocytes, and adipokines, myokines, and hepatokines should be considered for further researches.

References

1. Cadegiani FA, Kater CE. Novel insights of overtraining syndrome discovered from the EROS study. BMJ Open Sport Exerc Med. 2019;5(1):e000542. https://doi.org/10.1136/bmjsem-2019-000542.
2. William R, Hankinson SE, Sluss PM, Vesper HW, Wierman ME. Challenges to the measurement of estradiol: an endocrine society position statement. Clin Endocrinol Metab. 2013;98:1376–87. https://doi.org/10.1210/jc.2012-3780.
3. Fiers T, Casetta B, Bernaert B, Vandersypt E, Debock M, Kaufman JM. Development of a highly sensitive method for the quantification of estrone and estradiol in serum by liquid chromatography tandem mass spectrometry without derivatization. J Chromatogr B Analyt Technol Biomed Life Sci. 2012;893–4:57–62. https://doi.org/10.1016/j.jchromb.2012.02.034.
4. Stanczyk FZ, Jurow J, Hsing AW. Limitations of direct immunoassays for measuring circulating estradiol levels in postmenopausal women and men in epidemiologic studies. Cancer Epidemiol Biomarkers Prev. 2010;19:903–6. https://doi.org/10.1158/1055-9965.EPI-10-0081.
5. Dorgan JF, Fears TR, McMahon RP, Friedman LA, Patterson BH, Greenhut SF. Measurement of steroid sex hormones in serum: a comparison of radioimmunoassay and mass spectrometry. Steroids. 2002;67:151–8. https://doi.org/10.1016/S0039-128X(01)00147-7.
6. Christina W, Catlin DH, Demers LM, Starcevic B, Swerdloff RS. Measurement of total serum testosterone in adult men: comparison of current laboratory methods versus liquid chromatography-tandem mass spectrometry. J Clin Endocrinol Metab. 2004;89:534–43. https://doi.org/10.1210/jc.2003-031287.
7. Huhtaniemi IT, Tajar A, Lee DM, O'Neill TW, Finn JD, Bartfai G, et al. Comparison of serum testosterone and estradiol measurements in 3174 European men using platform immunoassay and mass spectrometry; relevance for the diagnostics in aging men. Eur J Endocrinol. 2012;166:983–91. https://doi.org/10.1530/EJE-11-1051.

Chapter 13
Clinical Hormonal Guidelines for the Research of the Endocrinology in Sports and Athletes: Beyond Overtraining Syndrome

Outline

Clinical hormonal guidelines for the research of the endocrinology in sports and athletes: beyond overtraining syndrome – in the current context of the growing importance of physical activity, while cardiovascular and psychological aspects have been extensively researched and its knowledge well-structured, the understanding of the endocrine adaptations to sport still lacks. The author presents guidelines for the standardization of the research on the endocrinology of the athletes, including the principles, design, classical and novel parameters, and key characteristics to be present in studies on the field.

13.1 Introduction

Physical activity has historically been key for human development, including preparation for activities demanding exhaustive physical efforts (battles, wars, and hunting), rituals, entertainment, competitions, and demonstration of political power and diplomatic reasons. Besides, the sense of competitiveness is a primitive feeling in humans, which led to the development of sports and exercises, from the Old Olympics, started in 776 BC, to the modern ones, which was launched in 1896, also in Athens, Greece [1–4]. Overall, sports and physical activity have a broad range of meanings in the human history.

In addition to the importance for human development, since the nineteenth century, it has been observed that those who practiced physical activity regularly showed improvement of overall health [5]. The first clear benefits from exercise were demonstrated in the cardiovascular system [5, 6], but have been extended for other systems and types of diseases [7–10], and have now been demonstrated to prevent up to 10% of the deaths [11–13].

© The Editor(s) (if applicable) and The Author(s), under exclusive license to Springer Nature Switzerland AG 2020
F. Cadegiani, *Overtraining Syndrome in Athletes*,
https://doi.org/10.1007/978-3-030-52628-3_13

Along time, physical activity demonstrated to be highly beneficial for a growing range of conditions at different levels of approaches, including primary and secondary prevention, treatment, and prognosis, as well as for improvement of life quality and expectancy, overall well-being, and psychological health [8]. The broadening range of conditions and levels of approaches benefited from exercising encouraged an active lifestyle not only for healthy but also for a range of clinical populations at higher risk of overall morbidity and mortality, including frail elderly [14, 15], those affected by multiple comorbidities, including chronic kidney disease [16], end-stage liver failure [17], and secondary prevention of cardiovascular diseases [6], and for several different types of cancer [18–24] in terms of being an active part of treatment, improvement of life quality and prevention of recurrence, as well as for supportive palliative care [25], postmenopause highly symptomatic women [26], and more than 200 other conditions [8, 27].

The level of importance of physical activity achieved levels that has been recognized by as a major world public health game changer, including worldwide plans including the Global Action Plan on Physical Activity, from the World Health Organization (WHO) [28], and "Exercise is Medicine," also a worldwide Global Health Initiative, from the American College of Sports Medicine (ACSM) [29].

The massive positive propaganda to engage people to exercise leaded to a substantial increase of physically active subjects, among which many have become amateur elite athletes and gym members [29–34], although structured and quality researches on the epidemiology of physical activity lack [28]. The number of elite amateur athletes is particularly growing in individual sports, rather than ball games, including half-marathons, triathlon, and high-intensity functional training (HIFT), like CrossFit, rather than ball and collective games [28–34].

Physical conditioning is defined as the ability to use a same amount of energy for more intense, exhaustive, and/or larger volume of training and to recover faster and more efficiently from physical demands [35–39]. Underlying conditioning processes and adaptive physiology that allow physical conditioning have been reported in the cardiovascular, musculoskeletal, neuromuscular, autonomic, pain, and neuropsychiatric systems [35–39]. While findings in these systems are consistent and conclusive, a comprehensive and conclusive understanding of the endocrine and metabolic adaptations to physical activity remains largely unclear [40], despite the findings that exercise has been shown to induce acute and chronic hormonal effects, including increase of insulin contrarregulatory hormones (glucagon, cortisol, growth hormone (GH), and catecholamines and increase of testosterone in males, respectively [41–43]). Increased hormonal responses to exercise have been described [44, 45], but an inherent hormonal conditioning process that occurs independently from physical effort has only been identified recently [40]. Consequently, the unclear hormonal adaptive changes to exercise preclude from consistent findings on sports-derived endocrine abnormalities, as identified in a systematic review on hormonal aspects on OTS [46], which is further explored.

The inconsistency, non-reproducible, and inadequate conclusions at times found in studies on the endocrine system in athletes and physically active subjects may be due to the lack of participation of experts, in this case, endocrinologists, in group researches of sports medicine [44].

Indeed, oppositely to cardiology and orthopedic fields, in the endocrinology field very few seem to have researched in the sports-related endocrine physiology and disorders thoroughly, while very few endocrine research groups apparently jointed sports medicine teams to perform comprehensive analyses of the endocrinological aspects of athletes and physical activity. Notable exceptions exist in the extensively explored range of aspects on the management of type 1 diabetes in sports and elucidation of the metabolic benefits from exercise, particularly in the glucose and lipid metabolism [47–49]. The lack of scientific interest of research groups of endocrinology, endocrinology societies, and endocrinologists on the field of endocrinology of the physical activity and sport is possibly due to the stigma associated with the correlation between hormones in athletes and anabolic androgenic steroids (AAS) abuse and misuse and the misperceptions of endocrinologists on the researchers of this field.

Apparently, since researches on sports endocrinology rarely had participation of endocrinology experts, and methodological aspects on research on hormones are naturally challenging, issues on the hormonal assessment methods during fasting and stimulation tests have been identified [38, 44] and resemble those identified in the studies on OTS. Some of the major flaws in the specific research on hormones in athletes, regardless of the objective and the studied population of athletes, include (1) the use of unvalidated and heterogeneous hormonal assessment methods; (2) being conducted in a highly heterogeneous populations of athletes, which seem to be randomly selected, according to the availability of the athletes; (3) lack of control for variables that could influence the hormonal and metabolic results, including specific eating and sleeping patterns; (4) misalignment with the correct timing to collect, as the appropriate period of the menstrual cycle; and (5) absence of a control group of healthy sedentary, matched for sex, age, BMI, and other baseline characteristics.

Consequently, despite the interesting findings on athletes in the endocrinology field, many of results were largely inconsistent, conflicting, and many times contradictory. Findings were conclusive. Because of the heterogeneity of the baseline characteristics between groups, as well as the lack of appropriated description of these findings, comparative analyses between studies were unfeasible, precluding from strong systematic reviews or any sort of meta-analysis [44]. Apparently, hormonal and metabolic adaptations to exercises are type of sport-, effort-, sex-, age-, intensity-, and nutritional-dependent. Hence, studies that did not minimally describe these characteristics may have results influenced by these factors, and not considered in the analyses.

All these factors prevented the building of sufficient and concordant evidence for a structured understanding of the endocrine changes in response to physical activity, according to each training pattern and athlete characteristics, in particular in regard with the hormonal adapted physiology to sports.

Once we are unfamiliar with what to physiologically expect from healthy athletes, the diagnosis of overt endocrine dysfunctions when in physically active subjects, and sports-related conditions, such as overtraining syndrome (OTS), gets compromised. Alteration observed in athletes, when their levels are compared to the normal range for general population, is not necessarily pathological, as the apparent altered levels could be actually resulted from a physiological adaptive process. Conversely, parameters that are within the normal range could be actually altered when compared to the expected levels for athletes. Since standardization of adaptations of normal ranges for athletes do not exist in the endocrinology field, accurate analyses of the biochemical parameters on athletes of any level is unfeasible.

The arguments and the *rationale* that reinforce the need for the development of a comprehensive guideline with standardization of endocrine, hormonal, and metabolic parameters and tests in exercise and sports are summarized in Table 13.1 and Fig. 13.1, respectively.

In this chapter, a guideline based on multiple sources, including updated guidelines published by endocrinology or related societies, reviews on the suggested parameters or tests, and original researches that unveiled new markers, new roles, uses, or clinical applications for already existing parameters or tests, or performed standardization of the tests is presented.

13.2 Principles for the Hormonal Research in Athletes

Studies on the endocrinology of physical activity and sport should be conducted with well-defined objectives and potential clinical applications and learnings from the findings and their applications in clinical practice.

The first principle is to delimitate the population of athletes to be studied, as well as the controls, since this is essential to have the questions of the research answered without major biases. The characteristics of the athletes to be defined include:

1. Whether athletes are healthy or affected by any condition, related (e.g., overtraining syndrome) or unrelated (e.g., type 1 diabetes *mellitus*) to sports.
2. Which sport and type of sport (endurance, resistance, explosion (stop-and-go), or mixed).

Table 13.1 Why should endocrine research in physical activity be standardized?

1. An increasing population is becoming physically active, among which many are becoming elite athletes, either amateur or professional
2. Hormones undergo multiple adaptive processes to regular exercises, which may lead to changes in the levels compared to the expected in general population
3. Ranges of hormones and metabolic and other biochemical parameters should be adapted and validated for active subjects and athletes and may be sports-, sex-, age-, and intensity-specific
4. It is urgent to understand the hormonal physiology of the athlete in order to properly identify and treat dysfunctions when in athletes and to diagnose sports-related diseases
5. Standardization of the methods for hormonal research on athletes still lacks
6. The indisputable importance of sports on health care on prevention, remission, and prognosis on a broader range of diseased than previously expected can be expanded to benefits on endocrine diseases, if properly assessed
7. The homogenization of the hormonal parameters and tests and control for modifiable variables is key to strengthen the findings on the endocrine findings in physical activity

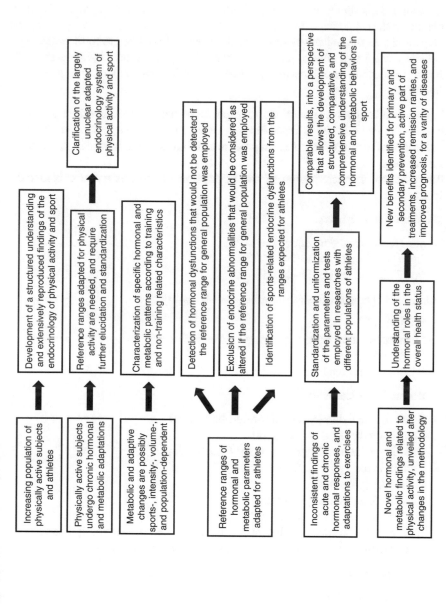

Fig. 13.1 Rationale for the development of guidelines for research on the endocrinology of Physical activity and sport

3. Training patterns: training volume, intensity, frequency of training sessions.
4. Conditioning and training level: sedentary (to be longitudinally followed during a conditioning program), leisure or everyday activities, irregular training, regular mild training, moderate and intense training, amateur athlete, amateur elite athlete, professional elite athlete, ultra-elite athlete, or athlete of an extreme sport.
5. Control groups should be of similar athletes with similar training and baseline characteristics.
6. Depending on the objectives of the study, a second control group of sedentary subjects, when two control groups, of control athletes or healthy athletes, and healthy sedentary should provide more comprehensive understanding of the findings.

The second but not less important principal is the determination of the baselines characteristics that must be assessed in every research, since a large number of variables have been identified as potential predictors of clinical and biochemical behaviors in athletes.

The essential characterization of the population of athletes to be considered includes:

1. Baseline characteristics:

   - Sex (female/male/undefined)
   - Age (y/o), body weight (kg)
   - Body mass index (BMI – kg/m^2)

2. Additional characteristics (desired but not obligatory):

   - Body fat (%)
   - Lean mass (kg/lb or %)
   - Body water (%)
   - Waist circumference (cm)

3. Nutritional characteristics:

   - 3-day to 7-day dietary record (preferably 7-day)
   - Caloric intake per day (kcal/day)
   - Caloric intake per kilogram per day (kcal/kg/day)
   - Protein intake (g/kg/day)
   - Fat intake (g/kg/day)
   - Carbohydrate intake (g/kg/day)
   - Daily intake of whey protein (yes/no)
   - Dietary patterns (intervals between meals, pre- and post-training caloric and macronutrient-specific intake, fasting periods, caloric variations)

4. Sleeping patterns:

   - Sleep duration (h/day)
   - Sleep quality (self-reported, quantified with standardized questionnaires)
   - Sleep tracking devices
   - Sleep hygiene

5. Social patterns:

   - Cognitive concurrent effort (work and/or study; h/day and level of responsi-
     bility and demand)
   - Smoking (yes/no, cigarettes/day, duration of smoking)
   - Alcohol intake (yes/no; estimated alcohol intake per week)
   - Current external stressors (personal, financial, social, familial)

6. Health status:

   - Presence of pre-existing conditions
   - Use of centrally acting drugs
   - Current or recent use of hormones (<6 months)
   - Current or healing orthopedic lesions (yes/no)

The use of medications of any sort may have important interferences in hormonal responsiveness and should be therefore criteria of exclusion.

The main characteristics that must be present in studies on the endocrinology of physical activity are depicted in Fig. 13.2.

The third principal is the design of the study that should be drawn strictly for accurate, applicable, and conclusive findings and must include:

1. Employment of the most suitable validated tests for the objective, for which the choice should result from a thorough examination of the range of possibilities.
2. Definition of the exact timing of the data collect, based on justifications and on previously standardized protocol.
3. Avoidance of unnecessary adaptations whose changes in responses are unknown, unless the development of new protocols is the primary objective. In this case, correlations with currently standardized tests are mandatory.
5. Determination of whether comparisons will be within the same subjects (before vs after intervention), transversally between groups, and/or longitudinally between groups (intervention vs control, intervention "a" vs intervention "b") and/or within the same subjects (response to intervention for each group), according to the questions to be answered by the research.
6. When studying a specific hormonal axis or system, employment of all represen- tative parameters of that axis/system, at all levels, instead of partial evaluation part of axis, since this should give more complete answers to the questions. An example: to test the hypothalamic-pituitary-gonadal (HPG) axis in males, instead of measuring testosterone, it is recommended to measure LH, FSH, SHBG, tes- tosterone, estradiol, and sperm analysis, optionally.
7. Slight differences in baseline characteristics between groups may affect the results, even when differences are not significant. Hence, this should be addressed by rerandomization, if necessary.

For more comprehensive and accurate researches, besides the employment of appropriate parameters and tests, the forth principle of the research on the endocri- nology and metabolism of athletes is the chronology and timing of the tests, which

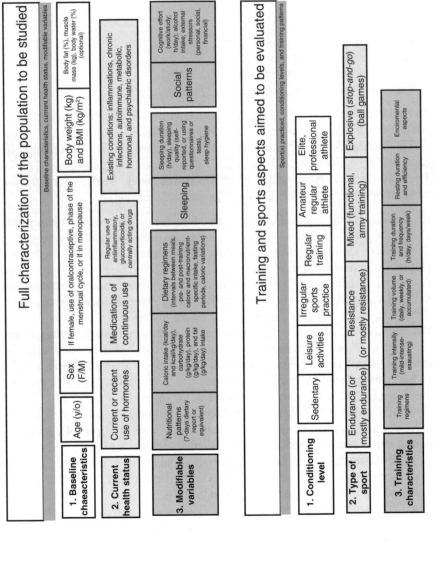

Fig. 13.2 Variables to be considered and characterized for the research on endocrinology of physical activity and sport

should be thoroughly designed. The chronology of these three protocols is based on the rationale, plausibility, existing data, and objectives.

These protocols should allow correlation of different methods, different moments, and therefore a more detailed explanation of the physiological adaptive processes or responses to interventions that occur in athletes.

The three major protocols for the chronology of the research in athletes are detailed in Fig. 13.3 and include:

- Protocol 1. Acute responses to exercise (immediate and direct hormonal and metabolic responses to physical effort)
- Protocol 2. Changes in acute responses to exercise in response to an intervention
- Protocol 3. Chronic hormonal and metabolic changes during a pre-season/pre-event program

In Protocol 1, the assessment of acute responses to exercise aims to unveil which and how hormones and other metabolic parameters respond to a certain training intensity and volume, according to the type of training and sport. Studies on Protocol 2 aim to evaluate how a specific intervention affects the acute responses to exercise, i.e., changes in the hormonal and metabolic behavior under physical effort. In this protocol, repeat the protocol for the acute responses to exercise twice: before and after the intervention. The types of intervention include (1) intensification or changes of trainings patterns; (2) pre-season or pre-event training program; (3) conditioning program (in previously sedentary or not); or (4) non-exercise intervention (nutritional, stress management, sleeping, etc.). Finally, in Protocol 3 hormonal and metabolic changes to be analyzed are not directly stimulated by exercise (>48h after last training session), but may present chronic changes in response to chronic and regular training. For Protocol 3, the timing of the collect should preferably be at least 48 hours after the last training session, to avoid influences from responses to exercise. The time since last training should be similar between participants.

13.3 Standardized Research of the Endocrinology of Physical Activity and Sport

13.3.1 Essays

Accurate and reproductible hormonal and metabolic assessments are central to good quality research. Current commercially available assays exhibit sufficient sensitivity and acceptable intra-assay variability (<20%). However, some hormones, including estradiol (E2), total testosterone (T), growth hormone (GH), freeT4 (fT4), and gonadotropins (LH and FSH, at a lower extent), have large inter-assay variability, which reinforces the importance of assessing the assay quality when interpreting hormone levels. When employing or testing using an assay, researchers and clinicians should be aware of the accuracy, intra- and inter-assay variabilities, and

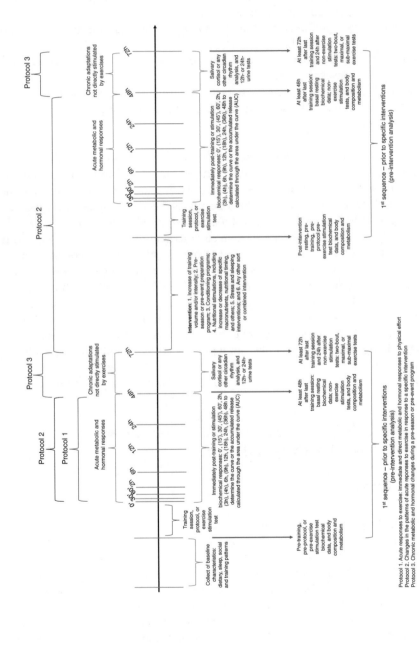

Fig. 13.3 Protocols for the chronology of the research in athletes

limitations of the proposed assay. The most common clinical assays used in research and clinical practice include radioimmunoassay (RIA), chemiluminescence (CLIA), and liquid chromatography by tandem mass spectrometry (LC-MS/MS).

13.3.1.1 Radioimmunoassay (RIA)

In radioimmunoassay (RIA), the use of antibodies against an antigen, which is the specific parameter aimed to be detected, is the basic principle of RIA. It has been initially employed for insulin in the late 1950s and brought a revolution to the biochemical analyses [50].

Along time, RIA underwent sensible improvements, from the detection of autoantibodies until for use of a triple layer ("sandwich") manner (antibody-antigen-antibody), which has been demonstrated to be more accurate than previous assays. However, the use of two antibodies to detect the presence of hormones, particularly steroids, is imprecise and progressively less employed [50–67].

13.3.1.2 (Electro) Chemiluminescence (CLIA)

Chemiluminescence (CLIA) is an immunoassay technique where the emission of visible or near-visible radiation in the form of light is generated when an electron transition from an excited state to ground state occurs, resulted from the analytic reaction between the antibody and the targeted antigen (in which the antigen is the parameter to be quantified) [51].

It has some advantages over RIA, but still lacks accuracy for the detection of steroids. However, for clinical use, once limitations of CLIA are considered, precision for diagnoses and monitoring is sufficient.

The two most important limitations in the use of RIA and CLIA are in low levels of testosterone and low levels of estradiol. In these two cases, variability between tests may be as high as 200% to 400%, which make diagnostic approaches unfeasible. Therefore, in contexts in which testosterone are expectedly low, including women, castrated men (for prostate cancer), and pre-pubertal children, the employment of RIA or CLIA is discouraged.

13.3.1.3 Liquid Chromatography by Tandem Mass Spectrometry (LC-MS/MS)

Clinicians are increasingly using liquid chromatography-mass spectrometry (LC-MSMS), the combination of two techniques, chromatography and mass spectrometry, which delivers high accuracy and reduced interference. The chromatography separates the molecules by their time of retention (min), which occurs according to their molecular weight and divides part of the molecules to be analyzed. The mass spectrometry separates the molecules according to the ratio between their molecular weight and charge levels (mass-to-charge ratio) using different types of ionization processes. The combined analysis of the intact and fragments of the molecules also improves the accuracy of the measurements. The previous separation by the chromatography allows a more precise differentiation between structurally similar molecules that may be similar in mass and which the only difference may be on their charge.

Besides LC, chromatography may also be generated by gas (GC). For steroids, LC is superior in terms of precision and turnaround time, when compared to GC. For untargeted metabolomics, direct infusion (DI), i.e., without separation by any method prior to the mass spectrometer, is an alternative (DI-MS/MS).

Hence, in contrast to RIA/CLIA that measure intact hormones, LC-MS/MS separates molecules twice and also measures molecular fragments of these hormones. The double separation and fragmentation eliminates the problem with interferences in RIA/CLIA when autoantibodies are present. An example is the differentiation between cortisol and prednisolone, which is unfeasible using other methods than LC-MS/MS.

In addition, LC-MS/MS improves can analyze more than one molecule simultaneously and increases comparability between different laboratories.

13.3.2   Parameters, Tests, and Procedures

The present guideline presents the most standardized and reproduced parameters and tests, those with clinical applications and those with specific roles in athletes. For an easier comprehension, parameters are divided according to the gland, axis, or system they belong.

Figure 13.4 summarizes the parameters and tests recommended for research on the field of endocrinology of physical activity and sport and were divided into primary, secondary, and tertiary parameters and tests, according to the level of priority for further researches. Tables depict the parameters and tests of each axis or system, the summary of their main roles and their potential for use in the research on endocrinology of physical activity and sport, according to the existing literature and biological plausibility and mechanisms of their influences on exercises. For steroids and adrenal axis, preferred essays have also been proposed.

Below is the summary of the areas and axes to be analyzed:

- Hypothalamic-pituitary-gonadal (HPG) axis and sex steroids (Table 13.2)
- Hypothalamic-pituitary-adrenal (HPA) axis (Table 13.3)
- GH-IGF-1 axis (Table 13.4)
- Hypothalamic-pituitary-thyroid (HPT) axis (Table 13.5)
- Prolactin (Table 13.5)
- Renin-angiotensin-aldosterone system (RAAS) (Table 13.6)

- Catecholamines and adrenal medulla (Table 13.7)
- Bone metabolism (Table 13.8)
- Water metabolism (Table 13.9)
- Glucose metabolism (Table 13.10)
- Lipid metabolism (Table 13.11)

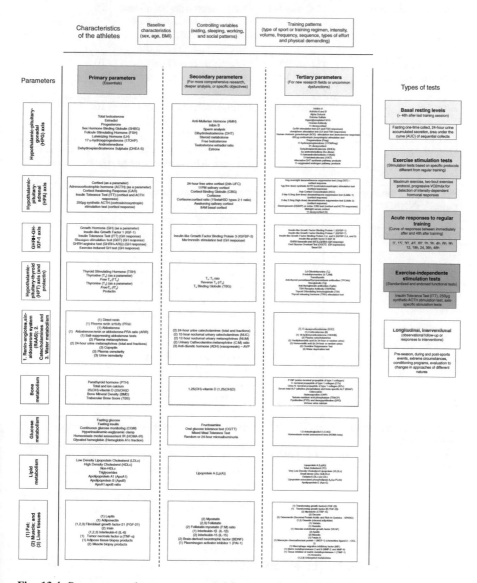

Fig. 13.4 Parameters and tests recommended for research on the field of endocrinology of physical activity and sport

Table 13.2 Parameters of the hypothalamic-pituitary-gonadal axis (HPG) axis and sexual steroids

Hypothalamic-pituitary-gonadal axis (HPG) axis and sexual steroids	Preferred assay(s), if any	Roles (summary)	Potential use for research on endocrinology of physical activity and sport (0–10)*
Steroid metabolome	LC-MS/MS	Complete profiling of all sex steroid metabolites; digital fingerprint of sex steroid secretion and metabolism	10
Total testosterone *Free testosterone*	LC-MS/MS LC-MS/MS	Multiple roles diffusely, improvement of performance and strength	10 7
Estradiol	LC-MS/MS	Multiple roles in women; synergistic effects with testosterone in men	10
Progesterone	LC-MS/MS	Multiple roles in women; very few known biological actions in men	10 (women) / 0 (men)
Sex hormone binding globulin (SHBG)	GC-MS/MS; LC-MS/MS	Stabilize and regulate steroids actions; sex steroid-independent actions?	8
Luteinizing hormone (LH)	–	Oocyte maturation through androgen subtract for estrogen synthesis, optimization of maturation, and other complex functions in women; stimulation of testosterone secretion by Leydig cells in men	8
Follicle-stimulating hormone (FSH)	–	Recruitment, growth, and development of immature ovarian follicles in women; stimulation of spermatogenesis in Sertoli cells in men	8
Dehydroepiandrosterone sulfate (DHEAS)	LC-MS/MS	Androgenic actions in women	7 (women) / 1 (men)
17α-Hydroxyprogesterone (17OHP)	LC-MS/MS	Intermediate product of adrenal steroidogenesis; its increase is related to congenital adrenal hyperplasia (CAH), a hyperandrogenic disease, more preventive in female athletes than general population	6 (women; for screening purposes) / 0 (women, when for other purposes) / 0 (men)
Testosterone/estradiol ratio	LC-MS/MS	Predictor of muscle and fat mass, moods, libido, and performance in men. Roles in women?	2 (women) / 8 (men)

Table 13.2 (continued)

Hypothalamic-pituitary-gonadal axis (HPG) axis and sexual steroids	Preferred assay(s), if any	Roles (summary)	Potential use for research on endocrinology of physical activity and sport (0–10)*
Androstenedione	LC-MS/MS	Intermediate product of the adrenal and gonadal steroidogenesis; indirect marker of hyperandrogenism, polycystic ovary syndrome (PCOS), and CAH in women	5 (women) / 0 (men)
Dihydrotestosterone (DHT)	LC-MS/MS	Highly-potent androgen; main roles in men, although actions in women have been reported	3 (women) / 6 (men)
Inhibin B	–	Inhibition of FSH in both sexes	3
Sperm analysis	Morphology	Characterization of male fertility; marker of testicles health	5 (men)
Estrone	LC-MS/MS	Weak estrogenic effect; produced peripherally	1
Anti-Mullerian hormone (AMH)	–	Oocyte maturation; marker of ovarian viable reserve, premature menopause and PCOS; investigation of secondary amenorrhea	2 (women) / 0 (men)

*0 not important, 10 extremely important
LC-MS/MS liquid chromatography by tandem mass spectrometry

- Adipose tissue metabolism and endocrinology (Table 13.12)
- Muscle metabolism and endocrinology (Table 13.12)
- Inflammatory, muscular, and immunologic parameters (Table 13.12 for other cytokines)
- Body composition and metabolism

13.3.2.1   Hypothalamic-pituitary-gonadal (HPG) axis and sex steroids
(Table 13.2)

The most remarkable hormones and related parameters comprised in the HPG axis include total, bioavailable, and free testosterone, estradiol, progesterone, sex hormone binding globulin (SHBG), luteinizing hormone (LH), follicle-stimulating hormone (FSH), androstenedione, and dihydrotestosterone (DHT). In addition, the testosterone/estradiol ratio may independently predict specific clinical and metabolic outcomes in males.

In addition, the most clinically relevant adrenal sex hormones, which are regulated in an independent manner from HPG, include dehydroepiandrosterone sulfate

Table 13.3 Parameters of the hypothalamic-pituitary-adrenal (HPA) axis

Hypothalamic-pituitary-adrenal (HPA) axis	Preferred assay(s), if any	Roles (summary)	Potential use for research on endocrinology of physical activity and sport (0–10)*
Steroid metabolome	LC-MS/MS	Complete profiling of glucocorticoid metabolites; digital fingerprint of glucocorticoids secretion and metabolism	10
Cortisol (as a parameter)	LC-MS/MS	Fight-or-flight mechanism; stress-driven; energy recruitment	10
Adrenocorticotropic hormone (ACTH) (as a parameter)	Immunoassays	Regulation of cortisol secretion	9
Cortisol binding globulin (CBG)	GC-MS/MS; LC-MS/MS	Stabilize and regulate cortisol actions; glucocorticoid-independent actions?	6
Cortisone	LC-MS/MS	Marker of cortisol inactivation	6
Insulin tolerance test (ITT) (cortisol and ACTH responses)	LC-MS/MS	Responsiveness of the HPA axis; requires integrity of all levels of the HPA axis for normal response	8
250µg synthetic ACTH (cosyntropin - corticotropin) test (cortisol response)	LC-MS/MS and immunoassays	Responsiveness of the fasciculata layer of the adrenal cortex	8
Cortisol awakening response (CAR)	LC-MS/MS	Marker of fatigue and sleep quality; correlated with energy levels after awakening	7
24-hour free urine cortisol (24h-UFC)	LC-MS/MS	Accumulated cortisol secretion	5
Cortisone/cortisol ratio	LC-MS/MS	Balance between 11β-hydroxysteroid dehydrogenase (11β-HSD) types 1 and 2; indirect marker of level of cortisol inactivation	5
Testosterone/cortisol (T/C) ratio	LC-MS/MS	Historical marker of anabolic/catabolic ratio; lacks standardization	5
11PM salivary cortisol	LC-MS/MS	Integrity of the cortisol circadian rhythm	4

Table 13.3 (continued)

Hypothalamic-pituitary-adrenal (HPA) axis	Preferred assay(s), if any	Roles (summary)	Potential use for research on endocrinology of physical activity and sport (0–10)*
Low-dose (1µg) synthetic ACTH (cosyntropin – corticotropin) test (cortisol response)	LC-MS/MS	Responsiveness of the fasciculata layer of the adrenal cortex	3
Awakening salivary cortisol	LC-MS/MS	Potential marker of fatigue and sleep quality	1
8AM basal cortisol	LC-MS/MS	Restricted to investigation of hypocortisolism	1
Hair cortisol concentration (HCC)	-	Chronic cortisol release; marker of chronic exposure to stress?	1
1mg overnight dexamethasone suppression test (1mg-DST) (cortisol response)	LC-MS/MS	Investigation of hypercortisolism	0
2-day low-dose dexamethasone suppression test (cortisol response)	LC-MS/MS	Investigation of hypercortisolism	0
Desmopressin (DDAVP) or ovine-CRH test (cortisol and ACTH responses)	LC-MS/MS and immunoassays	Investigation of the etiology of endogenous hypercortisolism	0
Midnight serum cortisol	LC-MS/MS	Integrity of the cortisol circadian rhythm	0

*0 not important, 10 extremely important
LC-MS/MS liquid chromatography by tandem mass spectrometry

(DHEAS), 17α-hydroxyprogesterone (17OHP), and androstenedione, which is also present in the regulation of the HPG axis. Multiple additional sex steroids can be obtained by steroid metabolome, including more than 300 sex steroid metabolites.

Functional tests in the HPG axis can be employed, including the GnRH-stimulation test, used to elucidate the etiology of male hypogonadism and female hyperandrogenism, and the clomiphene and hCG stimulation tests to determine the etiology sex steroid and fertility abnormalities in both males and females.

Testosterone (T) is the major circulating androgen in males that is also present in females, although at concentrations 10 to 15 times lower than those in men. Given the important actions in both sexes, testosterone should no longer be termed as the "male hormone." Total testosterone includes testosterone bound to SHBG (30–40% of total testosterone), weakly bound to albumin (55–65% of total testosterone), and in its free form (only 2% of total testosterone) [64, 68–74] and varies according to

Table 13.4 Parameters of the GH-IGF-1 axis

GH-IGF-1 axis	Roles (summary)	Potential use for research on endocrinology of physical activity and sport (0–10)*
Growth hormone (GH) (as a parameter)	Bone, muscular, and overall growth; biological cycle; glucose metabolism	10
Insulin-like growth factor 1 (IGF-1)	Overall growth; glucose metabolism; mediates most of GH actions	10
Insulin-like growth factor binding protein 3 (IGFBP-3)	Stabilizes and modulates IGF-1 actions; independent actions on inhibitory and stimulatory effects on cell proliferation (dual effects)	5
Insulin-like growth factor binding protein 1 (IGFBP-1)	Regulates IGF-I and IGF-II; inhibition of tumorigenesis and promotion of apoptosis	3
Insulin-like growth factor binding protein 2,4, and 5 (IGFBP-2,4, and 5)	IGFs-independent effects on metabolism, growth, and cell regulation; confirmatory findings lack	2
Insulin-like growth factor-II (IGF-II)	Growth during fetal development; effects on adults unclear	1
Insulin tolerance test (ITT)	Integrity of the GH responsiveness	10
Glucagon-stimulation test (GST)	Integrity of the GH responsiveness	10
GHRH-arginine test (GHRH+ARG)	Integrity of the GH responsiveness	10
Exercise-induced GH test	Integrity of the GH responsiveness	10
Macimorelin stimulation test	Integrity of the GH responsiveness	7
GHRH-hexarelin test (HEX+GHRH)	Integrity of the GH responsiveness	4
Oral glucose overload test (OGOT)	Suppression of GH secretion in suspected acromegaly	0
Basal GH	Screening for acromegaly; no other roles	0

*0 not important, 10 extremely important

age, health status, diet, and physical activity. In males, testosterone levels demonstrate a circadian rhythm with maximum values between 6h and 8h00 AM, and nadir levels approximately 12 hours later [70, 75], while its patterns in females are uncertain. Testosterone release is negatively affected in males with aging, acute and subacute illnesses [70, 76], systemic diseases, use of medications or recreational drugs, hyperprolactinemia, severe caloric reduction, and eating disorders. In athletes, exhaustive training or caloric restriction may reduce testosterone production, while an appropriate conditioning program increases T levels.

Estradiol (E2) is a predominantly female hormone, although its role in males has been increasingly recognized. In females, its levels vary along menstrual cycle;

Table 13.5 Parameters of the hypothalamic-pituitary-thyroid (HPT) axis and prolactin

Hypothalamic-pituitary-thyroid (HPT) axis (and prolactin).	Preferred assay(s), if any	Roles (summary)	Potential use for research on endocrinology of physical activity and sport (0–10)*
Thyroid-stimulating hormone (TSH)	Immunoassays	Stimulation of the thyroid hormones (T4 and T3)	10
Thyroxine (T_4) (as a parameter) *FreeT4 (fT4)*	LC-MS/MS LC-MS/MS	Pro-hormone; its conversion into T3 by deiodinase (DIO) types 1 and 2 is required for biological actions	10 10
Thyronine (T_3) (as a parameter) *FreeT3 (fT3)*	LC-MS/MS LC-MS/MS	Regulation of overall metabolism, metabolic speed, energy demand, cognitive functions and hemodynamic patterns	10 10
Prolactin	CLIA/RIA	Lactation; stress-regulated; interactions with other axes	8
T4:T3 ratio	LC-MS/MS	Level of conversion from T4 into T3	5
Reverse T3 (rT3)	LC-MS/MS	Level of thyroid hormone inactivation	4
T_4 Binding Globulin (TBG)	GC-MS/MS; LC-MS/MS	Regulation of thyroid hormone actions	3
3,5-Diiodothyronine (T2)	LC-MS/MS	"Hot" thyroid hormone metabolite, with recently described actions in metabolism	2
3-iodothyronamine (3-T1AM)	LC-MS/MS	Brain metabolism	1
Transthyretin	–	Stabilization of thyroid hormones	1
Anti-thyroid peroxidase/ thyroperoxidasis antibodies (TPOAb)	–	Detection of Hashimoto's thyroiditis and other autoimmune thyroiditis	0
Tireoglobulin (Tg)	–	Thyroid activity; follow-up post-total thyroidectomy	0
Anti-thyroglobulin antibodies (TgAb)	–	Detection of autoimmune thyroiditis	0
TSH receptor antibody (TSHRAb)	–	Detection of Graves's disease	0
Thyroid-stimulating immunoglobulin (TSI)	–	Detection non Grave's disease hyperthyroidism	0
Thyroid-releasing hormone (TRH) stimulation test	–	Diagnosis of panhypopituitarism, secondary hypothyroidism, and TSH-producing pituitary adenoma (TSHoma)	0

*0 not important, 10 extremely important
LC-MS/MS liquid chromatography by tandem mass spectrometry
CLIA/RIA chemiluminescence assay and radioimmunoassay

Table 13.6 Parameters of the renin-angiotensin-aldosterone system (RAAS)

Renin-angiotensin-aldosterone system (RAAS)	Preferred assay(s), if any	Roles (summary)	Potential use for research on endocrinology of physical activity and sport (0–10)*
Direct renin	LC-MS/MS	Stimulation of angiotensin and aldosterone; indirect regulation of blood pressure and serum potassium	10
Plasma renin activity (PRA)	LC-MS/MS	Stimulation of aldosterone; indirect regulation of blood pressure and serum potassium	10
Aldosterone	LC-MS/MS	Regulation of blood pressure and serum potassium; indirect regulation of water content	10
Aldosterone to renin ratio (ARR)	LC-MS/MS	Screening for primary hyperaldosteronism	5 (10 for the diagnosis of primary hyperaldosteronism)
Salt-suppressing aldosterone tests	LC-MS/MS	Confirmatory test for the diagnosis of primary hyperaldosteronism	0

*0 not important, 10 extremely important
LC-MS/MS liquid chromatography by tandem mass spectrometry

Table 13.7 Parameters of the catecholamines and adrenal medulla

Catecholamines and adrenal medulla	Roles (summary)	Potential use for research on endocrinology of physical activity and sport (0–10)*
Plasma metanephrines	Metabolites of catecholamines; inactivation of catecholamines	10
24-hour urine metanephrines (total and fractions)	Accumulated secretion of metabolites of catecholamines; level of inactivation of catecholamines	10
24-hour urine catecholamines (total and fractions)	Accumulated secretion of catecholamines; fight-or-flight reaction; endocrine adrenergic functions	8
12-hour nocturnal urinary catecholamines (NUC)	Accumulated secretion of catecholamines; fight-or-flight reaction; endocrine adrenergic functions	7
12-hour nocturnal urinary metanephrines (NUM)	Accumulated secretion of metabolites of catecholamines; level of inactivation of catecholamines	7
Plasma catecholamines	Fight-or-flight reaction; endocrine adrenergic functions	3
Catecholamine/metanephrine (C:M) ratio	Level of inactivation of catecholamines	2
Vanillylmandelic acid (in 24-hour or random urine)	Norepinephrine metabolite	0
Homovanillic acid (in 24-hour or random urine)	Dopamine metabolite	0
Clonidine suppression test	Suppression of catecholamines; exclusion of the diagnosis of pheochromocytoma	0

*0 not important, 10 extremely important

Table 13.8 Parameters of the bone metabolism

Bone metabolism	Roles (summary)	Potential use for research on endocrinology of physical activity and sport (0–10)*
Parathyroid hormone (PTH)	Maintenance of calcium in the serum	8
Total and ion calcium	Multiple actions	8
25(OH)-vitamin D (25(OH)D)	Prohormone; it is conversion into 1,25(OH)D, the active isoform of vitamin D, by 1alpha-hydroxilase enzyme, mostly located in the kidneys	7
Bone mineral density (BMD)	Level of mineralization of trabecular and cortical (long) bones	7
Trabecular bone score (TBS)	Quality of the microarchitecture of trabecular bones	5
1,25(OH)-vitamin D (1,25(OH)D)	Stimulation of intestinal calcium absorption, renal reabsorption, and calcium and other minerals absorption from the skeleton; genomic regulation	3
P1NP (amino-terminal propeptide of type 1 collagen)	Bone reabsorption	2
C- terminal propeptide of type 1 collagen (CTx)	Bone reabsorption	1
Urine N- terminal propeptide of type 1 collagen (NTx)	Bone reabsorption	1
Serum total ALP (alkaline-phosphatase) and bone-specific ALP (BSAP)	Bone formation	1
Osteocalcin	Bone formation	1
Hydroxyproline (OHP)	Bone formation and reabsorption	0
Tartrate-resistant acid phosphatase (TRACP)	Bone reabsorption	0
Pyridinoline (PYD) and deoxypyridinoline (DPD)	Bone reabsorption	0
24-hour urine calcium	Secondary effects of primary hyperparathyroidism; hypercalciuric states; investigation of etiology of nephrolithiasis	3

*0 not important, 10 extremely important

therefore the phase of the cycle should be known when interpreting estradiol levels. Women under oral contraceptives and peri- and post-menopause will also present unreliable estradiol levels. For research purposes, the phase of the cycle should be questioned and aligned with the estradiol measurement. In males, estradiol should be proportional to the levels of testosterone, as a mechanism toward a proper

Table 13.9 Parameters of the water metabolism

Water metabolism	Roles (summary)	Potential use for research on endocrinology of physical activity and sport (0–10)*
Copeptin	More stable marker of ADH secretion	10
Plasma osmolarity	Marker of level of hydration and water balance	10
Urine osmolarity	Marker of level of hydration and water balance	10
Anti-diuretic hormone (ADH) (vasopressin – AVP)	Regulation of plasma osmolarity	7
Urine and plasma osmolarity responses to desmopressin (DDAVP) intake	Kidney sensitivity to ADH (theoretical)	1
Water deprivation test	Diagnosis of diabetes insipidus	0

*0 not important, 10 extremely important

balance between these two hormones, since increased testosterone has better effects on body composition, energy levels, and libido when accompanied by increased estradiol, when compared to testosterone alone, while excessive estradiol in relation to the testosterone levels may be a sign of metabolic, inflammatory, or energy-balancing dysfunction [77].

Progesterone is almost inexistent in men, except in those affected by congenital adrenal hyperplasia (CAH), while it is present and highly variable in women, according to the phase of the cycle, use of oral contraceptives, and menopause. Its roles in a variety of systems and its effects on sports performance are yet to be determined, but the dosage of progesterone in women is desired, as it may address some missing gaps on the study findings. In female athletes, both estradiol and progesterone should be assessed together when psychological evaluation is desired, since neither estradiol nor progesterone predicted physical or mental fatigue alone [78, 112].

Sex hormone binding globulin (SHBG) is the major carrier globulin that stabilizes while temporarily inactivates sex steroids. Several factors increase SHBG, including aging (1% per year), nephrotic syndrome, cirrhosis and hepatitis, HIV, hyperthyroidism, use of estrogens and anticonvulsivants, estrogen excess, severe androgen deficiency, and polymorphisms in the SHBG gene [68]. Oppositely, decrease of SHBG occurs in aging metabolic syndrome and other metabolic conditions, obesity, type 2 diabetes mellitus (T2DM), secondary hypogonadism, hypothyroidism, use of glucocorticoids, and other polymorphisms in the SHBG gene [68]

Gonadotropin-releasing hormone (GnRH), the key regulator of LH and FSH release by the pituitary, is modulated by hypothalamic neuropeptides such as kiss-peptins, neurokinin-B, dynorphin-A, and phoenixins. Hypothalamic neuropeptide

Table 13.10 Parameters of the glucose metabolism

Glucose metabolism	Roles (summary)	Potential use for research on endocrinology of physical activity and sport (0–10)*
Fasting glucose	Glucose levels without the interference of meals or carbohydrate intake; glucose behavior in a prolonged absence of exogenous glucose intake	10
Fasting insulin	Insulin levels without the acute response to meals or carbohydrate intake	10
Continuous glucose monitoring (CGM)	Glucose behavior in response to multiple physiological processes	10
Hyperinsulinemic-euglycemic clamp	Gold standard test for the evaluation of the level of insulin sensitivity	5 (lack of feasibility)
Homeostasis model assessment IR (HOMA-IR)	Prediction of the level of sensitivity to insulin for glucose uptake	4
Glycated hemoglobin (hemoglobin A1c)	Estimation of average glucose level in the past 2 to 3 months	4
Fructosamine	Estimation of average glucose level in the past two to three weeks	3
1,5-Anhydroglucitol (1,5-AG)	Estimation of average glucose level in the past seven to 14 days	2
Mixed Meal Tolerance Test	Glucose behavior (optionally: C-peptide) in response to a standardized meal	1
Homeostasis model assessment-beta (HOMA-beta)	Prediction of the level of beta-mass activity	0
Oral glucose tolerance test (OGTT)	Glucose behavior in response to acute glucose intake	0
Random or 24-hour microalbuminuria	Initial renal protein loss; present in early stages of diabetic nephropathy, or physiologically, in response to exercise.	0

*0 not important, 10 extremely important

expressions are dependent on metabolic status, but little is known regarding responses to exercises. Gonadotropin-inhibitory hormone (GnIH) is another hypothalamic that is released in response to androgens and in a lesser extent to estrogens.

The two gonadotrophic hormones (gonadal-stimulatory hormones), LH and FSH, are simultaneously produced by the same population of cells in the pituitary (gonadotrophs), although they have partially different regulatory mechanisms. The amount of LH and FSH released from the pituitary depends on the frequency and width of GnRH release pulses, whereas continuous GnRH release paradoxically inhibits these hormones.

Table 13.11 Parameters of the lipid metabolism

Lipid metabolism	Roles (summary)	Potential use for research on endocrinology of physical activity and sport (0–10)*
Low-density lipoprotein cholesterol (LDLc)	Transport of cholesterol molecules	10
High-density cholesterol (HDLc)	Reverse transport of cholesterol from atheroma plaques	10
Non-HDLc	Sum of all atherogenic particles	10
Triglycerides	Molecules with three fatty acids and a glycerol	10
Apolipoprotein A1 (ApoA1)	Quantification of the number of anti-atherogenic particles	10
Apolipoprotein B (ApoB)	Quantification of the number of atherogenic particles	10
ApoA1:apoB ratio	Most accurate predictor of cardiovascular disease (according to the INTER-HEART study)	10
Lipoprotein A (Lp(A))	An atherogenic particle, independent from LDLc and HDLc	7
Total cholesterol (TC)	Sum of all particles that contain cholesterol	5
Very Low Density Cholesterol Lipoprotein (VLDLc)	Transport of triglycerides	4
Small dense LDLc (sdLDLc)	Most atherogenic LDLc particles	3
oxidized LDLc (ox-LDL)	Dubious actions – marker of LDLc capacity to protect against oxidative stress, but also atherogenic	2
Lipoprotein-associated phospholipase A_2 (Lp-PLA2)	Independent risk factor for cardiovascular diseases; negative synergistic effects with C-reactive protein (CRP)	2
Apolipoprotein E (Apo-E)	Brain cholesterol metabolism	1

*0 not important, 10 extremely important

In females, LH is essential to provide androgen substrate for estrogen synthesis, which contributes to oocyte maturation, and play relevant roles in the optimization of fertilization and embryo quality [119–121]. In men, LH is similarly regulated by GnRH and acts in the Leydig cells in the testicles to produce testosterone [68, 69, 79].

FSH actions in women include stimulation of the recruitment, growth, and development of immature ovarian follicles in the ovary, while in men FSH stimulates the spermatogenesis process, although indirect functions of FSH on the intratesticular testosterone production have also been reported.

Table 13.12 Adipokines, myokines, and hepatokines

Cytokines	Adipokines, myokines, and/ or hepatokines?	Roles (summary)	Potential use for research on endocrinology of physical activity and sport (0–10)*
Leptin	Adipokine	Anorexigenic; male and female fertility; bone mass gain; positive energy homeostasis	10
Adiponectin	Adipokine	Improvement of glucose and lipid metabolism; thermogenic (upregulation of UCP-1); anti-inflammatory	10
Fibroblast growth factor-21 (FGF-21)	Adipokine, myokine, adipokine	Improvement of glucose and lipid metabolism; cardioprotection; muscle atrophy; bone mass loss	10
Irisin	Myokine	Thermogenesis; browning of white adipose tissue (WAT)	10
Interleukin-6 (IL-6)	Adipokine, myokine, hepatokine	Acutely, as an energy sensor; chronically, as pro-inflammatory and negative regulator of muscle growth	10
Tumor necrosis factor α (TNF-α)	Adipokine	Pro-inflammatory	8
Myostatin	Myokine	Negative regulator of muscle growth	8
Follistatin	Myokine, hepatokine	Myostatin inhibitor; muscle hypertrophy	6
Follistatin/myostatin (F:M) ratio	Myokine-derived ratio	Marker of muscle anti-atrophy/ atrophy balance	5
Interleukin-1beta (IL-1beta)	Adipokine	Pro-inflammatory	5
Interleukin-15 (IL-15)	Myokine	Improvement of glucose and lipid metabolism; conflicting findings on actions on muscle	4
Brain-derived neurotrophic factor (BDNF)	Myokine	Fat oxidation; positive metabolic changes; increase of brain volume	3
Plasminogen activator inhibitor 1 (PAI-1)	Adipokine	Pro-inflammatory; thrombogenic	3
Transforming growth factor-β (TGF- β)	Adipokine	Pro-inflammatory	3
Myonectin (CTRP-15)	Myokine	Muscle hypertrophy; fatty acid utilization for muscle contraction	2
Decorin	Myokine	Myostatin inhibitor; muscle hypertrophy	2

(continued)

Table 13.12 (continued)

Cytokines	Adipokines, myokines, and/or hepatokines?	Roles (summary)	Potential use for research on endocrinology of physical activity and sport (0–10)*
Osteonectin (Secreted Protein Acidic and Rich in Cysteine - SPARC)	Myokine	Apoptosis of cancer cells; cancer prevention; muscle regeneration; bone mass gain	2
Omentin	Apipokine (visceral fat); hepatokine;	Anti-inflammatory; positive metabolic outcomes; prevention of type 2 diabetes mellitus	2
Visfatin	Adipokine	Pro-inflammatory mediator	2
Resistin	Adipokine	Pro-inflammatory mediator	2
Vascular endothelial growth factor (VEGF)	Adipokine		2
Apelin	Myokine	Improvement of glucose metabolism; prevention of type 2 diabetes mellitus and sarcopenia	2
Musclin	Myokine	Prevention of muscle atrophy and metabolic dysfunctions	1
Fetuin-A	Adipokine	Carrier of free fatty acids in circulation; anti-inflammatory; anti-atherogenic	1
Monocyte chemoattractant protein 1 (MCP-1) (chemokine ligand-2 – CCL-2)	Adipokine	Immunologic regulator; pro-autoimmune	1
Macrophage migration inhibitory factor (MIF)	Adipokine	Innate immunity; pro-inflammatory	1
Matrix metalloproteinase 2 and 9 (MMP-2 and MMP-9)	Adipokine	Cancer progression and worse prognosis; pathological angiogenesis and neovascularization	1
Tissue inhibitor of matrix metalloproteinase-1 (TIMP-1)	Adipokine	Cancer progression and worse prognosis	1
Annexins	Adipokine	Prevention of cancer and type 2 diabetes mellitus, triggering of autoimmune disorders	1
Untargeted metabolome	–	Full profile of all intermediate and end products of all metabolic pathways and reactions within an organism or tissue	9

Table 13.12   (continued)

Cytokines	Adipokines, myokines, and/or hepatokines?	Roles (summary)	Potential use for research on endocrinology of physical activity and sport (0–10)*
Adipose tissue biopsy	–	Intra- and inter-adipose macrophage infiltration and accumulation; local and regional cytokine and morphological profiling; adipokines with paracrine, autocrine, and intacrine actions	8
Muscle biopsy	–	Intra- and inter-muscular fat accumulation; local myokine production; myokines with paracrine, autocrine, and intacrine actions	8

*0 not important, 10 extremely important

Compared to LH, FSH has a longer half-life and alters earlier in pathological states and is therefore more sensitive to detect early alterations and to provide adequate results [69].

Dehydroepiandrosterone sulfate (DHEAS) is the final product of the reticular layer of the adrenal cortex and has androgenic effects, which are particularly important in women [70], and is a precursor of sex steroid hormones and is converted to testosterone and estradiol. Besides the adrenal and gonadal steroidogenesis, recent reports revealed that conversion of DHEA into androgens and estrogens occurs in peripheral tissues including brain, skin, liver, kidney, bone, and skeletal muscles, using circulating DHEA as the substrate. Although the use of DHEA may elevate testosterone and DHT, particularly in women [80], the idea of DHEA as being a "supplement" may explain its wide use, particularly among amateur-athletes and non-doping controlled sports and in training centers. Thus, the use of DHEA should be assessed in participants of studies on the field of sports endocrinology, prior to their inclusion in the study.

Spermatogenesis is the process of maturation of spermatozoids, from spermatogonia into spermatozoa, and plays important roles to reflect the male gonadal health as menstrual cycles in females. Methodologies of sperm analysis, they still need sensitive improvement, although current assays already allow that sperm analysis, including account, analysis of the dead sperms rate, morphological normality, and multi-directional mobility, can be a more sensitive marker of testicles health than testosterone production, as 85% of the testicle mass is constituted of Sertoli cells. In clinical practice, semen analysis should be performed when testosterone replacement is recommended and fertility required [68, 70].

Studies on the interference of sports on male fertility are sparse and could add important data to the currently existing understanding and should include semen analysis but go beyond [81–83]. Since we have now learned that metabolic disorders in males affect fertility rate and miscarriage, in an independent manner from normal sperm analysis, possibly due to increased oxidative stress, athletes should also be evaluated for characteristics of sperm other than those described in the classic sperm analysis, since elite sports and exhaustive competitions are known to increase oxidative stress status, as observed in increased DNA fragmentation in semen in ultra-endurance exercises, affecting fertility [82]. Also, intense physical activities may temporarily affect the semen concentration and the number of motile and morphologically normal spermatozoa, in a linear manner [81, 83].

In female athletes, the employment of progesterone, estradiol, testosterone, androstenedione, DHEA-S, LH, FSH, and SHBG is recommended, while in male athletes, testosterone, estradiol, LH, FSH, and SHBG are the preferred markers of the HPG axis. Noteworthy, hyperandrogenism is particularly prevailing in female elite athletes, which should be considered when assessing this population. All female participants should be dosed within the same phase of the menstrual cycle, as comparisons become unfeasible otherwise.

13.3.2.2 Hypothalamic-Pituitary-Adrenal (HPA) Axis (Table 13.3)

The major markers of the HPA axis include cortisol and adrenocorticotropic hormone (ACTH). However, these parameters without specific testing protocols provide limited data. Functional tests aiming to evaluate adrenal reserve in cases of suspect of adrenal insufficiency include low-dose (1μg) or regular-dose (250μg) synthetic ACTH (cosyntropin – corticotropin) stimulation test (CST) that directly stimulates cortisol secretion and insulin tolerance test (ITT), leading to ACTH and cortisol responses to a hypoglycemic episode.

On the opposite direction, tests to detect cortisol pathological overexpression include 1mg overnight dexamethasone suppression test (1mg-DST) (cortisol response), 2-day low-dose dexamethasone suppression test (cortisol response), and desmopressin (DDAVP) or ovine-CRH test (cortisol and ACTH responses). This last test can also be employed to determine the etiology of adrenal insufficiency (whether primary, adrenal, or secondary, pituitary).

For the detection of abnormalities in the cortisol secretion patterns, a 24-hour free urinary cortisol can be performed as a manner to measure the accumulated secretion in a 24-hour period, and the late-night salivary cortisol, that should be physiologically undetectable, and which detectable levels may suggest impaired cortisol circadian rhythm.

Additional tests include the cortisol wakening response (CAR) using salivary cortisol, an indirect marker of fatigue and sleep quality (although does not demonstrate causality), cortisol binding globulin (CBG), and cortisone, a marker of cortisol inactivation by 11β-hydroxysteroid dehydrogenase (11β-HSD) type 2 (11β-HSD type 1 actives cortisol from cortisone).

Circulating cortisol is 80% bound to CBG, 10–15% to albumin and a very low percentage circulates unbound [84]. Cortisol is produced in substantial amounts, and so its measurement in serum, urine, and saliva by different assays is not a critical issue. In healthy states, cortisol acutely increases in response to exercise, as an insulin contrarregulatory hormone, to avoid hypoglycemia and to indirectly provide prompt energy availability under high demands.

Testosterone/cortisol ratio and hair cortisol concentration (HCC) are alleged markers of anabolic/catabolic ratio and stress profile in the past months, respectively, although both require further confirmatory data and standardization.

The adrenal gland encompasses two almost independent glands of distinct embryogenic origin: the adrenal medulla, located centrally in the gland, originated from the neuroectodermal embryogenic cells, and the adrenal cortex, originated in the mesoderm. Despite virtually distinct, both adrenal cortex and medulla have a similar greater function as the key gland to respond to stress, by releasing cortisol and adrenaline, respectively, as the major hormones that induce the fight-or-flight response. While the cromafin cells of the adrenal medulla are the main source of catecholamines produced to act as endocrine hormones, the adrenal cortex is where adrenal steroidogenesis occurs. The adrenal cortex has three zones: glomerulosa, *fasciculata* and *reticularis/reticulata* layers, and their end bioactive steroidal products are aldosterone, cortisol, and DHEA-S, respectively. However, multiple metabolites from each intermediate steroid produced the adrenal cortex provide a vast amount of steroids, measurable through steroid metabolome. Among all these steroids, the major products, their respective metabolites, and the location of their production include:

All cortical-adrenal layers:

- Pregnenolone (Preg) – one metabolite: Pregnenediol (5PD)
- Progesterone (Prog) – one metabolite: Pregnanediol (PD)

Fasciculata and *reticulata* layers:

- 17 Hydroxypregnenolone (17OHPreg) – one metabolite: Pregnenetriol (5PT)
- 17-Hydroxyprogesterone (17OHP) – two metabolites: Pregnanetriol (PT) and 17-hydroxyregnanolone (17HP)
- 21-Deoxycortisol – one metabolite: Pregnanetriolone (PTONE)

Fasciculata layer (end product = cortisol):

- 11-Deoxycortisol (S) – one metabolite: Tetrahydro-11-deoxycortisol (THS)
- Cortisol (F) - six metabolites: 6β-Hydroxycortisol (6β-OHF), tetrahydrocortisol (THF), 5α-tetrahydrocortisol (5α-THF), α-cortol, β-cortol, 11β-hydroxy etiocholanolone (11β-OHEt)
- Cortisone (E) - five metabolites: E, tetrahydrocortisone (THE), α-cortolone, β-cortolone, 11-ketoetioclolanolone (11-ketoEt)

Glomerulosa layer (end product = aldosterone):

- 11-Deoxycorticosterone   (DOC)   –   one   metabolite: Tetrahydro-11-deoxycorticosterone (THDOC)
- Corticosterone (B) – four metabolites: Tetrahydro-11-dehydrocorticosterone (THA), 5α-tetrahydro-11-dehydrocorticosterone (5α-THA), tetrahydrocorticosterone (THB), 5α-tetrahydrocorticosterone (5α-THB)
- 18-Hydroxycorticosterone   (18OHB)   –   one   metabolite: 18-Hydroxy-tetrahydro-11-dehydrocorticosterone (18OHTHA)
- Aldosterone (Aldo) – one metabolite: 3α5β-Tetrahydroaldosterone (THAldo)

Reticularis/reticulata layer (end product = DHEAS):

- Dehydroepiandrosterone (DHEA) – two metabolites: DHEA, 16α-hydroxy-DHEA (16α-OHDHEA)
- Dehydroepiandrosterone sulfate (DHEAS) – one metabolite: DHEA
- Androstenedione (A4) – two metabolites: Androsterone (An), etiocholanolone (Et)
- 5α-androstenedione (5α-dione)
- Testosterone (T) – two metabolites: Androsterone (An), etiocholanolone (Et)
- 5α-Dihydrotestosterone (DHT) – one metabolite: Androsterone (An)
- 11β-Hydroxyandrostenedione   (11OHA4)   –   one   metabolite: 11β-Hydroxyandrosterone (11β-OHAn)
- 11-Ketoandrostenedione (11KA4) – no metabolite
- 11-Ketotestosterone (11KT) – no metabolite

Alternative pathways in addition to the classic previously described pathways have also been uncovered and include:

Alternative DHT synthesis pathway:

- Prog → 5α-dihydroprogesterone (DHProg)
- DHProg → allopregnanolone (allopreg)
- DHProg → 17OH-dihydroprogesterone (17OHDHP)
- Allopreg → 17OH-allopregnanolone (5α-17HP)
- 17OHDHP → 17OH-allopregnanolone (5α-17HP)
- 5α-17HP → androsterone (An)
- An → androstanediol (3α-diol)
- 3α-diol → DHT

11-Oxygenated androgen pathway:

- A4 → 11β-hydroxyandrostenedione (11OHA4)
- 11OHA4 → 11β-hydroxytestosterone (11OHT)
- 11OHA4 → 11-ketoandrostenedione (11KA4)
- 11OHT → 11-ketotestosterone (11KT)
- 11KA4 → 11-ketotestosterone (11KT)

Collectively, the analysis of the multiple recently described steroids allowed the identification of each different condition as disclosing a unique, fingerprint metabolomic profile [85]. However, unlike LC-MS/MS, which although expensive is available for clinical use, the analysis of steroid metabolome is still restricted to research purposes, owing to its multiple technical complexities.

Under exhaustive sports, androsterone, DHEA, and estradiol are reduced, whereas cortisol, tetrahydrocortisol, and tetrahydrocortisone are increased, in an overall hormonal metabolomic profile toward catabolism [86]. Strength training in male athletes leads to a pro-catabolic hormonal metabolome state including reductions in epitestosterone, androstenedione, androsterone, etiocholanolone, estrone, and tetrahydrocortisone (THE), which may last for up to 48 hours after training [87] that occurs similarly in trainings focusing in both concentric and eccentric contractions [88]. Strength training in female athletes leads to reduction in estradiol during follicular phase and reduction in progesterone during luteal phase, reinforcing the differences that occur in hormonal responsiveness related to performance according to the phase of menstrual cycle [89].

The evaluation of athletes should not include basal serum cortisol alone; on the contrary, cortisol response to stimulations or salivary cortisol rhythm better reflects the correlations between the HPA axis and sports performance and related parameters.

13.3.2.3 GH-IGF-1 Axis (Table 13.4)

The major hormones encompassed in the somatotropic axis include growth hormones (GH), insulin-like growth factor 1 (IGF-1), and IGF binding globulin-3 (IGFBP-3). IGF-2 and IGFBPs 1, 2, 4, and 5 have been described, although several of their biological effects remain uncertain [90–93].

GH is a hormone released by the somatotrophs in the pituitary in response to GH-releasing hormone (GHRH) released by the hypothalamus, and its secretion is highly pulsatile, remaining undetectable in 95% of the time. Hence, its measurement in basal states finds substantiation to detect excessiveness, even when in normal range, in case levels are detectable in three or more occasions, particularly when IGF-1 is elevated. For unexplained reasons, athletes demonstrate increased GH release even when not in peak.

While basal GH provides limited data, functional tests to detect either excessive GH secretion or GH deficiency yields more reliable data regarding the biochemical confirmation or exclusion of the suspected diagnosis. Whenever basal GH is persistently detectable, a suppression test using oral glucose can be performed, since under physiological circumstances glucose suppresses GH release. In case GH deficiency is suspected, different GH stimulation tests can be employed, including the ITT, glucagon (GST), GHRH-arginine, macimorelin, and GHRH-hexarelin stimulation tests. For athletes, the ability to release GH under stress can predict performance, which can be estimated by any of these stimulation tests. Among these, ITT would provide the most reliable results, but can bring ethical concerns from the risk of hypoglycemia.

IGF-1 is a hormone produced in the liver and the most important bioactive hormone released in response to GH, a log-linear relationship. Circulating IGF-1 half-life is approximately 15 to 30 hours, which can be extended by binding to IGF binding globulins (IGFBPs). Only 2% of IGF-1 circulates freely, while 20 to 25% is bounded to IGFBP-3, and approximately 75% is doubly bounded, to IGFBP-3 and its acid labile subunit (ALS).

Although there are differences between the liver-derived and local IGF-I actions [94–103], generally IGF-1 release leads to protein synthesis and muscle hypertrophy, neural growth, reduction of inflammation, and several other physiological actions.

Factors that reduce IGF-1 levels include malnutrition, chronic liver disease, and renal diseases, severe infection, poorly controlled diabetes mellitus, use of oral contraceptives, fasting for more than 12 hours/day, and reduction of more than 50% of the previous daily caloric intake, while increase of IGF-1 occurs with metformin use, increase of vitamin D, and increase in protein intake [104–111].

From the IGFBP family of six proteins, IGFBP-3 is the most described and elucidated. IGFBP-3 is the most important binding globulin of IGF-1, is regulated by GH in a similar manner than IGF-1 [112, 113], and acts stabilizing and modulating IGF-1 actions, exhibiting a complex apparently paradoxical concurrent inhibitory and stimulatory effects on cell proliferation. IGFBP-3 has also been proposed to be an indirect marker of liver health, as its production is compromised in steatohepatitis, liver fibrosis, and cirrhosis. IGFBP-3 is reduced in response to estrogens, while it is consistently and acutely increased with exercise, equally in males and females, which may help attenuate the effects of aging-related IGF-1 decline.

Similarly to IGFBP-3, IGFBP-1 is an endocrine factor which modulates serum IGF-I and IGF-II bioavailability and also exerts paracrine and autocrine effects, particularly in the inhibition of IGF-II, leading to inhibition of tumorigenesis and promotion of apoptosis, while IGFBP-1 may enhance IGF-1 activity [114, 115]. IGFBP-1 protects from hypoglycemia, has pro-erythrocyte effects, and plays multifunctional roles in the female reproductive tract.

IGFBP-1 expression is strongly and acutely stimulated under catabolic conditions, including prolonged fasting and hypocaloric diets, hypoxia, and stress, use of glucocorticoids, T1DM, polycythemia *vera*, and endurance training while is suppressed by body fat, insulin levels, and acute exercises (<45 minutes of duration).

IGFBP-2 is a less characterized IGFBP that has demonstrated both inhibitory and stimulatory effects on IGF-mediated functions, including weak intensification of IGF-I actions, fetal development, transport of IGF-II to central nervous system (CNS), and the major IGFBP in CNS. IGFBP-4 is a IGFBP with inhibitory effects on IGF actions, inhibition of cancer development and progression, and inhibition of atherosclerosis while can restore cardiomyocytes and heart function [112, 116, 117]. IGFBP-5 is the prevailing IGFBP in the bone; has atherogenic effects, particularly when bound to IGFs; enhances ovarian function; is upregulated in cancer, although its functions in cancer cells are conflicting and dependent on IGF-I expression; and has important roles in glucose metabolism and within vessels. Like other IGFBPs, IGFBP-5 is increased in some autoimmune disorders, particularly systemic sclerosis [112, 117]. IGFBP-6 is the least known IGFBP, with apparent inhibitory actions in the HPG axis [112, 117].

Insulin-like growth factors (IGFs) are key growth-promoting peptides that act both as endocrine hormones and as autocrine and paracrine growth factors; IGF-II has been classically reported to act as a growth hormone during pregnancy and in cancer cells, while recent reports show IGF-1 to be an emerging metabolic regulator, as a fine-tuning of body composition and metabolism, with metabolic roles in different tissues, including skeletal muscle, adipose tissue, bone, and ovary. In addition, skeletal muscle has high levels of locally produced IGF-II, which is essential for myoblast differentiation. IGF-II paracrine actions on muscle increases exercise capacity and time to fatigue [118, 119].

While IGF-I receptor mediates the mitogenic and metabolic effects of both IGF-I and IGF-II, the specific IGF-II receptor (IGF-II/M-6-P) has distinct actions, primarily responsible for reducing IGF-II levels during fetal development, and acts antagonistically against IGF-I actions through IGF-I receptor [120, 121].

For athletes, GH responses to stimulations, IGF-1, IGFBP-3, and optionally IGFBP-1 are the most indicated parameters to be employed, particularly for the evaluation of well-being, performance, and balance between training, nutrition, and resting.

However, in opposition to the lack of association between exercise performance and circulating IGF-I, exercise exerts muscle strength by paracrine and autocrine actions of locally produced IGF-1, rather than diffuse circulating IGF-1 [109–111, 122–132], and evaluation of tissue IGF-1 may provide more useful information than serum IGF-1.

13.3.2.4 Hypothalamic-Pituitary-Thyroid (HPT) Axis (Table 13.5)

The parameters of the thyrotropic (HPT) axis more commonly measured with clinical relevance include thyroid-stimulating hormone (TSH), produced by the thyrotropes in the pituitary to stimulate the release of thyroid hormones (TH) by the thyroid; thyroxine (T_4) in its total (tT_4) and free (fT_4) isoforms, which is the prohormone that represents 80–90% of the total hormones released by the thyroid and that requires conversion into thyronine (T_3) for its biological actions; total (tT_3) and free T_3 (fT_3), the most bioactive thyroid hormone, including 10–20% released by and 80–90% converted from T_4 in almost all tissues, by a group of enzymes named *deiodinase* types I and II ("*de-*" that means "remove" and "*-ionidase*" that means "iodine") (DIO1 and DIO2, respectively); reverse T_3 (rT_3), that is the bioinactive product from T_4 converted by the *deiodinase* type III (DIO3), with clinical relevance in the evaluation of severe illnesses that may lead to a state of euthyroid-sick syndrome; T_4 binding blobulin (TBG), the main globulin carrier of thyroid hormones; transthyretin, a secondary TH binding carrier; thyroglobulin (Tg), a marker of thyroid activity mostly used for post-thyroidectomy follow-up in thyroid cancer; anti-thyroglobulin (TgAb) and anti-thyroid peroxidase (TPOAb) that are the two suppressing antibodies; TSH receptor (TSHRAb) and thyroid stimulating (TSI) antibodies that are the two major stimulating antibodies; and T_4:T_3 ratio, an indirect marker of *deiodinase* types 1 and 2 activity.

Other thyroid metabolites are being increasingly employed with a variety of clinical meanings, including the groups of "hot" (that stimulate metabolism) and "cold" (that do not stimulate metabolism) metabolites.

TSH levels vary diurnally by up to approximately 50% throughout the day, are lowest in late afternoon and highest during sleep, and are extremely sensitive to minor changes in serum free form of T_4. Thus, variations in the serum TSH of up to 40% when within the range do not necessarily reflect a change in thyroid status and function [133]. Factors that increase TSH secretion include use of dopamine receptor blockers, ritonavir, increased norepinephrine, adrenal insufficiency, sleep, cold ambient temperature, and emotion-physical stress, which is falsely elevated by the presence of heterophilic or interfering antibodies [134–136], while decrease of TSH secretion can be induced by dopamine and use of glucocorticoids, metformin, somatostatin analogues, opiates, bromocriptine, and oral estrogens and is falsely reduced with the use of biotin as a supplement (interferences in the TSH assay). Short-duration moderate-to-intense (>60% of maximal oxygen uptake – VO_{2max}) aerobic exercises results in elevated blood TSH levels, whereas more prolonged submaximal exercise (>60 minutes) effects on TSH are inconsistent. Usually, TSH progressively increases until 40 minutes of exercise and then stabilizes.

T_4 (3,5,3′,5′-tetraiodo-L-thyronine), although being a prohormone, has critical roles in development, growth, metabolism, and neuron health mediated through its active isoform, T_3. T_3 (3,5,3′-triiodothyronine) is the biologically active isoform of the thyroid hormones, mostly generated in peripheral tissues, and has circadian variations, following the circadian cycle of TSH, and may drive metabolic status, including total cholesterol and SHBG, in a stronger manner than TSH. Only a small minority of T_4 and T_3 circulates unbound (0.03% and 0.3%, respectively), the most biologically form of thyroid hormones.

Despite the recommendation against the use of rT3 in clinical practice, the rT_3 (3,3′,5′-triiodothyronine), an inactive isomer of T_3, shows promising roles to help unveil metabolic and endocrine adaptations to elite sports. In athletes under extreme situations, including ultra-elite endurance sports, rapid weight loss for fights, strict diets together with intense training, and those athletes affected by OTS, measurement of rT_3 may be useful as a signal of dysfunctional metabolism on starvation or a hypometabolism-like mode (316–321). In sports, while immediately after long training session lead to evaluation of both rT_3 and fT_3, rT_3 tends to increase and fT_3 decreases in the 12 hours following a short high-intensity interval training (HIFT) session, which does not occur in steady-state endurance exercise. The prolonged reduction in fT_3 means that HIIT may require longer periods of resting compared to moderate longer steady-state trainings (316–321).

Other TH metabolites include 3,5-diiodothyronine (T_2) [137–155], an active hot metabolite that rapidly increases resting metabolic rate; elicits short-term beneficial hypolipidemic effects; stimulates oxygen consumption in the liver; enhances mitochondrial activity; increases resting metabolic rate (RMR); stimulates mitochondrial activity and uncoupling in skeletal muscle; increases the energy capacity of the heart, skeletal muscle, liver, and brown adipose tissue (BAT); consequently improves survival in the cold; increases lipid β-oxidation with subsequent decrease in fat content in liver and peripheral fat tissue; enhances glucose-induced insulin secretion; and induces thermogenesis in the brown adipose tissue (BAT) via enhanced uncoupling C protein-1 (UCP1) activity.

Conversely, 3-iodothyronamine (3-T_1AM) is a "cool" thyroid metabolite with a single atom of iodine located in the carbon 3, with reported actions in metabolism, brain function, dementia, degenerative brain disease, cancer [137, 138, 156], inhibition of insulin, and stimulation of glucagon in the pancreas.

Besides 3,5-T_2 and 3-T_1AM, other thyroid metabolites have been identified and described as having potential biological actions [137, 138], including iodothyroacetic acids, such as 3,5,3',5'-thyroacetic acid (TA_4), 3,5,3'-thyroacetic acid (TA_3), and 3-thyroacetic acid (TA_1).

In summary, the uncovered pathways of the TH metabolism, including the novel TH metabolites, and their respective confirmed or likely enzymes responsible for the conversion (in parenthesis) are the following:

- T4 → T3 (DIO1 and DIO2)
- T4 → rT3 (DIO3)
- T3 → 3,5-T2 (DIO1? DIO2?)
- rT3 → 3,3-T2 (DIO1 and DIO2)
- 3,5-T2 → 3-T1 (DIO1 and DIO2)
- 3,3-T2 → 3-T1AM (3-iodothyronamine) (DIO1 and DIO2)
- 3-T1→ 3-T1AM (ornithine decarboxylase - ODC)
- 3-T1AM → TA1 (3-T1AC) (amino-oxidase - AO)
- 3-T1AM → T0AM (thyronamine) (DIO3?)
- T3 → TRIAC (amino-oxidase - AO)
- 3,5-T2 → DIAC (amino-oxidase - AO)
- T0AM → T0AC (amino-oxidase - AO)
- TRIAC → DIAC (DIO1?)
- DIAC → T1AC (3-iodothyroacetic acid) (DIO3?)
- T1AC → T0AC (DIO3?)

For research in athletes, the employment of TSH, fT4, and fT3 should be the preferred parameters, while rT3 may provide key information in exhaustive and extreme sports. TgAb and APOAb may help the investigation of impaired athletes.

13.3.2.5 Prolactin

Prolactin is the sole marker of the lactotrophic axis, secreted by the lactotrophs in the anterior pituitary, since there is not any specific hormone produced by the hypothalamus to stimulate its production, while prolactin itself does not stimulate the secretion of any hormone. While the other hormones in the anterior pituitary are only released when stimulated by their respective releasing hormones in the hypothalamus, lactotrophs need to be constitutively inhibited by dopamine to avoid prolactin production. The loss of the continuous inactivation by dopamine leads to uncontrolled prolactin release.

Prolactin is also secreted at lesser extents by the breast, adipose tissue, parts of the central nervous system (CNS), and some components of the immune system. Release of prolactin and its biological actions are linked to emotional and physical stress response, immune system activation, reproductive function, water balance regulation, and development of fetal surfactant [157–159].

Besides the constitutive inhibitory action of dopamine from the hypothalamus, other prolactin inhibitors include GABA, caloric deficiency, and non-classical inflammatory markers (Il-1bet, IL-2, IL-4), while physiological increase of prolactin occurs in response to estrogens, TRH, GnRH, serotonin, oxytocin insulin, catecholamines, opiates, sleep, stress, lactation, and orgasm [157–159].

Stimulatory functional tests of prolactin include TRH, L-dopa, nomifensine, and domperidone stimulation tests, but were demonstrated to provide no further information than single serum prolactin.

During endurance exercise, prolactin levels increase proportionally to the training intensity, at maximal increase during HIIT regimens. Conversely, longer exercise duration results in graduate increase of prolactin, which seems to be strongly driven by the elevation in core temperature that occurs during trainings, as cooling mitigates prolactin increase, although prolactin increase after training can partially persist for the rest of the training day. In contrast to the extensively described effects of endurance training on prolactin, response to resistance or strength exercise is poorly studied, and findings are contradictory [160–163].

13.3.2.6 Renin-Angiotensin-Aldosterone System (RAAS) (Table 13.6)

The RAAS comprises a sequence of (renin)-(angiotensin I)-(angiotensin II)-(aldosterone) sequence of hormones released by the kidneys, lungs, and adrenals, respectively, that regulates the intravascular volume, blood pressure, and serum potassium. The exacerbation of the full RAAS, leading to a hyperreninemic-hyperaldosteronemic state, is the responsible for essential hypertension, which comprises 90% of the cases of hypertension. The main actors of the RAAS include renin measured directly (direct renin), plasma renin activity (PRA) measured by angiotensinogen, aldosterone, and aldosterone-to-renin ratio.

Given the large variability in the water availability and distribution, hydration status, and kidney function, the RAAS metabolism likely plays important roles during exercise and requires further elucidation, even for the understanding of the sport-adapted RAAS physiology. Physiologically, both PRA and aldosterone increase in response to exercise and decrease with recovery and water intake [164], although in acute exercises that achieve 90% of VO2max and ultra-endurance sports lead to exceeding increase of aldosterone release, without proportional increase of PRA, leading to a sort of renin-independent hyperaldosteronism [165]. For athletes, the measurement of acute and chronic aldosterone and renin levels may provide, particularly together with the parameters of the water metabolism, early signs of dehydration, insufficient salt intake, or any sort of imbalance between blood volume and solutes.

13.3.2.7   Catecholamines and Adrenal Medulla (Table 13.7)

Catecholamines are produced by neuroectodermic-originated chromaffin cells located in the adrenal medulla (the inner part of the adrenal glands) and the postganglionic fibers of the sympathetic nervous system (SNS) [166]. The major types of active catecholamines include dopamine, norepinephrine (noradrenaline), and epinephrine (adrenaline), in this exact sequence of biosynthesis pathways of catecholamines, whereas the inactive products from the metabolism of catecholamines are metanephrines, vanillylmandelic, and homovanillic acids. While adrenal medulla secretes adrenaline and at lesser extension noradrenaline and dopamine, postganglionary cells of the SNS lack phenylethanolamine N-methyltransferase (PNMT) that converts noradrenaline into adrenaline and therefore secrete only dopamine and noradrenaline.

The major parameters to be measured include total catecholamines and their fractions (dopamine, noradrenaline, and adrenaline) and total and fractioned metanephrines (metanephrine and normetanephrine), in the plasma or in the 24-hour urine. Plasma metanephrines are the most accurate marker to detect pheochromocytomas or paragangliomas, but has limited relevance for athletes. In athletes, the 12-hour nocturnal urinary catecholamines, as an adaptation of the 24-hour urine, are by far the parameter most measured in athletes [167]. However, catecholamine-to-metanephrine ratio that is an indirect marker of level of inactivation of catecholamines has shown clinical relevance in some sport-related dysfunctions, particular in overtraining syndrome (OTS).

13.3.2.8   Bone Metabolism (Table 13.8)

The bone metabolism comprises the parathyroid glands that produce the parathyroid hormone (PTH), the kidneys that convert vitamin D into its bioactive isoform, and the bones. The most important biochemical parameters for clinical practice and research include PTH, total and ion calcium, 25(OH)-vitamin D (25(OH)D), 1,25(OH)-vitamin D (1,25(OH)D), and the bone mineral density (BMD) and more recently the trabecular bone score (TBS) that accurately estimates the quality of the bone microarchitecture. Markers of bone formation including osteocalcin, bone-specific alkaline-phosphatase (BS-ALP/BSAP), and hydroxyproline (OHP), and those of bone reabsorption, including C-terminal (CTx), N-terminal (NTx), and amino-terminal (P1NP) propeptides of the type 1 collagen, tartrate-resistant acid phosphatase (TRACP), pyridinoline (PYD), and deoxypyridinoline (DPD), can be employed in specific situations to measure the current bone metabolic state, since bone metabolism encompasses dynamic, continuous, and simultaneous anabolic and catabolic pathways.

In healthy populations, intermittent running decreased CTx concentrations acutely when running interval was below 5–10 seconds, while it remained in the same levels in longer intervals [168]. P1NP did not change in any of the training programs. In high-impact resistance exercise, P1NP and osteocalcin were increased, while CTx was unaltered, with positive correlations between increased bone markers and quality of life (QoL) [169].

Military training may result in increase of trabecular and cortical volumes, and cortical thickness and area, and decrease of βCTX, while no changes in P1NP were observed. The osteogenic alterations in bone density and geometry in military training are likely due to its inherent unaccustomed, dynamic, and high-impact loading characteristics [170].

Low energy availability (LEA) achieved through dietary energy restriction resulted in a significant decrease in bone formation but no change in bone resorption, whereas LEA achieved through exercise energy expenditure did not significantly influence bone metabolism [171].

In summary, for athletes, metabolic parameters including PTH, 25(OH)D, 1,25(OH)D, serum and urine calcium, phosphate and magnesium, and bone markers including CTx, urinary NTx, P1NP, and osteocalcin should be employed to evaluate circumstances of deprivation, strong utilization of strength, and ultra-endurance sports. The collective analysis of all these parameters may better answer to the highly complex and dynamic bone metabolism.

13.3.2.9   Water Metabolism (Table 13.9)

Water metabolism comprises the regulation of the blood and urine osmolarity and includes the anti-diuretic hormone (ADH, also termed as vasopressin – AVP), released by the termination of the hypothalamic neurons located in the posterior pituitary, as its major actor, by keeping water within the vessels, leading to decrease of plasma and increase of urine osmolarity. ADH is tightly regulated by blood osmolarity, causing an exponential release of ADH when it reaches 292 mOsm/kg or above (reference range 285–295 mOsm/kg). Plasma and urine osmolarity are the two major parameters that indicate the level of secretion and action of ADH. Conversely, the concurrent measurement of plasma and urine osmolarity and ADH can indicate the level of ADH activity, since high ADH levels is expected to be associated with decreased plasma and increased urine osmolarity, as per its inherent antidiuretic actions. ADH is mainly used in clinical practice to diagnose and follow up cases of diabetes *insipidus* (DI), a disease characterized by impaired ability to retain water within the organism. The etiology of DI can be neurogenic, when the abnormality is located centrally in the neurohypophisis, and is caused by impaired ADH secretion, nephrogenic, when kidney response to ADH is impaired due to decreased sensibility, abnormalities, or downregulation of aquaporins, which are the ADH receptors in the kidneys, or can be of both neurogenic and nephrogenic etiologies [172, 173].

While ADH has a short half-life and reflects the momentaneous state with a short delay compared to plasma and urine osmolarity, copeptin, a metabolite released together with ADH equimolarly, has more stability in the blood, indicates the hydration status from the previous hours, and is more precisely correlated with plasma and urine osmolarity [174, 175]. Besides dysfunctions of ADH release and actions, copeptin is among the earliest signs of acute coronary syndrome and has important prognostic values in heart failure and cardiogenic shock.

Exercise induces a continuous rise of copeptin in healthy male athletes, independently of regulations from sodium levels and fluid intake [176–178]. For research in sports, the use of copeptin can provide more reliable information on the effects of exercise on water metabolism, due to its longer half-life, compared to ADH.

The assessment of copeptin, ADH, plasma, and urine osmolarity in athletes may contribute to the elucidation of the role of water status during sports, by measuring these parameters in response to acute exercises, exhausting sports in response to specific interventions (increased water intake, salt intake, improvement of sleep, eating changes), and how water metabolism independently correlates and predicts clinical and biochemical behaviors.

13.3.2.10   Glucose Metabolism (Table 13.10)

Glucose is the major source of energy for metabolism and is obtained from direct carbohydrate digestion, glycogenolysis (breakdown of the glucose storage in liver and muscles), and gluconeogenesis (glucose generation from amino acids and fatty acids) [179]. Dysfunctions of glucose metabolism include diabetes mellitus, hypoglycemia, and inherent glucose metabolic dysfunctions.

Glucose metabolism goes beyond random glucose levels and includes fasting glucose, glucose response to meals, glucose patterns along the day, average glucose, fasting insulin, post-meal insulin, correlations between glucose and insulin, and insulin sensitivity. In addition, specific glucose responses to eating, including carbohydrates of different types and glycemic indexes, protein, fat and fibers, exercises, other physical efforts, cognitive demands, and sleeping characteristics, may also be assessed and provide additional and helpful information regarding the glucose metabolic patterns.

The major parameters to assess the glucose metabolism include fasting and post-meal glucose and insulin; continuous glucose monitoring (CGM) with new technological devices; assessment of insulin sensitivity by the prediction using the homeostasis model assessment-insulin resistance (HOMA-IR) and the gold standard parameters of the hyperinsulinemic-euglycemic clamp; assessment of the estimated average glucose by glycated hemoglobin (estimated period: 2 to 3 months), fructosamine (estimated period: 2 to 3 weeks), and 1,2-anydroglucitol (1,5-AG) (estimated period: seven to 14 days); and assessment of glucose homeostasis including the oral glucose tolerance test, as the gold standard test for the diagnosis of diabetes mellitus, and mixed meal tolerance test, particularly important for the detection of suspected hypoglycemia.

Among these tests, the CGM for the assessment of the exact glucose behavior in response to specific types, intensity, and other patterns of exercise, and possible correlations with hormonal responses to exercise, and the hyperinsulinemic-euglycemic clamp to determine the exact level of insulin sensitivity are the most promising tests to be employed in athletes.

For research purposes, plasma, capillary, and interstitial glucose simultaneously collected may disclose important differences, due to the delay between the glucose in the plasma within the principal veins and its flow to capillarity and interstitial space. The capillary or interstitial glucose may better reflect the glucose status in the brain than the plasma glucose, once the delay between the plasma and brain glucose tends to be similar to the differences between plasma and capillary glucose, due to the blood-brain barrier [180].

13.3.2.11   Lipid Metabolism (Table 13.11)

Lipid metabolism comprises all cholesterol and lipid particles and includes the following parameters: total cholesterol (TC); low (LDLc), high (HDLc), intermediate (IDLc), and very low (LDLc) density lipoproteins; non-HDLc; triglycerides; apolipoprotein A1 (apoA1) and B (apoB); lipoprotein A (Lp(a)); small dense LDLc (sdLSLc); oxidized LDLc (ox-LDLc); and lipoprotein-associated phospholipase A_2 (Lp-PLA2) [181–183]. The collective analysis of the changes in lipid patterns in athletes of different sports provides a more precise understanding of the dynamics of the lipid metabolism, rather them specific lipid particles alone, including utilization of lipids for a range of pathways, in special the adrenal and gonadal steroidogeneses. Hence, the employment of the majority of the parameters, in particular TC, LDLc, HDLc, triglycerides, apoA1, and apoB, is highly recommended.

Total cholesterol (TC) reflects the sum of all particles that contain cholesterol, including triglycerides, VLDLc, LDLc, IDLc, and HDLc. Although TC has been classically used as a marker of cardiovascular diseases and prognosis, more specific parameters have been shown to be more accurate.

LDLc encompasses a *spectrum* of lipoprotein particles with similar structure that differ in terms of size and density, with apo-B100 as being the prevailing apoprotein. Size and density of LDLc particles influence its atherogenic effect, since smaller and more dense LDLc tend to be more atherogenic, which is the predominant pattern of lipid profile in metabolic diseases. Increase in LDLc occurs inherently, due to increased hepatic production (genetic), or in hypothyroidism, nephrosis, anabolic steroid treatment, cholestatic diseases, and use of protease inhibitors.

Correspondingly to LDLc, HDLc encompasses a *spectrum* of HDLc sizes clustered into HDLc-1, HDLc-2, and HDLc-3 that are, in common, the smallest lipoprotein particles due to its high protein content, with apo-A1 as the prevailing apoprotein. The size of HDLc modulates its ability to remove cholesterol and fat from artery walls and reduce macrophage accumulation in atheroma plaques, although epidemiological studies have not provided conclusive evidence that measurement of HDL size contributes to risk prediction.

VLDLc are particles with high endogenous triglycerides content, as its main function is to transport cholesterol, while chylomicrons transport exogenous, i.e., dietary triglycerides. In the bloodstream, VLDLc is converted into LDLc and IDLc, by the lipoprotein lipase (LPL) enzyme. VLDLc is estimated as being 20% of triglyceride concentrations, although in slightly lower concentrations when triglycerides are higher than 500 mg/dL.

Unlike lipoproteins, triglycerides are esters constituted of glycerol and three fatty acids, and not a type of lipoprotein. Its measurement reflects the fat content in the bloodstream and may indirectly measure chylomicrons or VLDLc, in non-fasting and fasting states, respectively. Its breakdown into free fatty acids occurs via interaction with LPL and serves as an important source of energy. Non-fasting triglycerides are approximately 30 mg/dL higher than fasting samples (455). Hypertriglyceridemia results from increased triglyceride production, reduced triglyceride catabolism, or both. Factors that are known to induce hypertriglyceridemia include overweight and obesity, sedentarism, excessive alcohol intake, sugar craving, metabolic syndrome, insulin resistance, T2DM, rapid weight loss (due to intense release of triglycerides from adipocytes), chronic renal disease, hypothyroidism, use of oral contraceptives, use of thiazide diuretics, and genetic conditions.

Apolipoproteins are fat-binding components of lipoproteins shells. Apolipoprotein A1 (apo-A1) is the major protein component of HDLc, and its major action is the fat clearance, particularly cholesterol, from leukocytes within artery walls, preventing fat overload and transformation into foam cells, eventually preventing the progression of atherosclerosis, although these actions are apo-E dependent. While HDLc measures the sum of the volumes of HDLc particles, apo-A1 assesses the number of HDLc particles. Hence, low HDLc levels with normal apo-A1 suggests the existence of an adequate number of HDLc particles with lower cholesterol content. In this situation, low HDLc is not suggestive of increased cardiovascular risk.

Apolipoprotein B (apo-B) is the prevailing apolipoprotein of atherogenic lipoproteins, including chylomicrons, VLDLc, IDLc, and LDLc particles (456–458). Its major types include apo-B48 and apo-B100, produced by the small intestine and liver, respectively [184]. Apo-B48 is the major apoprotein from chylomicrons and represents less than 1% of all apo-B, while apo-B100 is the prevailing apoprotein of the other lipoproteins. Since there is one apo-B100 molecule per hepatic-derived lipoprotein and apo-B100 represents 99% of all apo-B, apo-B100 concentrations reflect the number of atherogenic particles in plasma (455–460) and can be superior to LDLc or even non-HDLc to predict cardiovascular risks and better reflect metabolic dysfunctions. Coexisting high apo-B levels (>mg/dL) and LDLc lower than 160 mg/dL reflects small dense LDLc, the main etiology of premature CVDs, since disproportional low LDLc means that particles have very low volume.

Lipoprotein(a) [Lp(a)] consists of a particle that resembles LDLc, with an apo-A covalently bound to apo-B contained in the outer shell of the particle. Lp(a) has large pro-atherogenic and pro-thrombogenic effects, including reduction of fibrinolysis capacity, stimulation of plasminogen activator inhibitor-1 (PAI-1) secretion, and carry of atherogenic-causing cholesterol bound to atherogenic pro-inflammatory oxidized phospholipids and is independently correlated with CVDs, atherosclerosis, thrombosis, and stroke. Lp(a) levels are less modifiable than other lipoproteins, and is only marginally affected by diet, exercise, and other environmental factors, as well as life-style changes [185–187].

Other major particles include oxidized LDLc (ox-LDL) that reflects oxidative stress within protein and lipid content in LDLc, with dual but prevailing pro-atherogenic effects [188, 189]; apolipoprotein E (Apo-E), an essential apoprotein for the metabolism of triglyceride-containing lipoproteins, produced in the liver and macrophages, and the major cholesterol carrier in the brain [190, 191]; and lipoprotein-associated phospholipase A_2 (Lp-PLA_2), an enzyme produced by inflammatory cells, with paradoxical anti-atherogenic and pro-atherogenic effects when linked to HDLc and LDLc, respectively, 80% linked to the LDLc, and with synergistic actions with C-reactive protein (CRP) to enhance CVD risk.

Other tests of the lipid metabolism include apo-B mutations, apo-E Genotype, lipoprotein fractionation by ion mobility, subclasses of HDLc, nonesterified fatty acids, omega-3 fatty acid, omega-6 fatty acid, phospholipids, very long chain fatty acids, and anti-LDLc antibodies.

13.3.2.12 Inflammatory and Immunologic Markers

Markers of classical and non-classical inflammatory pathways provide complementary information regarding the overall metabolism, once inflammatory paths are strong modulators of hormonal, immunologic, muscular, oxidative, glucose, lipid, and amino acid metabolism, as well as on the various systems. In addition, the immunologic system has a large complexity, since it comprises multiple types of responses that work inter-dependently and influences the majority of the metabolic pathways and systems, in different but equally important forms than inflammatory system [192–194].

Hence, for a more comprehensive analysis, or whenever specific characteristics is aimed to be fully elucidated, inflammatory, immunologic, and muscular markers should be assessed alongside with hormones, aiming to detect inter-correlations, associations, and predictions between hormones and non-hormonal biochemical parameters.

The two major classical inflammatory parameters include C-reactive protein (CRP) and erythrocyte sedimentation rate (ESR), while non-classical inflammatory parameters include interleukin-1 beta (IL-1beta), IL-6, tumor necrosis factor α (TNF-α), hepcidin, transforming growth factor-β (TGF- β), and others.

The major immunologic markers include neutrophils, lymphocytes, eosinophils, neutrophil/lymphocyte ratio, eosinophil/lymphocyte ratio, natural killer (NK) cells (levels and activity), cluster of differentiation-3 (CD3), CD4 and CD8 (T-lymphocytes), CD20 (B-lymphocytes), total immunoglobulin (IgG), IgG subclasses, and IgA.

Non-inflammatory non-immunologic markers that are directly related to inflammation and immunologic responses comprise muscular markers, including creatine kinase (CK), CK-MB (Muscle-Brain) fraction, lactate dehydrogenase (LDH), lactate, myoglobin, and aspartate transaminase (AST), and minerals and vitamins,

including magnesium, selenium, zinc, folate, ferritin, serum iron, ferritin, cobala-
min (vitamin B12), homocysteine, red blood cell folate (folate-RBC), methylmalo-
nic acid, thiamine (vitamin B1), riboflavin (vitamin B2), niacin (vitamin B3),
pantothenic acid (vitamin B5), pyridoxine (vitamin B6), total and isoforms of vitamin A, vitamin C, total and isoforms of vitamin E, cupper, ceruloplasmin,
and others.

Among these parameters, a broad and detailed profile using classical and non-
classical inflammatory, immunologic, and muscular parameters will certainly unveil
important information regarding the sophisticated web of inter- and external combined influences that these markers play. Inflammatory and immunologic panels
should not be limited to the basic parameters, since multiple studies have already
demonstrated that the most commonly employed markers do not yield convincing
understanding of these systems.

13.3.2.13 Adipose, Muscle, and Liver Tissue Metabolism and Endocrinology (Table 13.12)

Fat and muscle tissues have been increasingly recognized as real endocrine organs,
that together produce more than 100 hormones. The multiple cross talks ("conversa-
tions" through cell-to-cell communication using hormones and other signals)
between them and between each of these and other glands, brain, gut microbiota,
and likely all other tissues and organs highlight the complexity of the biological
actions generated by adipokines, myokines, and hepatokines. Actions promoted by
these cytokines may seem apparently contradictory, since anti- and pro-inflammatory responses may concurrently occur in response to a same cytokine. However,
cytokine-mediated responses are tissue-specific, resulted from a combination of
cytokines, and dependent on the overall tissue environment.

Adipose tissue encompasses two major types of fat, the white (WAT) and the
brown (BAT) adipose tissues. The WAT comprises the subcutaneous and visceral
(omental and mesenteric) subtypes, with almost distinct functions between them.
WAT produces a range of hormones, growth factors, cytokines, complement factors,
and matrix proteins, including TNF-α, IL-1ß, IL-6, visfatin, and resistin as pro-
inflammatory mediators, adiponectin as anti-inflammatory marker, leptin and
angiotensin for control of nutritional intake, and plasminogen activator inhibitor 1
(PAI-1), vascular endothelial growth factor (VEGF), and adipsin as pathways
[195–197].

Other biologically active adipokines include apelin; acylation-stimulating protein (ASP); calprotectin; cardiotrophin-1; chimerin; chemokine ligands 2, 3, 5, 8,
and 11; clusterin; fetuin A; ghrelin; hepcidin; lipocalin-2; monocyte chemoattrac-
tant protein 1 (MCP-1); macrophage migration inhibitory factor (MIF); matrix
metalloproteinase 2 and 9 (MMP-2 and MMP-9); PAI-1; pigment epithelium-
derived factor (PEDF); progranulin; six-transmembrane protein of prostate-2
(STAMP-2); transforming growth factor-β (TGF- β); tissue inhibitor of matrix
metalloproteinase-1   (TIMP-1);   vaspin;   Wnt1-inducible   signaling   pathway

protein-1; and, more recently, the annexins [198]. Collectively, these adipokines act diffusely in the organism, deteriorating tissues via inflammation and degeneration.

Leptin is a hormone predominantly produced by adipocytes, and its major biological action is the regulation of fat storage by anorexigenic effects. Secondary actions include positive cognitive functions and brain health, improvement of male and female fertility, increase of bone mass, positive energy homeostasis, and metabolic benefits, including suppression of glucagon and corticosterone production, increased glucose uptake, and inhibition of hepatic glucose output. However, chronic exposure to leptin leads to reduced hypothalamic sensitivity to leptin, with consequent loss of its anorexigenic effects, and leptin may play harmfully in a wide range of tissues, enhancing the metabolic dysfunctions related to obesity, increasing the pro-inflammatory adipokines, and as a growth factor for some types of cancer [199]. In response to regular exercising (more than two weeks), leptin reduces independently of body fat and has exacerbation decrease when body fat reduces. Disturbances of leptin may be particularly important in sports-related eating disorders and chronic deprivations.

Adiponectin is also a major hormone produced by the WAT in an inverse relation with adipocyte size and is enhanced by prolonged fasting states, ketogenic diets, and weight loss. Its major biological actions include regulation of glucose and lipid metabolism, including decrease of glucogenogenesis, increase of glucose uptake, sensitization to insulin, lipid catabolism, triglyceride clearance, and additional effects on weight loss, reduction of TNF-alpha, and upregulation of uncoupling proteins (UCPs) [200]. In addition, adiponectin levels also predict the ability to induce cold-induced BAT thermogenesis [201]. Overall exercises increase muscle-induced adiponectin, while more intense exercise also induces an exacerbation of the increase of HMW-adiponectin on the top of adiponectin overall increase, indirectly providing additional metabolic benefits through the exacerbation of adiponectin and HMW-adiponectin [202–204]. In addition, high-intensity interval training (HIIT) programs induce the adiponectin gene expressions, inducing further insulin sensitization [205].

Several substances have been demonstrated to act directly in the BAT, including beta3-agonists, indomethacin, and erythropoietin (EPO) [206, 207]. The major batokines, the bioactive cytokines produced by the BAT, are the insulin-like growth factor-1 (IGF-1), interleukin-6 (IL-6), and fibroblast growth factor-21 (FGF-21). Although IL-6 is classically a pro-inflammatory cytokine, for not fully clear reasons, its actions when released by the muscles and BAT have protective roles. Similarly, FGF-21 can act paradoxically, according to the site of expression, as further detailed. Other batokines that act auto- or paracrinally include local prostaglandins, T3, angiotensinogen, IL-1alpha, IL-6, VEGF, IGF-1, fibroblast growth factor-2 (FGF-2), FGF-21, retinol-binding protein-4, and lipocalin prostaglandin D synthase.

Fibroblast growth factor-21 (FGF-21) is likely the endocrine factor that mediates the most metabolic benefits associated with BAT. FGF-21 is expressed in several metabolically active organs and interacts with different tissues, with opposite effects depending on the source of its production. At the same time that FGF-21 have important roles for the metabolic benefits, including increased glucose uptake, fatty acid oxidation, improvement of insulin secretion, and cardioprotection, it has paradoxical harmful effects, causing muscle atrophy and reduction of bone mineral density. FGF-21 is likely the factor unanimously produced in the major metabolic-regulator sites, including adipose tissue (as an adipokine), muscle (as a myokine), and liver (as a hepatokine) [208, 209].

The most important myokines produced by myocytes are myostatin, irisin, IL-6, IL-15, brain-derived neurotrophic factor (BDNF), FGF-21 as a myokine, follistatin, brain-derived neurotrophic factor (BDNF), myonectin (CTRP15), decorin, osteonectin (SPARC), follistatin-like 1 (FSTL 1), apelin, and musclin [210–212].

Irisin is acutely released in response to skeletal muscle activity, and its major activities include promotion of browning process of WAT, inhibition of adipogenesis, and BAT-induced thermogenesis [210–215]. Irisin also induces muscle hypertrophy by increasing muscle protein synthesis and activating muscle satellite cells, particularly in the elderly exposed to training [216], and may show anti-cancer properties (512). Irisin increases equally to resistance trainings and HIITs, but only in approximately 50% of the subjects, which may explain the debate on its response to exercise [210–218]. Chronically, it may reduce under intense training regimens for explosive sports (ball games and fighting) [218, 219]. Also, irisin seems to be reduced under metabolic states, including T2DM and obesity [220]. For research purposes, a proposed marker would be the irisin/muscle mass ratio, i.e., the amount of insulin produced per kg of muscle mass, in a specific period of time [210–212, 220].

Myostatin, also termed as growth differentiation factor 8, and part of the TGF-beta superfamily, is a negative regulator of muscle growth by leading to inhibition of myogenesis [210–212] and has been proposed to be a reliable marker of sarcopenia and frailty [221, 222]. Its expression is inhibited by aerobic exercises and strength training for 24 hours, as well as in resistance training and HIIT regimens [223] while is increased after explosive (stop-and-go) exercises [219]. Exacerbation of myostatin is positively correlated with IL-6 and may lead to muscle atrophy, although myostatin alone does not cause decreased muscle.

Follistatin is considered a hepatokine and a myokine, since it is released by both tissues. It is an activin-binding protein expressed widely and which its primary function is to neutralize the members of the TGF-beta super family. The follistatin/myostatin (F:M) ratio may help explain changes in muscular size in response to resistance training, since follistatin are among the most important myogenic and anti-myogenic myokines, respectively.

Interleukin-15 (IL-15) is a cytokine structurally similar to IL-2, which main biological actions include the increase of fat oxidation, glucose tolerance, glucose uptake in the muscles, myofibrillar protein synthesis, and endurance capacity, although in healthy subjects IL-15 administration may paradoxically cause muscle atrophy [210–212, 224, 225]. IL-15 seems to increase more intensely in exhausting exercises, particularly when it reaches a peak higher than 90% of Vo2max.

Overall, owing to the highly complex correlations between different classical and non-classical inflammatory, immunological, and adipokines, myokines, and hepatokines, it is recommended that these parameters should not be assessed apart from each other, as comprehensive inflammatory, immunologic, and cytokine panels that encompass the major actors of each system should be employed instead.

13.3.2.14   Body Composition and Metabolism

Different parameters of body composition and metabolism that encompass more than just body fat, lean mass, and basal metabolic rate (BMR) can be obtained through different methods. Parameters that can be clinically relevant for athletes include body fat (%), visceral fat (%;cm²), muscle mass, lean mass, body water (% of total body weight), extracellular water (% of total body water), waist circumference, chest/waist circumference ratio (in males), regional body fat (%; % of expected), truncal body fat, overall visceral fat, omental visceral fat, mesenteric visceral fat, and intermuscular fat (IMAT) that can be assessed through the following methods: dual-energy X-ray absorptiometry (DEXA), air displacement plethysmography, hydrostatic scale, electrical bioimpedance (angle phase), magnetic resonance imaging (MRI) (for visceral fat compartments and IMAT), 3D scanning, skin-fold anthropometry, and skin ultrasound (for subcutaneous thickness) [226].

The major parameters of body metabolism include resting metabolic rate (RMR) (kcal/day), basal metabolic rate (BMR) (kcal/day), fat oxidation (kcal/day and % of RMR/BMR), and measured/predicted BMR ratio (% of expected BMR). Body metabolism can be assessed through indirect calorimetry or double-labeled water, or can be predicted from body composition analysis.

The potential use for physical activity of body composition and metabolism analyses include acute and chronic overall and fat metabolic behaviors in response to exercise.

13.4   Remarkable Aspects of the Research in Athletes

13.4.1   The Importance of Understanding the Healthy Athlete

The progressively increasing number of physically active subjects, among which many have become athletes, a population with many peculiarities resulted from the multiple adaptations that athletes undergo during the conditioning programs. Since athletes disclose multiple physiological particularities, the detection of pathological states from what is physiologically expects needs the elucidation and standardization of the specific reference ranges for all parameters. While reference ranges for athletes are not established, it is challenging to diagnose dysfunctions, sports-related or not, in this growing population, leading to misdiagnoses and lack of other diagnoses.

Two examples are testosterone for males and creatine kinase (CK) for all athletes. It is well characterized that total testosterone increases with regular training. Consequently, the reference range applied for general population (264–850 ng/dL) can be inappropriate for athletes, while an adapted range of approximately 400–900 ng/dL has been proposed from the findings on healthy athletes. Hence, a highly trained male athlete with 25 years of age with normal body weight, and without comorbidities, for instance, that presents total testosterone levels of 310 ng/dL, although considered as having 'normal' testosterone levels, could be attributed to can be actually considered as presenting hypogonadism, particularly in the presence of clinical manifestations that cannot be explained by other disorders. With an adapted range standardized, this same athlete could be reassessed and be appropriately diagnosed with hypogonadism and managed accordingly.

CK levels increase in response to changing of any of the training patterns, and its increase occurs exponentially with the level of the change. Athletes that undergo regular and frequent changes in their training regimens to optimize results may present high CK levels continuously, without necessarily meaning any muscle- or heart-related dysfunction. This is constantly confounded by health practitioners, in particular by those not familiarized with athletes, since muscle soreness is related with higher CK levels and is a reason for visit to a physician. With exacerbated increase of CK, athletes and physically active subjects can be over-investigated and over-treated. Therefore, the fact that CK levels may disclose substantially increased levels in those who exercise regularly should be highlighted within the results report, and all those who interpret biochemical exams should be educated for this peculiarity. However, whether there is an upper limit for athletes, even shortly after training, above which harmful effects can occur regardless, should be elucidated.

It is therefore recommended that whole physiology of the healthy athlete should be understood, including the metabolic and hormonal adaptive physiology, which goes beyond the acute responses to exercise. Researches should learn the adaptations that occur in the basic hormonal and metabolic parameters, through the performance of extensive and reproduced studies, using standardized methods, to consolidate the learning of the behaviors of these parameters. From the solid understanding of the basic parameters, further markers should be assessed, and correlations between the well-established parameters with the novel markers should be performed, in order to understand the context of the findings of the new parameters, regardless of which system or metabolism is being assessed. Moreover, changes to be steadily learned should be specified according to each major characteristic of the training patterns, since many of the observed changes can be sports-specific, intensity-mediated, and based on external characteristics.

13.4.2   Key Characteristics and Variables to Be Considered for a High-Quality Research in Sports Medicine

Compared to other research fields, sports science has additional confounding factors, since physical activity is not a single variable, but a cluster with several variables that interact between them, and which the external variables may have different effects on athletes, according to the interactions between these external effects and training patterns. Hence, besides the choice of the appropriately parameters to be evaluated and tests to be employed, researches should consider the employment of additional parameters that may influence the results and which are not considered in the general population, since the effects of these additional parameters may be dependent on the practice of exercises. In addition, the assessments of external variables that influence the outcomes are highly recommended.

Not only for athletes, but in all researches, exercising and physical activity should be considered as a major variable and should go beyond the simples answer whether yes or no, since responses, benefits, changes, and adaptations vary according to the type of physical effort, level of intensity, volume of training, regularity, and the characteristics of those who exercise.

Which exercise regimens and how optimized they should be for prevention of diseases are yet to be elucidated and are still based on poor data, mostly epidemiological and observational studies. With the acquisition of more specific data, recommendations of physical activity will be prescribed individually, based on the diseases aimed to be prevented or treated, objectives, and responses according to the genomics and metabolomics.

Since the specification of the assessment all training and external variables are key for results high-quality and consistent conclusions, baseline data regarding training patterns should be present in every study in the field of sports science. Figure 13.2 details the variables that should be considered in the research on athletes.

In particular, the assessment of the type of sport is of extreme importance, once responses are likely specific to each type of effort. Hence, besides the description of the sports practiced by the studied population, the categorization into type of sports can provide further clustering of results for comparative analysis [227].

Although the division into types of sports aims to understand the effects of each type of effort, sports are rarely a sole one type of export. One example is that currently endurance training regimens should be complemented by some resistance activities, which bring additional benefits to the endurance sports and prevent injuries and other harms. The four major types of sports include:

1. Endurance: triathlon, long distance running, cycling
2. Short intense explosions: short distance running (100–400m), swimming
3. Continuous explosive sports ("stop-and-go"): basketball, soccer, American football, rugby, handball, volleyball, high-intensity interval training (HIIT)
4. Strength: bodybuilding, weight lifting
5. Mixed sports: high-intensity functional training (HIFT – including "CrossFit"), army training

Additionally, mixed ones could be subdivided into mostly resistance, equally distributed, and mostly endurance, as illustrated in Fig. 13.5.

In summary, researches should know the variables and control for them. Although time-consuming, a full assessment of athletes helps to better understand the context of the findings and strengthen the conclusions, since the risk of biased findings becomes less likely. The summary of the proposal from the present guidelines for researches on endocrinology of physical activity and sport is presented in Fig. 13.6.

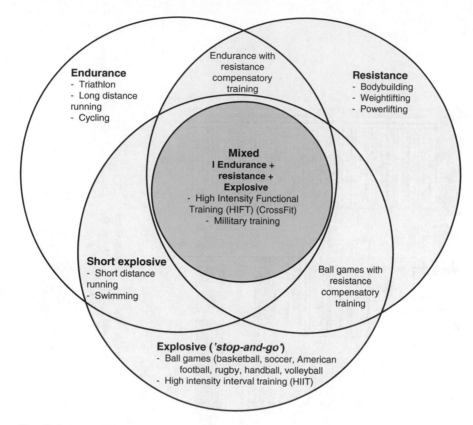

Fig. 13.5 Proposed division of sports into categories, for research purposes

13.4.3   CLIA/RIA or MS: Which to Use in Research?

Despite the clearly superior accuracy of LC-MS/MS compared to CLIA/RIA, some aspects should be considered before choosing the most appropriate method:

1. LC-MS/MS requires a highly trained technical team, has low turnaround time (compared to RIA/CLIA), and is more expensive. Not all centers, particularly in sports research centers, may have available LC-MS/MS.
2. Although there may be significant differences in absolute measurements between RIA/CLIA and LC-MS/MS, there is a high correlation between these methods for the levels of steroid hormone. Indeed, despite lack of absolute accuracy, particularly when in lower concentrations, RIA/CLIA has sufficient relative precision when the objective is to compare between groups, or before vs post-intervention within the same participants, without major loss of quality.

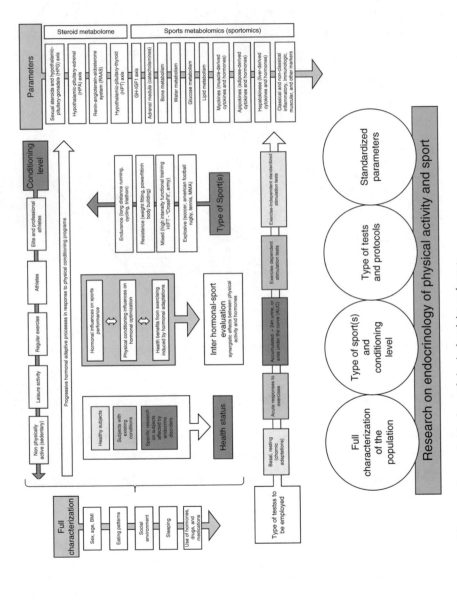

Fig. 13.6 Summary of the research on endocrinology of physical activity and sports

3. RIA/CLIA and LC-MS/MS yield similar results in epidemiologic analyses, which allows RIA/CLIA to be used in large population-based studies.
4. Incorporation of LC-MS/MS should be strongly encouraged to measure a more complete profile of androgen and estrogen metabolites, due to its high reproducibility and accuracy for these steroids, but cannot be mandatory due to the limitations in clinical practice at large scale.

13.4.4   The Importance of a Matched Sedentary Control Group

In research of adaptations to sports exercise, longitudinal protocols with specific interventions on the training regimens or the evaluation of external interventions on training or non-training outcomes have great importance in the pre- vs post-intervention comparisons within the same participants. Conversely, for a more comprehensive analysis of the findings, the inclusion of non-physically active participants as an additional control group may be highly relevant, depending on the objectives of the research. Some of the types of studies could be benefited from the inclusion of sedentary controls including (1) cross-sectional analysis of the effects of physical activity on hormones and metabolism, by the comparison between sex-, age-, and BMI-matched healthy athletes and healthy sedentary, and (2) biochemical and metabolic responses to an exercise test between matched conditioned and sedentary participants.

13.4.5   Correlations Between Exercise-Dependent and Exercise-Independent Tests

Traditionally, the evaluation of hormonal responses in athletes was performed through exercise-dependent stimulations, like maximum strength and the two-bout protocol exercise tests. However, hormonal responses to these tests are dependent on the performance of the athlete, which depends on the current conditioning level and health state. Moreover, in exercise-dependent tests, hormonal responses are secondary to external musculoskeletal and cardiovascular signaling, and differences in hormonal responses may be resulted from differences in the levels of signaling from these or other external systems, and not necessarily due to differences in glands responsiveness.

Indeed, responses to exercise-dependent tests result from an integrative model of assessment, in which the differences in outcomes are more important than the identification of the exact underlying etiology of these differences.

Distinctly from exercise-stimulatory tests, exercise-independent ones have specific targets to be assessed. The employment of an exercise-independent test is based on the specific parameters that are aimed to be determined. In opposition to the exercise-dependent tests, the focus is based on the detection of the responses of a specific parameter or cluster of parameters, in which differences in performance and conditioning levels are mitigated and athletes of different levels and sedentary are evened.

If one aims to evaluate the roles of the hormones in sport, it is crucial that hormones should be assessed without confounding causality biases, including the influence of the level of physical capacity on hormonal responsiveness. In this context, direct stimulation of hormonal axes may be superior to tests that require sequential responses from different systems, and therefore in this particular case exercise-independent tests are preferred.

Conversely, when the level of hormonal responsiveness in real-life stimulation is the main objective, irrespective of the exact biological location from which differences occur, exercise-based stimulation tests are better representatives.

Ideally, correlations between the hormonal and metabolic responses to exercise and non-exercise stimulation tests may provide additional information regarding whether any stimulation, or only physical stimulation, leads to the differences identified. The analysis of the differences between the results also suggests the biological sites in which differences are generated.

13.4.6   Unvalidated Hormonal Tests Commonly Employed in Researches

One of the aims of the present guideline is to standardize and uniformize the research of hormones and metabolism in athletes and in response to physical activity, i.e., in the "sports endocrinology" field, because it has been identified that parameters assessed in this research field are highly heterogeneous, which precludes from further comparative and integrative analyses, and not always fully validated.

Indeed, unsubstantiated tests are still commonly employed in athletes, particularly regarding hormonal assessments. The lack of standardization and reliability of the tests employed may lead to misleading interpretations and conclusions. The rationale for the choice of tests to be employed should be based on which parameters and tests are validated as assessment methods for the specific objectives and whether these parameters will adequately answer the questions of the research. Some examples of unvalidated tests that may have confounding interpretations include:

1. Testosterone/cortisol (T:C) ratio: although widely used, as T:C is typically seen as an anabolic/catabolic ratio, which a reduced T:C ratio is correlated with poorer outcomes, the T:C ratio has some major flaws. First, testosterone and cortisol are hormones from completely different axes, independently regulated, with few factors that simultaneously influence testosterone and cortisol expression. Second, there is a premise that higher cortisol levels are necessarily harmful, based on the correct assumption that chronically cortisol leads to muscle atrophy that is not entirely true, because chronic exposure to chronic excessive cortisol also leads to visceral fat accumulation, which is an undesired but an anabolic effect of corti-

sol, and because acute cortisol responses are desired, since suboptimized cortisol responses to exercise tend to be correlated with worse performance. In this case, for steady testosterone levels, an elevation of the T:C ratio is paradoxically undesirable, in opposition to what is preconized by the common sense. Third, there is a lack of sustained scientific validation for this ratio. For the standardization of the T:C cortisol, increase of testosterone and decrease of cortisol should be independently and consistently correlated with anabolism in a linear matter, while decrease of testosterone and increase of cortisol should yield the exact opposite correlations. The linear correlations should be consistent along different studies to reinforce and validate the method. Instead, the use of testosterone/estradiol (T:E) ratio, since they are regulated by a single enzyme (aromatase) and reaction and are intrinsically regulated, may disclose more plausibility.

2. Post-DST ACTH stimulation test: the combination of cortisol suppression by dexamethasone with further stimulation with synthetic ACTH is based on the idea that the suppression provided by cortisol provides a strict "control" for "unwanted and confounding ACTH response." However, this test lacks any validation.

Noteworthy, not always the parameters and tests themselves are unsubstantiated, but the interpretation and conclusions from the results are. For example, basal cortisol is commonly but inappropriately interpreted as an identifier of stress levels, which is not based in any scientific literature; on contrary, cortisol levels are influenced on a wide variety of factors, since pre-analytical factors (painful puncture), until punctual factors, including sleeping quality in the night before the collect, and presence of current infections or inflammations, even when subclinical.

The knowledge on the factors that may influence the results for each parameter and test is highly desired for the detection of confounding biases in a participant that may interfere in the statistical analysis, particularly because the usual sample of subjects in studies on the sports science field is not large and biased results from one single participant may lead to changes in the final conclusions, even when the sample size tests are employed.

The standardization of specific tests for athletes, from consistent correlations between these and validated tests, may strengthen the findings from previous researches that employed these tests and provide reliable alternatives that suit more adequately for athletes.

13.4.7   The Underappreciated Importance of Water Metabolism

The understanding of the water metabolism and the roles of its major direct players, including vasopressin (ADH), serum sodium, and urine and plasma osmolarity, as well as factors that indirectly regulate the water content, including the RAAS (aldosterone, renin, angiotensinogen I and II), is critical, but highly challenging, due to the large number of factors that directly or indirectly influence the water balance, in manners that are dependent or independent of the direct regulators of the water metabolism.

In result, the assessment methods of the hydration status are still poor and largely influenced by physiological adaptations in athletes. Furthermore, mild dehydration states are challenging to be identified, as compensatory mechanisms tend to be highly efficient and effective, which precludes from decompensation of the intracellular and intravascular water content and consequently from identifiable markers within the plasma or urine.

The joint analysis of the main parameters related to the water balance (ADH, copeptin, plasma and urine osmolarity, aldosterone, renin, sodium, potassium, plasma glucose) with the assessment of overall and extracellular water content using angle phase in electrical bioimpedance may provide precise information on water status, although even when analyzed together, biochemical parameters and body composition-derived markers may not always be capable to identify milder water deprivation.

Since ADH and copeptin changes occur in an earlier stage of water metabolic dysfunctions and that copeptin has larger half-life and better stability, the measurement of copeptin in athletes, particularly under any sort of water restriction or extreme circumstance, may add important information to understand the underlying dysfunctional mechanisms that may occur in athletes.

13.5 Trends in the Research on Endocrinology of Physical Activity and Sport

Improvements on the hormonal and metabolic parameters assays and learnings from previous studies allowed the proposal of novel methodologies for the analysis of the results, which may result in novel insights and understanding of the hormonal behaviors and patterns of the metabolic responses.

A sequential analysis of a parameter to determine its behavior during a specific period allows the estimation of the accumulated release of this parameter, through the calculation of the area under the curve (AUC). In many circumstances, particularly when a "dose-exposure" effect plays a more important role than isolated concentrations, the estimated AUC may provide more accurate data for the understanding the mechanism investigated. Besides the calculated AUC, the analysis of the rhythm of the secretion of the analyzed parameter, by the calculations of percentage of the loop, decrease, increase, and other changes, between different collected times, may also yield useful information, since several effects of hormones are not due to their increase or decrease, but resulted from the speed that these hormones change instead.

A more precise way to measure the hormonal behavior and response throughout the process investigated is to perform a 24h-urine analysis. Unlike the measured AUC, the 24-hour sample provides precise results of the accumulated secretion during the period. In addition, in the context of the emerging employment of LC-MS/MS and other mass spectrometer assays, in which not only the active hormones but also its pre- and post-receptor metabolites are expressed in the urine, a more complete profiling of the hormones analyzed, including the metabolization, activation, and inactivation speed, and clearance, has become feasible with the use of these assays.

Several studies have demonstrated that ratios between different hormones, rather than hormones alone, are more accurate predictors of biochemical or clinical behaviors and outcomes. With the advent of the mass spectrometer-based assays, a massive number of new steroids have been identified, among which some of the ratios between these steroids, including active hormones and hormonal metabolite, have been identified as being the most accurate markers for the diagnosis of certain hormonal-related conditions, clinical outcomes, and prognostic purposes. In additional, non-hormonal ratios, including the neutrophil/lymphocyte and platelet/lymphocyte ratios, have been proposed as being markers of overall prognosis in the presence of chronic and severe diseases.

Hormones that are regulated by a single reaction and enzyme, including the testosterone/estradiol and catecholamine/metanephrine (norepinephrine/normetanephrine and epinephrine/metanephrine) ratios, may indirectly reflect the regulation of anabolic and adrenergic effects on the organism, respectively.

Hence, the inclusion of ratios, in addition to the hormones alone, may bring insightful information to elucidate mechanisms to explain the hypothesis investigated.

Owing to the emerging importance of sports for overall health, and the fact that the population of physically active subjects has become highly relevant, newly standardized tests and parameters should be promptly applied for research on physical activity, alongside with the research on general populations.

13.6 Muscle as an Endocrine Organ: Implications for Research

Muscle tissue has been increasingly recognized as an active endocrine organ that directly and indirectly regulates most of the hormonal axes and metabolic profile. In addition, the characteristics of the satellite cells, myotubes, and progenitor muscular cells are highly dependent on the conditioning and metabolic patterns, which demonstrated the intimate correlations between muscles and metabolism.

With the production of myokines with essential roles in metabolism, including irisin, myostatin, IL-6, IL-15, and BDNF, the understanding of the muscle tissue as not only being an endocrine organ but also an active protagonist in the promotion of health benefits of physical activity is imperative.

Additionally, the intermuscular fat tissue (IMAT) has been demonstrated to be a highly harmful type of adipose tissue, even when compared to visceral fat, and can impair muscle hyperplasia and hypertrophy and decrease muscle strength, power, and training capacity. Moreover, IMAT has an important role in the pathogenesis of sarcopenia, since this tissue plays actively to replace the lieu of the muscle tissue. Hence, the study of the muscles should go beyond the muscle tissue and should encompass the surrounding tissues and the multiple cross talks with distant organs.

Given the newly recognition of muscle as an active tissue that goes beyond orthopedic functions that the endocrine muscle has multiple tissue-to-tissue communications, and the fact that muscle is a major player in the regulation of the metabolism, sports-related researches that evaluate metabolic and hormonal aspects should preferably include analysis of myokines or other muscle-related parameters and correlate with biochemical profile.

13.7   Contribution to the Field of Endocrinology of Physical Activity and Sport

The substantial and exponential growing number of athletes and physically active subjects demonstrates that active population has become highly relevant in the medical field. Together with the increasing number of discoveries of novel health benefits related to sports and the extensively described physiology of the cardiovascular and musculoskeletal systems, the understanding of the hormonal physiology and metabolism of physical activity has become critical for further elucidation of underlying mechanisms identified from the findings of future researches. The present practical guidelines for the standardization of the research of endocrinology of physical activity and sports will be highly useful for the uniformization of the parameters and tests employed and validation of the findings and conclusions of upcoming researches, which will collectively help build a structured understanding of the adapted physiology of the endocrinology system in athletes and physically active subjects. These guidelines will also help researchers improve the level of hormonal assessment, which will likely lead to novel findings, hypothesis, and understanding of the hormonal and metabolic roles in the athlete.

13.8   Conclusion

The number of athletes and physically active subjects as well as the discovery of novel health benefits related to sports is continuously growing, and the physiology of the cardiovascular and musculoskeletal systems in athletes is extensively described. Conversely, the understanding of the hormonal physiology and metabolism of physical activity, which is critical for further elucidation of underlying mechanisms identified from the findings of future researches, still needs further clarification.

The proposal of guidelines for the standardization of the research of endocrinology of physical activity and sports can be useful for the uniformization of the parameters and tests to be employed and validation of the findings and conclusions of upcoming researches. Collectively, the standardization of the research on sports endocrinology may help build a structured understanding of the adapted physiology of the endocrinology system in athletes and physically active subjects.

Finally, the present guidelines should help researchers to improve the level of hormonal assessment, which will likely lead to novel findings, hypothesis, and understanding of the hormonal and metabolic roles in the athlete.

References

1. Bellis M. "A Brief History of Sports." ThoughtCo, Oct. 16, 2019, thoughtco.com/history-of-sports-1992447 (last access: 20th April 2020).
2. https://en.wikipedia.org/wiki/History_of_sport (last access: 20th April 2020).
3. Crowther NB. Sport in Ancient Times. Praeger series on the ancient world, ISSN 1932-1406. Westport, Connecticut: Greenwood Publishing Group. p. xxii. ISBN 9780275987398. Retrieved 30 May 2018. People in the ancient world rarely practiced sports for their own sake, especially in the earliest times, for physical pursuits had strong links with ritual, warfare, entertainment, or other external features. 2007.
4. Blanchard K. The anthropology of sport: an introduction. ABC-CLIO. 1995. p. 99. ISBN: 978-0-89789-330-5.
5. Karjalainen JJ, Kivinicmi AM, Hautala AJ, Piira OP, Lcpojärvi ES, Perkiömäki JS, Junttila MJ, Huikuri HV, Tulppo MP. Effects of physical activity and exercise training on cardiovascular risk in coronary artery disease patients with and without type 2 diabetes. Diabetes Care. 2015;38(4):706–15.
6. Piepoli MF, Villani GQ. Lifestyle modification in secondary prevention. Eur J Prev Cardiol. 2017;24(3_suppl):101–7.
7. McKenzie F, McKenzie F, Biessy C, Ferrari P, Freisling H, Rinaldi S, Chajès V, Dahm CC, Overvad K, Dossus L, Lagiou P, Trichopoulos D, Trichopoulou A, Bueno-de-Mesquita HB, May A, Peeters PH, Weiderpass E, Sanchez MJ, Navarro C, Ardanaz E, Ericson U, Wirfält E, Travis RC, Romieu I. Healthy lifestyle and risk of cancer in the European prospective investigation into cancer and nutrition Cohort Study. Medicine (Baltimore). 2016;95(16):e2850.
8. Warburton DER, Bredin SSD. Health benefits of physical activity: a systematic review of current systematic reviews. Curr Opin Cardiol. 2017;32(5):541–56.
9. Loprinzi PD, Addoh O, Wong Sarver N, Espinoza I, Mann JR. Cross-sectional association of exercise, strengthening activities, and cardiorespiratory fitness on generalized anxiety, panic and depressive symptoms. Postgrad Med. 2017;129(7):676–85.
10. Machado S, Filho ASS, Wilbert M, Barbieri G, Almeida V, Gurgel A, Rosa CV, Lins V, Paixão A, Santana K, Ramos G, Neto GM, Paes F, Rocha N, Murillo-Rodriguez E. Physical exercise as stabilizer for alzheimer's disease cognitive decline: current status. Clin Pract Epidemiol Ment Health. 2017;13:181–4.
11. Liu Y, Shu XO, Wen W, Saito E, Rahman MS, Tsugane S, Tamakoshi A, Xiang YB, Yuan JM, Gao YT, Tsuji I, Kanemura S, Nagata C, Shin MH, Pan WH, Koh WP, Sawada N, Cai H, Li HL, Tomata Y, Sugawara Y, Wada K, Ahn YO, Yoo KY, Ashan H, Chia KS, Boffetta P, Inoue M, Kang D, Potter JD, Zheng W. Association of leisure-time physical activity with total and cause-specific mortality: a pooled analysis of nearly a half million adults in the Asia Cohort Consortium. Int J Epidemiol. 2018;27.
12. Carlson SA, Adams EK, Yang Z, Fulton JE. Percentage of deaths associated with inadequate physical activity in the United States. Prev Chronic Dis. 2018;15:E38.
13. Kraus WE, Powell KE, Haskell WL, et al. Physical activity, all-cause and cardiovascular mortality, and cardiovascular disease. Med Sci Sports Exerc. 2019;51(6):1270–81.

14. Zhao R, Bu W, Chen X. The efficacy and safety of exercise for prevention of fall- related injuries in older people with different health conditions, and differing intervention protocols: a meta-analysis of randomized controlled trials. BMC Geriatr. 2019;19(1):341.
15. Silva RB, Aldoradin-Cabeza H, Eslick GD, Phu S, Duque G. The effect of physical exercise on frail older persons: a systematic review. J Frailty Aging. 2017;6(2):91–6.
16. Taryana AA, Krishnasamy R, Bohm C, et al. Physical activity for people with chronic kidney disease: an international survey of nephrologist practice patterns and research priorities. BMJ Open. 2019;9(12):e032322.
17. Duarte-Rojo A, Ruiz-Margáin A, Montaño-Loza AJ, et al. Exercise and physical activity for patients with end-stage liver disease: Improving functional status and sarcopenia while on the transplant waiting list. Liver Transpl. 2018;24(1):122–39.
18. Rezende LFM, Sá TH, Markozannes G, et al. Physical activity and cancer: an umbrella review of the literature including 22 major anatomical sites and 770000 cancer cases. Br J Sports Med. 2018;52(13):826–33.
19. Oruç Z, Kaplan MA. Effect of exercise on colorectal cancer prevention and treatment. World J Gastrointest Oncol. 2019;11(5):348–66.
20. Arthur R, et al. The combined association of modifiable risk factors with breast cancer risk in the Women's Health Initiative. Cancer Prev Res (Phila). 2018: canprevres.0347.2017.
21. McKenzie F, et al. Healthy lifestyle and risk of breast cancer among postmenopausal women in the European Prospective Investigation into Cancer and Nutrition cohort study. Int J Cancer. 2015;136(11):2640–8.
22. Loughney LA, et al. Exercise interventions for people undergoing multimodal cancer treatment that includes surgery. Cochrane Database Syst Rev. 2018;12:CD012280.
23. Romero SAD, et al. The association between fatigue and pain symptoms and decreased physical activity after cancer. Support Care Cancer. 2018;26(10):3423–30.
24. Witlox L, et al. Four-year effects of exercise on fatigue and physical activity in patients with cancer. BMC Med. 2018;16(1):86.
25. Navigante A, Morgado PC. Does physical exercise improve quality of life of advanced cancer patients? Curr Opin Support Palliat Care. 2016;10(4):306–9.
26. Stojanovska L, Apostolopoulos V, Polman R, Borkoles E. To exercise, or, not to exercise, during menopause and beyond. Maturitas. 2014;77(4):318–23.
27. Part   D.   https://health.gov/paguidelines/second-edition/pdf/Physical_Activity_Guidelines_2nd_edition.pdf. (last access: 20th April 2020).
28. Global Action Plan on Physical Activity, from the World Health Organization. https://www.who.int/ncds/governance/who-discussion-paper-gappa-9april2018.pdf?ua=1. (last access: 20th April 2020).
29. https://www.exerciseismedicine.org/assets/page_documents/Exercise%20Profes   sionals'%20Action%20Guide.pdf. (last access: 20th April 2020).
30. https://www.premierhealth.com/news-and-events/news/adults-become-amateur-athletes-in-effort-to-stay-healthy. (last access: 20th April 2020).
31. https://qz.com/1230097/new-cdc-report-shows-americans-exercise-more-than-ever-but-the-obesity-rate-is-growing/. (last access: 20th April 2020).
32. https://www.statista.com/statistics/236123/us-fitness-center%2D%2Dhealth-club-memberships/. (last access: 20th April 2020).
33. https://euathletes.org. (last access: 20th April 2020).
34. https://www.worldathletics.org. (last access: 20th April 2020).
35. Alex C, Lindgren M, Shapiro PA, McKinley PS, Brondolo EN, Myers MM, Zhao Y, Sloan RP. Aerobic exercise and strength training effects on cardiovascular sympathetic function in healthy adults: a randomized controlled trial. Psychosom Med. 2013;75(4):375–81.
36. Lindgren M, Alex C, Shapiro PA, McKinley PS, Brondolo EN, Myers MM, Choi CJ, Lopez-Pintado S, Sloan R. Effects of aerobic conditioning on cardiovascular sympathetic response to and recovery from challenge. Psychophysiology. 2013;50(10):963–73.

37. McGlory C, Phillips SM. Exercise and the regulation of skeletal muscle hypertrophy. Prog Mol Biol Transl Sci. 2015;135:153–73.
38. Szuhany KL, Bugatti M, Otto MW. A meta-analytic review of the effects of exercise on brain-derived neurotrophic factor. J Psychiatr Res. 2015;60:56–64.
39. Douglas JA, King JA, McFarlane E. Appetite, appetite hormone and energy intake responses to two consecutive days of aerobic exercise in healthy young men. Appetite. 2015;92:57–65.
40. Cadegiani FA, Kater CE. Enhancement of hypothalamic-pituitary activity in male athletes: evidence of a novel hormonal mechanism of physical conditioning. BMC Endoc Dis. 2019.
41. Crewther B, Keogh J, Cronin J, Cook C. Possible stimuli for strength and power adaptation: acute hormonal responses. Sports Med. 2006;36(3):215–38.
42. Durand RJ, Castracane VD, Hollander DB, Tryniecki JL, Bamman MM, O'Neal S, Hayes LD, Grace FM, Baker JS, Sculthorpe N. Exercise-induced responses in salivary testosterone, cortisol, and their ratios in men: a meta-analysis. Sports Med. 2015;45(5):713–26.
43. Shaner AA, Vingren JL, Hatfield DL, Budnar RG Jr, Duplanty AA, Hill DW. The acute hormonal response to free weight and machine weight resistance exercise. J Strength Cond Res. 2014;28(4):1032–40.
44. Meeusen R, Piacentini MF, Busschaert B, Buyse L, De Schutter G, Stray-Gundersen J. Hormonal responses in athletes: the use of a two bout exercise protocol to detect subtle differences in (over)training status. Eur J Appl Physiol. 2004;91(2-3):140–6.
45. Urhausen A, Gabriel HH, Kindermann W. Impaired pituitary hormonal response to exhaustive exercise in overtrained endurance athletes. Med Sci Sports Exerc. 1998;30(3):407–14.
46. Cadegiani FA, Kater CE. Hormonal aspects on overtraining syndrome: a systematic review. Med & Rehab: BMC Sports Sci; 2017.
47. Riddell MC, Gallen IW, Smart CE, et al. Exercise management in type 1 diabetes: a consensus statement. Lancet Diabetes Endocrinol. 2017;5(5):377–90.
48. Kirwan JP, Sacks J, Nieuwoudt S. The essential role of exercise in the management of type 2 diabetes. Cleve Clin J Med. 2017;84(7 Suppl 1):S15–21.
49. Sylow L, Richter EA. Current advances in our understanding of exercise as medicine in metabolic disease. Curr Opin Physiol. 2019;12:12–9.
50. Hsing AW, Stanczyk FZ, Bélanger A, Schroeder P, Chang L, Falk RT, Fears TR. Reproducibility of serum sex steroid assays in men by RIA and mass spectrometry. Cancer Epidemiol Biomark Prev. 2007;16(5):1004–8.
51. Cinquanta L, Fontana DE, Bizzaro N. Chemiluminescent immunoassay technology: what does it change in autoantibody detection? Auto Immun Highlights. 2017;8(1):9.
52. Huayllas MKP, Netzel BC, Singh RJ, Kater CE. Serum cortisol levels via radioimmunoassay vs liquid chromatography mass spectrophotometry in healthy control subjects and patients with adrenal incidentalomas. Lab Med. 2018;49(3):259–67.
53. Amed S, Delvin E, Hamilton J. Variation in growth hormone immunoassays in clinical practice in Canada. Horm Res. 2008; https://doi.org/10.1159/000114860.
54. Carrozza C, Lapolla R, Canu G, Annunziata F, Torti E, Baroni S, Zuppi C. Human growth hormone (GH) immunoassay: standardization and clinical implications. Clin Chem Lab Med. 2011;49(5):851–3. https://doi.org/10.1515/CCLM.2011.138.
55. Saleem M, Martin H, Coates P. Prolactin biology and laboratory measurement: an update on physiology and current analytical issues. The Clin Biochemist. 2018;39(1):3–16.
56. Hsing AW, Stanczyk FZ, Bélanger A, Schroeder P, Chang L, Falk RT, Fears TR. Reproducibility of serum sex steroid assays in men by RIA and mass spectrometry. Cancer Epidemiol Biomark Prev. 2007;16(5):1004–8.
57. Handelsman DJ, Jimenez M, Singh GKS, Spaliviero J, Desai R, Walters KA. Measurement of testosterone by immunoassays and mass spectrometry in mouse serum, testicular, and ovarian extracts. Endocrinology. 2015;156(1):400–5.

58. Handelsman DJ, Wartofsky L. Requirement for mass spectrometry sex steroid assays in the Journal of Clinical Endocrinology and Metabolism. J Clin Endocrinol Metab. 2013;98(10):3971–3.
59. Wang C, Catlin DH, Demers LM, Starcevic B, Swerdloff R. Measurement of total serum testosterone in adult men: comparison of current laboratory methods versus liquid chromatography-tandem mass spectrometry. J Clin Endocrinol Metab. 2004;89(2):534–43.
60. Rosner W, Auchus RJ, Azziz R, Sluss PM, Raff H. Utility, limitations, and pitfalls in measuring testosterone: an endocrine society position statement. J Clin Endocrinol Metab. 2007;92(2):405–13.
61. Dorgan JF, Fears TR, McMahon RP, Friedman LA, Patterson BH, Greenhut SF. Measurement of steroid sex hormones in serum: a comparison of radioimmunoassay and mass spectrometry. Steroids, March. 2002;67(3–4):151–8.
62. Wang C, Catlin DH, Demers LM, Starcevic B, Swerdloff R. Measurement of total serum testosterone in adult men: comparison of current laboratory methods versus liquid chromatography-tandem mass spectrometry. J Clin Endocrinol Metab. 2004;89(2):534–43.
63. Huhtaniemi IT, Tajar A, Lee DM, O'Neill TW, Finn JD, Bartfai G, Boonen S, Casanueva FF, Giwercman A, Han TS, Kula K, Labrie F, Lean ME, Pendleton N, Punab M, Silman AJ, Vanderschueren D, Forti G, Wu FC. EMAS Group. Comparison of serum testosterone and estradiol measurements in 3174 Europea n menusing platform immunoassay and mass spectrometry; relevance of the diagnostics in aging men. Eur J Endocrinol. 2012;166(6):983–91.
64. Rosner W, Hankinson SE, Sluss PM, Vesper HW, Wierman ME. Challenges to the measurement of estradiol: an endocrine society position statement. J Clin Endocrinol Metab. 2013;98:1376–87.
65. Fiers T, Casetta B, Bernaert B, Vandersypt E, Debock M, Kaufman JM. Development of a highly sensitive method for the quantification of estrone and estradiol in serum by liquid chromatography tandem mass spectrometry without derivatization. J Chromatogr B Analyt Technol Biomed Life Sci. 2012;893-894:57–62.
66. Stanczyk FZ, Jurow J, Hsing AW. Limitations of direct immunoassays for measuring circulating estradiol levels in postmenopausal women and men in epidemiologic studies. Cancer Epidemiol Biomark Prev. 2010;19(4):903–6.
67. Hammes SR, Levin ER. Impact of estrogens in males and androgens in females. J Clin Invest. 2019;129(5):1818–26.
68. Bhasin S, Brito JP, Cunningham GR, Hayes FJ, Hodis HN, Matsumoto AM, Snyder PJ, Swerdloff RS, Wu FC, Yialamas MA. Testosterone therapy in men with hypogonadism: an endocrine society clinical practice guideline. J Clin Endocrinol Metab. 2018;103(5):1715–44.
69. Petak SM, Nankin HR, Spark RF, Swerdloff RS, Rodriguez-Rigau LJ. American Association of Clinical Endocrinologists Medical Guidelines for clinicalpractice for the evaluation and treatment of hypogonadism in adult male patients--2002 update. Endocr Pract. 2002;8(6):440–56.
70. Fleseriu M, Hashim IA, Karavitaki N, Melmed S, Murad MH, Salvatori R, Samuels MH. Hormonal replacement in hypopituitarism in adults: an endocrine society clinical practice guideline. J Clin Endocrinol Metab. 2016;101(11):3888–921.
71. Spratt DI, O'Dea LS, Schoenfeld D, Butler J, Rao PN, Crowley WF Jr. Neuroendocrine-gonadal axis in men: frequent sampling of LH, FSH, and testosterone. Am J Physiol. 1988;254(5 Part 1):E658-E666.
72. Santen RJ. The testes. In: Felig P, Baxter J, Frohman L, editors. Endocrinology and metabolism. New York: McGraw-Hill; 1995. p. 885–972.
73. Carani C, Zini D, Baldini A, Della Casa L, Ghizzani A, Marrama P. Effects of androgen treatment in impotent men with normal and low levels of free testosterone. Arch Sex Behav. 1990;19:223–34.
74. Steinberger E, Ayala C, Hsi B, et al. Utilization of com- mercial laboratory results in management of hyperandro- genism in women. Endocr Pract. 1998;4:1–10.

75. Brambilla DJ, Matsumoto AM, Araujo AB, McKinlay JB. The effect of diurnal variation on clinical measurement of serum testosterone and other sex hormone levels in men. J Clin Endocrinol Metab. 2009;94:907–13.

76. Spratt DI, Bigos ST, Beitins I, Cox P, Longcope C, Orav J. Both hyper- and hypogonadotropic hypogonadism occur transiently in acute illness: bio- and immunoactive gonadotropins. J Clin Endocrinol Metab. 1992;75:1562–70.

77. Cadegiani FA, Kater CE. Basal Hormones and Biochemical Markers as Predictors of Overtraining Syndrome in Male Athletes: The EROS-BASAL Study. J Athl Train. 2019;54(8):906–14.

78. Li SH, Lloyd AR, Graham BM. Physical and mental fatigue across the menstrual cycle in women with and without generalised anxiety disorder. Horm Behav. 2019;31:104667.

79. Silveira LF, Latronico AC. Approach to the patient with hypogonadotropic hypogonadism. J Clin Endocrinol Metab. 2013;98:1781–8.

80. Collomp K, Buisson C, Gravisse N, et al. Effects of short-term DHEA intake on hormonal responses in young recreationally trained athletes: modulation by gender. Endocrine. 2018;59(3):538–46.

81. Safarinejad MR, Azma K, Kolahi AA. The effects of intensive, long-term treadmill running on reproductive hormones, hypothalamus-pituitary-testis axis, and semen quality: a randomized controlled study. J Endocrinol. 2009;200(3):259–71.

82. Vaamonde D, Algar-Santacruz C, Abbasi A, García-Manso JM. Sperm DNA fragmentation as a result of ultra-endurance exercise training in male athletes. Andrologia. 2018;50(1).

83. Jóźków P, Rossato M. The impact of intense exercise on semen quality. Am J Mens Health. 2017;11(3):654–62.

84. Bornstein SR, Allolio B, Arlt W, Barthel A, Don-Wauchope A, Hammer GD, Husebye ES, Merke DP, Murad MH, Stratakis CA, Torpy DJ. Diagnosis and treatment of primary adrenal insufficiency: an endocrine society clinical practice guideline. J Clin Endocrinol Metab. 2016;101(2):364–89.

85. Storbeck KH, Schiffer L, Baranowski ES, Chortis V, Prete A, Barnard L, Gilligan LC, Taylor AE, Idkowiak J, Arlt W, Shackleton CHL. Steroid metabolome analysis in disorders of adrenal steroid biosynthesis and metabolism. Endocr Rev. 2019;40(6):1605–25.

86. Marcos-Serrano M, Olcina G, Crespo C, Brooks D, Timon R. Urinary steroid profile in ironman triathletes. J Hum Kinet. 2018;61:109–17.

87. Timon R, Olcina G, Muñoz D, Maynar JI, Caballero MJ, Maynar M. Determination of urine steroid profile in untrained men to evaluate recovery after a strength training session. J Strength Cond Res. 2008;22(4):1087–93.

88. Timon R, Olcina G, Tomas-Carus P, Muñoz D, Toribio F, Raimundo A, Maynar M. Urinary steroid profile after the completion of concentric and concentric/eccentric trials with the same total workload. J Physiol Biochem. 2009;65(2):105–12.

89. Timon R, Corvillo M, Brazo J, Robles MC, Maynar M. Strength training effects on urinary steroid profile across the menstrual cycle in healthy women. Eur J Appl Physiol. 2013;113(6):1469–75.

90. Carrozza C, Lapolla R, Canu G, Annunziata F, Torti E, Baroni S, Zuppi C. Human growth hormone (GH) immunoassay: standardization and clinical implications. Clin Chem Lab Med. 2011;49(5):851–3. https://doi.org/10.1515/CCLM.2011.138.

91. Clemmons DR. Consensus statement on the standardization and evaluation of growth hormone and insulin-like growth factor assays. Clin Chem. 2011;57:555–9.

92. Biller BM, Samuels MH, Zagar A, et al. Sensitivity and specificity of six tests for the diagnosis of adult GH deficiency. J Clin Endocrinol Metab. 2002;87:2067–79.

93. Pagani S, Cappa M, Meazza C, et al. Growth hormone isoforms release in response to physiological and pharmacological stimuli. J Endocrinol Investig. 2008;31(6):520–4.

94. Barkan AL, Beitins IZ, Kelch RP. Plasma insulin-like growth factor-I/somatomedin-C in acromegaly: correlation with the degree of growth hormone hypersecretion. J Clin Endocrinol Metab. 1988;67:69–73.

95. Lewitt MS, Saunders H, Cooney GJ, Baxter RC. Effect of human insulin-like growth factor-binding protein-1 on the half-life and action of administered insulin-like growth factor-I in rats. J Endocrinol. 1993;136:253–60.
96. Dimaraki EV, Jaffe CA, DeMott-Friberg R, Chandler WF, Barkan AL. Acromegaly with apparently normal GH secretion: implications for diagnosis and follow-up. J Clin Endocrinol Metab. 2002;87:3537–42.
97. Bidlingmaier M, Friedrich N, Emeny RT, et al. Reference intervals for insulin-like growth factor-1 (igf-i) from birth to senescence: results from a multicenter study using a new automated chemiluminescence IGF-I immunoassay conforming to recent international recommendations. J Clin Endocrinol Metab. 2014;99:1712–21.
98. Frystyk J, Freda P, Clemmons DR. The current status of IGF-I assays–a 2009 update. Growth Hormon IGF Res. 2010;20:8–18.
99. Pokrajac A, Wark G, Ellis AR, Wear J, Wieringa GE, Trainer PJ. Variation in GH and IGF-I assays limits the applicability of international consensus criteria to local practice. Clin Endocrinol (Oxf) . 2007;67:65–70.
100. Kord-Varkaneh H, Rinaldi G, Hekmatdoost A, et al. The influence of vitamin D supplementation on IGF-1 levels in humans: A systematic review and meta-analysis. Ageing Res Rev. 2020;57:100996.
101. Yang X, Varkaneh HK, Talaei S, et al. The influence of metformin on IGF-1 levels in humans: A systematic review and meta-analysis. Pharmacol Res. 2019;6:104588.
102. Rahmani J, Kord Varkaneh H, Clark C, et al. The influence of fasting and energy restricting diets on IGF-1 levels in humans: A systematic review and meta-analysis. Ageing Res Rev. 2019;53:100910.
103. Kazemi A, Speakman JR, Soltani S³ Djafarian K. Effect of calorie restriction or protein intake on circulating levels of insulin like growth factor I in humans: A systematic review and meta-analysis. Clin Nutr. 2019: S0261-5614(19)30312-7.
104. Fontana L, Weiss EP, Villareal DT, Klein S, Holloszy JO. Long-term effects of calorie or protein restriction on serum IGF-1 and IGFBP-3 concentration in humans. Aging Cell. 2008;7(5):681–7.
105. Fontana L, Villareal DT, Das SK, et al. Effects of 2-year calorie restriction on circulating levels of IGF-1, IGF-binding proteins and cortisol in nonobese men and women: a randomized clinical trial. Aging Cell. 2016;15(1):22–7.
106. Aleidi SM, Shayeb E, Bzour J, et al. Serum level of insulin-like growth factor-I in type 2 diabetic patients: impact of obesity. Horm Mol Biol. Clin Investig. 2019;9:39(1).
107. Puche JE, Castilla-Cortazar I. Human conditions of insulin-like growth factor-I (IGF-I) deficiency. J Transl Med. 2012;10:224.
108. Fontana L, Villareal DT, Das SK, et al. Effects of 2-year calorie restriction on circulating levels of IGF-1, IGF-binding proteins and cortisol in nonobese men and women: a randomized clinical trial. Aging Cell. 2016;15(1):22–7.
109. Hagmar M, Berglund B, Brismar K, Hirschberg AL. Body composition and endocrine profile of male Olympic athletes striving for leanness. Clin J Sport Med. 2013;23(3):197–201.
110. Nemet D, Connolly PH, Pontello-Pescatello AM, Rose-Gottron C, Larson JK, Galassetti P, Cooper DM. Negative energy balance plays a major role in the IGF-I response to exercise training. J Appl Physiol (1985). 2004;96(1):276–82.
111. Nindl BC, Alemany JA, Rarick KR, Eagle SR, Darnell ME, Allison KF, Harman EA. Differential basal and exercise-induced IGF-I system responses to resistance vs. calisthenic-based military readiness training programs. Growth Hormon IGF Res. 2017;32:33–40.
112. Ho-Seong Kim HS, Rosenfeld RG, Oh Y. Biological roles of insulin-like growth factor binding proteins(IGFBPs)
113. Dig H, Wu T. Insulin-like growth factor binding proteins in autoimmune diseases. Front Endocrinol. 2018.
114. Lagundžin D, Vucić V, Glibetić M, Nedić O. Alteration of IGFBP-1 in soccer players due to intensive training. Int J Sport Nutr Exerc Metab. 2013;23(5):449–57.

115. Aïssa Benhaddad A, Monnier JF, Fédou C, Micallef JP, Brun JF. Insulin-like growth factor-binding protein 1 and blood rheology in athletes. Clin Hemorheol Microcirc. 2002;26(3):209–17.

116. Pilitsi E, Peradze N, Perakakis N, Mantzoros CS. Circulating levels of the components of the GH/IGF-1/IGFBPs axis total and intact IGF-binding proteins (IGFBP) 3 and IGFBP 4 and total IGFBP 5, as well as PAPPA, PAPPA2 and Stanniocalcin-2 levels are not altered in response to energy deprivation and/or metreleptin administration in humans. Metabolism. 2019;97:32–9.

117. Allard JB, Duan C. IGF-binding proteins: why do they exist and why are there so many? Front. Endocrinol. 2018.

118. Florini JR, Magri KA, Ewton DZ, James PL, Grindstaff K, Rotwein PS. "Spontaneous" differentiation of skeletal myoblasts is dependent upon autocrine secretion of insulin-like growth factor-II. J Biol Chem. 1991;266(24):15917–23.

119. Regué L, Ji F, Flicker D, Kramer D, Pierce W, Davidoff T, Widrick JJ, Houstis N, Minichiello L, Dai N, Avruch J. IMP2 increases mouse skeletal muscle mass and voluntary activity by enhancing autocrine insulin-like growth factor 2 production and optimizing muscle metabolism. Mol Cell Biol. 2019;19:39(7).

120. Cianfarani S. Insulin-like growth factor-II: new roles for an old actor. Front. Endocrinol. 2012.

121. Holly JMP, Biernacka K, Perks CM. The neglected insulin: IGF-II, a metabolic regulator with implications for diabetes, obesity, and cancer. Cells. 2019;8(10):E1207.

122. Gatti R, De Palo EF, Antonelli G, Spinella P. IGF-I/IGFBP system: metabolism outline and physical exercise. J Endocrinol Investig. 2012;35(7):699–707.

123. Adams GR. Role of insulin-like growth factor-I in the regulation of skeletal muscle adaptation to increased loading. Exerc Sport Sci Rev. 1998;26:31–60.

124. Nindl BC, Pierce JR. Insulin-like growth factor I as a biomarker of health, fitness, and training status. Med Sci Sports Exerc. 2010;42(1):39–49.

125. Nindl BC, Kraemer WJ, Marx JO, Arciero PJ, Dohi K, Kellogg MD, Loomis GA. Overnight responses of the circulating IGF-I system after acute, heavy-resistance exercise. J Appl Physiol (1985). 2001;90(4):1319–26.

126. Nindl BC, Urso ML, Pierce JR, Scofield DE, Barnes BR, Kraemer WJ, Anderson JM, Maresh CM, Beasley KN, Zambraski EJ. IGF-I measurement across blood, interstitial fluid, and muscle biocompartments following explosive, high-power exercise. Am J Phys Regul Integr Comp Phys. 2012;303(10):R1080–9.

127. Florini JR, Ewton DZ, Coolican SA. Growth hormone and the insulin-like growth factor system in myogenesis. Endocr Rev. 1996;17(5):481–517.

128. Kraemer WJ, Ratamess NA. Hormonal responses and adaptations to resistance exercise and training. Sports Med. 2005;35(4):339–61.

129. Copeland JL, Verzosa ML. Endocrine response to an ultra-marathon in pre- and post-menopausal women. Biol Sport. 2014;31(2):125–31.

130. Berg U, Enqvist JK, Mattsson CM, Carlsson-Skwirut C, Sundberg CJ, Ekblom B, Bang P. Lack of sex differences in the IGF-IGFBP response to ultra endurance exercise. Scand J Med Sci Sports. 2008;18(6):706–14.

131. Manetta J, Brun JF, Fedou C, Maïmoun L, Prefaut C, Mercier J. Serum levels of insulin-like growth factor-I (IGF-I), and IGF-binding proteins-1 and -3 in middle-aged and young athletes versus sedentary men: relationship with glucose disposal. Metabolism. 2003;52(7):821–6.

132. Manetta J, Brun JF, Maïmoun L, Fédou C, Préfaut C, Mercier J. The effects of intensive training on insulin-like growth factor I (IGF-I) and IGF binding proteins 1 and 3 in competitive cyclists: relationships with glucose disposal. J Sports Sci. 2003;21(3):147–54.

133. Karmisholt J, Andersen S, Laurberg P. Variation in thyroid function tests in patients with stable untreated subclinical hypothyroidism. Thyroid. 2008;18:303–8.

134. Spencer CA, LoPresti JS, Patel A, et al. Applications of a new chemiluminometric thyrotropin assay to subnormal measurement. J Clin Endocrinol Metab. 1990;70:453–60.

135. Wartofsky L, Dickey RA. The evidence for a narrower thyrotropin reference range is compelling. J Clin Endocrinol Metab. 2005;90:5483–8.

136. Halsall DJ, English E, Chatterjee VK. Interference from heterophilic antibodies in TSH assays. Ann Clin Biochem. 2009;46:345–6.
137. Zucchi R, Rutigliano G, Saponaro F. Novel thyroid hormones. Endocrine. 2019;66(1):95–104.
138. Köhrle J. The colorful diversity of thyroid hormone metabolites. Eur Thyroid J. 2019;8(3):115–29.
139. Pietzner M, Köhrle J, Lehmphul I, Budde K, Kastenmüller G, Brabant G, Völzke H, Artati A, Adamski J, Völker U, Nauck M, Friedrich N, Homuth G. A thyroid hormone-independent molecular fingerprint of 3,5-Diiodothyronine suggests a strong relationship with coffee metabolism in humans. Thyroid. 2019;29(12):1743–54.
140. Langouche L, Lehmphul I, Perre SV, Kohrle J, Van den Berghe G. Circulating 3-T1AM and 3,5-T2 in critically ill patients: a cross-sectional observational study. Thyroid. 2016;26:1674–80.
141. Pietzner M, Lehmphul I, Friedrich N, Schurmann C, Ittermann T, Dorr M, et al. Translating pharmacological findings from hypothyroid rodents to euthyroid humans: is there a functional role of endogenous 3,5-T2? Thyroid. 2015;25:188–97.
142. Antonelli A, Fallahi P, Ferrari SM, Di Domenicantonio A, Moreno M, Lanni A, et al. 3,5-diiodo-L-thyronine increases resting metabolic rate and reduces body weight without undesirable side effects. J Biol Regul Homeost Agents. 2011;25:655–60.
143. Padron AS, Neto RA, Pantaleao TU, de Souza dos Santos MC, Araujo RL, de Andrade BM, et al. Administration of 3,5-diiodothyronine (3,5-T2) causes central hypothyroidism and stimulates thyroid-sensitive tissues. J Endocrinol. 2014.
144. Jonas W, Lietzow J, Wohlgemuth F, Hoefig CS, Wiedmer P, Schweizer U, et al. 3,5-Diiodo-L-thyronine (3,5-t2) exerts thyromimetic effects on hypothalamus-pituitary-thyroid axis, body composition, and energy metabolism in male diet-induced obese mice. Endocrinology. 2015;156:389–99.
145. Lanni A, Moreno M, Lombardi A, de Lange P, Silvestri E, Ragni M, et al. 3,5-diiodo-L-thyronine powerfully reduces adiposity in rats by increasing the burning of fats. FASEB J. 2005;19:1552–4.
146. Coppola M, Glinni D, Moreno M, Cioffi F, Silvestri E, Goglia F. Thyroid hormone analogues and derivatives: actions in fatty liver. World J Hepatol. 2014;6:114–29.
147. Lombardi A, Senese R, De Matteis R, Busiello RA, Cioffi F, Goglia F, et al. 3,5-Diiodo-L-thyronine activates brown adipose tissue thermogenesis in hypothyroid rats. PLoS One. 2015;10:e0116498.
148. Cimmino M, Mion F, Goglia F, Minaire Y, Geloen A. Demonstration of in vivo metabolic effects of 3,5-di-iodothyronine. J Endocrinol. 1996;149:319–25.
149. Senese R, de Lange P, Petito G, Moreno M, Goglia F, Lanni A. 3,5-Diiodothyronine: a novel thyroid hormone metabolite and potent modulator of energy metabolism. Front Endocrinol (Lausanne). 2018;9:427.
150. Cioffi F, Senese R, Lanni A, Goglia F. Thyroid hormones and mitochondria: with a brief look at derivatives and analogues. Mol Cell Endocrinol. 2013;379:51–61.
151. Goglia F. The effects of 3,5-diiodothyronine on energy balance. Front Physiol. 2014;5:528.
152. Lombardi A, Lanni A, de Lange P, Silvestri E, Grasso P, Senese R, et al. Acute administration of 3,5-diiodo-L-thyronine to hypothyroid rats affects bioenergetic parameters in rat skeletal muscle mitochondria.
153. Lombardi A, de Lange P, Silvestri E, Busiello RA, Lanni A, Goglia F, et al. 3,5-Diiodo-L-thyronine rapidly enhances mitochondrial fatty acid oxidation rate and thermogenesis in rat skeletal muscle: AMP-activated protein kinase involvement. Am J Physiol Endocrinol Metab. 2009;296:E497–502.
154. Horst C, Rokos H, Seitz HJ. Rapid stimulation of hepatic oxygen consumption by 3,5-di-iodo-L-thyronine. Biochem J. 1989;261:945–50.
155. Lanni A, Moreno M, Lombardi A, Goglia F. 3,5-Diiodo-L-thyronine and 3,5,3'-triiodo-L-thyronine both improve the cold tolerance of hypothyroid rats, but possibly via different mechanisms. Pflugers Arch. 1998;436:407–14.

156. Biebermann H, Kleinau G. 3-Iodothyronamine induces diverse signaling effects at different aminergic and non-aminergic G-protein coupled receptors [published online ahead of print, 2019 Nov 7]. Exp Clin Endocrinol Diabetes. 2019; https://doi.org/10.1055/a-1022-1554.

157. Melmed S, Casanueva FF, Hoffman AR, Kleinberg DL, Montori VM, Schlechte JA, Wass JA. Diagnosis and treatment of hyperprolactinemia: an Endocrine Society clinical practice guideline. J Clin Endocrinol Metab. 2011;96(2):273–88.

158. Casanueva FF, Molitch ME, Schlechte JA, et al. A 2006 Guidelines of the Pituitary Society for the diagnosis and management of prolactinomas. Clin Endocrinol. 2006;65:265–73.

159. Mancini T, Casanueva FF, Giustina A 2008 Hyperprolactinemia and prolactinomas Endocrinol Metab Clin N Am 37:67–99, viii

160. Ansley L, Marvin G, Sharma A, Kendall MJ, Jones DA, Bridge MW. The effect of head cooling on endurance and neuroendocrine responses to exercise in warm conditions. Physiol Res. 2008;57:863–72.

161. Radomski MW, Cross M, Buguet A. Exercise-induced hyperthermia and hormonal responses to exercise. Can J Physiol Pharmacol. 1998;76:547–52.

162. Ben-Jonathan N, Hugo ER, Brandebourg TG, LaPensee CR. Focus on prolactin as a metabolic hormone. Trends Endocrinol. 2006;17:110–6.

163. Hackney AC. Characterization of the prolactin response to prolonged endurance exercise. Acta Kinesiol. 2008;13:31–8.

164. Patlar S. Effect of acute and chronic submaximal exercise on plasma renin and aldosterone levels in football players. Isokinet Exerc Sci. 2011;19(3):227–30.

165. Luger A, Deuster PA, Debolt JE, Loriaux DL, Chrousos GP. Acute exercise stimulates the renin-angiotensin-aldosterone axis: adaptive changes in runners. Horm Res. 1988;30(1):5–9.

166. Eisenhofer G, Kopin IJ, Goldstein DS. Catecholamine metabolism: a contemporary view with implications for physiology and medicine. Pharmacol Rev. 2004;56:331–49.

167. Mazzeo RS. Catecholamine responses to acute and chronic exercise. Med Sci Sports Exerc. 1991;23(7):839–45.

168. Evans W, Nevill A, Mclaren SJ, Ditroilo M. The effect of intermittent running on biomarkers of bone turnover. Eur J Sport Sci. 2019;16:1–11.

169. Pasqualini L, Ministrini S, Lombardini R, et al. Effects of a 3-month weight-bearing and resistance exercise training on circulating osteogenic cells and bone formation markers in postmenopausal women with low bone mass. Osteoporos Int. 2019;30(4):797–806.

170. O'Leary TJ, Izard RM, Walsh NP, Tang JCY, Fraser WD, Greeves JP. Skeletal macro- and microstructure adaptations in men undergoing arduous military training. Bone. 2019;125:54–60.

171. Papageorgiou M, Martin D, Colgan H, Cooper S, Greeves JP, Tang JCY, Fraser WD, Elliott-Sale KJ, Sale C. Bone metabolic responses to low energy availability achieved by diet or exercise in active eumenorrheic women. Bone. 2018;114:181–8.

172. Hughes F, Mythen M, Montgomery H. The sensitivity of the human thirst response to changes in plasma osmolality: a systematic review. Perioper Med (Lond). 2018;7:1.

173. Spasovski G, Vanholder R, Allolio B, et al. Clinical practice guideline on diagnosis and treatment of hyponatraemia. Eur J Endocrinol. 2014;170(3):G1–47.

174. Thomsen CF, Dreier R, Goharian TS, Goetze JP, Andersen LB, Faber J, Ried-Larsen M, Grøntved A, Jeppesen JL. Association of copeptin, a surrogate marker for arginine vasopressin secretion, with insulin resistance: Influence of adolescence and psychological stress. Peptides. 2019;115:8–14.

175. Stacey MJ, Delves SK, Britland SE, Allsopp AJ, Brett SJ, Fallowfield JL, Woods DR. Copeptin reflects physiological strain during thermal stress. Eur J Appl Physiol. 2018;118(1):75–84.

176. Popovic M, Timper K, Seelig E, Nordmann T, Erlanger TE, Donath MY, Christ-Crain M. Exercise upregulates copeptin levels which is not regulated by interleukin-1. PLoS One. 2019;14(5):e0217800.

177. Nolte HW, Nolte K, Hew-Butler T. Ad libitum water consumption prevents exercise-associated hyponatremia and protects against dehydration in soldiers performing a 40-km route-march. Mil Med Res. 2019;6(1):1.
178. Tyler CJ, Reeve T, Hodges GJ, Cheung SS. The effects of heat adaptation on physiology, perception and exercise performance in the heat: a meta-analysis. Sports Med. 2016;46(11):1699–724.
179. Aronoff SL, Berkowitz K, Shreiner B, Want L. Glucose metabolism and regulation: beyond insulin and glucagon. Diabetes Spectrum. 2004;17(3):183–90.
180. Piccoli F, Degen L, MacLean C, et al. Pharmacokinetics and pharmacodynamic effects of an oral ghrelin agonist in healthy subjects. J Clin Endocrinol Metab. 2007;92:1814–20.
181. Mach F, Baigent C, Catapano AL, et al. 2019 ESC/EAS Guidelines for the management of dyslipidaemias: lipid modification to reduce cardiovascular risk: The Task Force for the management of dyslipidaemias of the European Society of Cardiology (ESC) and European Atherosclerosis Society (EAS) Atherosclerosis. Novum. 2019;290:140–205.
182. Jellinger PS, Handelsman Y, Rosenblit PD, Bloomgarden ZT, Fonseca VA, Garber AJ, Grunberger G, Guerin CK, Bell DSH, Mechanick JI, Pessah-Pollack R, Wyne K, Smith D, Brinton EA, Fazio S, Davidson M. American Association of Clinical Endocrinologists and American College of Endocrinology Guidelines for Management of Dyslipidemia and Prevention of Cardiovascular Disease. Endocr Pract. 2017;23(Suppl 2):1–87.
183. Di Angelantonio E, Gao P, Pennells L, et al. Lipid-related markers and cardiovascular disease prediction. JAMA. 2012;307:2499–506.
184. Boren J, Williams KJ. The central role of arterial retention of cholesterol-rich apolipoprotein-B-containing lipoproteins in the pathogenesis of atherosclerosis: a triumph of simplicity. Curr Opin Lipidol. 2016;27:473–83.
185. Utermann G. The mysteries of lipoprotein(a). Science. 1989;246:904–10.
186. van der Valk FM, Bekkering S, Kroon J, Yeang C, Van den Bossche J, van Buul JD, Ravandi A, Nederveen AJ, Verberne HJ, Scipione C, Nieuwdorp M, Joosten LA, Netea MG, Koschinsky ML, Witztum JL, Tsimikas S, Riksen NP, Stroes ES. Oxidized phospholipids on lipoprotein(a) elicit arterial wall inflammation and an inflammatory monocyte response in humans. Circulation. 2016;134:611–62.
187. Marcovina SM, Albers JJ. Lipoprotein (a) measurements for clinical application. J Lipid Res. 2016;57:526–37.
188. Oliveira CL, Santos PR, Monteiro AM, Figueiredo Neto AM. Effect of oxidation on the structure of human low- and high-density lipoproteins. Biophys J. 2014;106(12):2595–605.
189. Quinn MT, Parthasarathy S, Fong LG, Steinberg D. Oxidatively modified low density lipoproteins: a potential role in recruitment and retention of monocyte/macrophages during atherogenesis. Proc Natl Acad Sci USA. 1987;84:2995–2998. [PMC free article] [PubMed] [Google Scholar].
190. Marais AD. Apolipoprotein E in lipoprotein metabolism, health and cardiovascular disease. Pathology. 2019;51(2):165–76.
191. Chen DW, Shi JK, Li Y, Yang Y, Ren SP. Association between ApoE polymorphism and type 2 diabetes: a meta-analysis of 59 studies. Biomed Environ Sci. 2019;32(11):823–38.
192. Bongiovanni T, Pintus R, Dessì A, et al. Sportomics: metabolomics applied to sports. The new revolution? Eur Rev Med Pharmacol Sci. 2019;23(24):11011–9.
193. Heaney LM, Deighton K, Suzuki T. Non-targeted metabolomics in sport and exercise science. J Sports Sci. 2019;37(9):959–67.
194. Sonkasen PH, et al. Why do endocrine profiles in elite athletes differ between sports? Clin Diabetes Endocrinol. 2018;4:3.
195. Kwok KH, Lam KS, Xu A. Heterogeneity of white adipose tissue: molecular basis and clinical implications. Exp Mol Med. 2016;48:e215.
196. Coelho M, Oliveira T, Fernandes R. Biochemistry of adipose tissue: an endocrine organ. Arch Med Sci. 2013;9(2):191–200.
197. Kershaw EE, Flier JS. Adipose tissue as an endocrine organ. J Clin Endocrinol Metab. 2004;89(6):2548–56.

198. Grewal T, Enrich C, Rentero C, Buechler C. Annexins in adipose tissue: novel players in obesity. Int J Mol Sci. 2019;20(14):E3449.
199. D'souza AM, Neumann UH, Glavas MM, Kieffer TJ. The glucoregulatory actions of leptin. Mol Metab. 2017;6(9):1052–65.
200. Woodward L, Akoumianakis I, Antoniades C. Unravelling the adiponectin paradox: novel roles of adiponectin in the regulation of cardiovascular disease. Br J Pharmacol. 2017;174(22):4007–20.
201. Sun L, Yan J, Goh HJ. Fibroblast Growth Factor-21, Leptin and Adiponectin Responses to Acute Cold-induced Brown Adipose Tissue Activation. J Clin Endocrinol Metab. 2020;8.
202. Rahimi GRM, Bijeh N, Rashidlamir A. Effects of exercise training on serum preptin, under-carboxylated osteocalcin and high molecular weight-adiponectin in adults with metabolic syndrome. Exp Physiol. 2019;23.
203. Martinez-Huenchullan SF, Tam CS, Ban LA, Ehrenfeld-Slater P, Mclennan SV, Twigg SM. Skeletal muscle adiponectin induction in obesity and exercise. Metabolism. 2020;102:154008.
204. He Z, Tian Y, Valenzuela PL, Huang C, Zhao J, Hong P, He Z, Yin S, Lucia A. Myokine/adipokine response to "aerobic" exercise: is it just a matter of exercise load? Front Physiol. 2019;10:691.
205. Asilah Za'don NH, Amirul Farhana MK, Farhanim I, Sharifah Izwan TO, Appukutty M, Salim N, Farah NMF, Arimi Fitri ML. High-intensity interval training induced PGC-1α and AdipoR1 gene expressions and improved insulin sensitivity in obese individuals. Med J Malaysia. 2019;74(6):461–7.
206. Kodo K, Sugimoto S, Nakajima H, Mori J, Itoh I, Fukuhara S, Shigehara K, Nishikawa T, Kosaka K, Hosoi H. Erythropoietin (EPO) ameliorates obesity and glucose homeostasis by promoting thermogenesis and endocrine function of classical brown adipose tissue (BAT) in diet-induced obese mice. PLoS One. 2017;12(3):e0173661.
207. Hao L, Kearns J, Scott S, Wu D, Kodani SD, Morisseau C, Hammock BD, Sun X, Zhao L, Wang S. Indomethacin enhances brown fat activity. J Pharmacol Exp Ther. 2018;365(3):467–75.
208. Tezze C, Romanello V, Sandri M. FGF21 as modulator of metabolism in health and disease. Front Physiol. 2019;10:419.
209. Lee SY, Burns SF, Ng KKC, Stensel DJ, Zhong L, Tan FHY, Chia KL, Fam KD, Yap MMC, Yeo KP, Yap EPH, Lim CL. Fibroblast growth factor 21 mediates the associations between exercise, aging, and glucose regulation. Med Sci Sports Exerc. 2020;52(2):370–80.
210. He Z, Tian Y, Valenzuela PL, Huang C, Zhao J, Hong P, He Z, Yin S, Lucia A. Myokine response to high-intensity interval vs. resistance exercise: an individual approach. Front Physiol. 2018;9:1735.
211. Lee JH, Jun HS. Role of myokines in regulating skeletal muscle mass and function. Front Physiol. 2019;10:42.
212. Son JS, Chae SA, Testroet ED, Du M, Jun HP. Exercise-induced myokines: a brief review of controversial issues of this decade. Expert Rev Endocrinol Metab. 2018;13(1):51–8.
213. Tavassoli H, Heidarianpour A, Hedayati M. The effects of resistance exercise training followed by de-training on irisin and some metabolic parameters in type 2 diabetic rat model. Arch Physiol Biochem. 2019;7:1–8.
214. Ma EB, Sahar NE, Jeong M, Huh JY. Irisin exerts inhibitory effect on adipogenesis through regulation of wnt signaling. Front Physiol. 2019;10:1085.
215. Korta P, Pocheć E, Mazur-Biały A. Irisin as a multifunctional protein: implications for health and certain diseases. Medicina (Kaunas). 2019;55(8).
216. Planella-Farrugia C, Comas F, Sabater-Masdeu M, Moreno M, Moreno-Navarrete JM, Rovira O, Ricart W, Fernández-Real JM. Circulating irisin and myostatin as markers of muscle strength and physical condition in elderly subjects. Front Physiol. 2019;10:871.
217. Maalouf GE, El Khoury D. Exercise-induced irisin, the fat browning myokine, as a potential anticancer agent. J Obes. 2019;2019:6561726.
218. Dundar A, Kocahan S, Sahin L. Associations of apelin, leptin, irisin, ghrelin, insulin, glucose levels, and lipid parameters with physical activity during eight weeks of regular exercise training. Arch Physiol Biochem. 2019;10:1–5.

219. Kabak B, Belviranli M, Okudan N. Irisin and myostatin responses to acute high-intensity interval exercise in humans. Horm Mol Biol Clin Invest. 2018;20:35(3).
220. Martínez Muñoz IY, Camarillo Romero EDS, Correa Padilla T et al. Association of irisin serum concentration and muscle strength in normal-weight and overweight young women. Front Endocrinol (Lausanne). 2019;10:621.
221. Laurent MR, Dupont J, Dejaeger M, Gielen E. Myostatin: a powerful biomarker for sarcopenia and frailty? Gerontology. 2019;65(4):383–4.
222. Arrieta H, Rodriguez-Larrad A, Irazusta J. Myostatin as a biomarker for diagnosis or prognosis of frailty and sarcopenia: current knowledge. Gerontology. 2019;65(4):385–6.
223. Wessner B, Ploder M, Tschan H, Ferunaj P, Erindi A, Strasser EM, Bachl N. Effects of acute resistance exercise on proteolytic and myogenic markers in skeletal muscles of former weightlifters and age-matched sedentary controls. J Sports Med Phys Fitness. 2019;59(11):1915–24.
224. Fujimoto T, Sugimoto K, Takahashi T, Yasunobe Y, Xie K, Tanaka M, Ohnishi Y, Yoshida S, Kurinami H, Akasaka H, Takami Y, Takeya Y, Yamamoto K, Rakugi H. Overexpression of Interleukin-15 exhibits improved glucose tolerance and promotes GLUT4 translocation via AMP-Activated protein kinase pathway in skeletal muscle. Biochem Biophys Res Commun. 2019;509(4):994–1000.
225. Nadeau L, Aguer C. Interleukin-15 as a myokine: mechanistic insight into its effect on skeletal muscle metabolism. Appl Physiol Nutr Metab. 2019;44(3):229–38.
226. Cadegiani FA, Kater CE. Body composition, metabolism, sleep, psychological and eating patterns of overtraining syndrome: results of the EROS study (EROS-PROFILE). J Sports Sci. 2018;36(16):1902–10.
227. Sonkasen PH, et al. Why do endocrine profiles in elite athletes differ between sports? Clin Diabetes Endocrinol. 2018;4:3.

Chapter 14
Endocrinology of Physical Activity and Sport in Practice: How to Research? Models for the Chronology and Tests of Studies on Athletes

Outline

Models of research on athletes: the author presents the major characteristics that must be specified in researches in the endocrinology of physical activity and sport.

Proposals of studies on athletes: important gaps in the knowledge of how the adaptations of the endocrine system occur according to the conditioning level, as well as the adaptive endocrine physiology of the athlete. The author proposes some protocols of researches that could help fill these missing gaps.

14.1 The Endocrinology of Physical Activity and Sport in Practice

The additional variable of training levels makes the research on athletes more complex compared to general population, since several different aspects should be considered, whereas each of these aspects may modify hormonal and metabolic behaviors independently [1–4]. The peculiarities of this population are one of the main reasons for the lack of consistency on the hormonal research on athletes, as well as may justify the unclear pathophysiology of overtraining syndrome (OTS), relative energy deficiency of the sport (RED-S), and other related conditions [5–11].

A structured thinking of the design and methodology of the research on the endocrinology of physical activity and sport is key for the collective improvement of the researches on the field. For this reason, an intuitive step-by-step and well-organized guide is provided to standardize the research on the endocrinology of sports.

F. Cadegiani, *Overtraining Syndrome in Athletes*, https://doi.org/10.1007/978-3-030-52628-3_14

14.2   Models of Research on Athletes

14.2.1   Characteristics to Be Present in Researches

Besides the additional confounding variable of training characteristics in athletes, this population may be even more sensitive to slight differences in all other variables, particularly habits and other modifiable aspects. Correspondingly, baseline characteristics also influence on overall results at higher extension in athletes, compared to other populations. The manner which parameters are evaluates, i.e., whether basal or during resting, after exercise stimulation, or in response to non-exercise stimulations tend to yield different meaning of the findings, and must be therefore explicitly be described together with the parameters to be measured, and a rationale of the reasons why the specific type of tests for the proposed parameters will be employed [1–4, 12–15]. Parameters to be tested should be classified according to which axis or system they belong, as depicted in Chap. 13. Figure 14.1 illustrates the key characteristics that must be specified in studies on the field.

14.2.2   Types of Comparisons to Be Employed

Eight major types of comparisons employed in studies with athletes. Making explicit which types of tests will be performed may help researches that will conduct the proposed study to review from a bigger picture if the proposed tests and comparisons are in accordance with the objectives [4]. The eight types of comparative analyses are put into an illustrative context in Fig. 14.2 and described below:

- Test 1. Analysis within the same group of affected athletes for resting/fasting markers (pre vs post intervention)
- Test 2. Analysis within the same group of affected athletes in response to stimulation (pre vs post stimulation) – non-longitudinal studies
- Test 3. Analysis within the same group of affected athletes of the difference of responses to stimulation between pre and post intervention (response to stimulations pre versus post intervention)
- Test 4. Between-group analysis (affected vs control(s) group(s)) of resting/fasting markers (pre vs post intervention)
- Test 5. Between-group analysis (affected vs control(s) group(s)) in response to stimulation (pre vs post stimulation)
- Test 6. Between-group analysis (affected vs control(s) group(s)) of the difference of responses to stimulation between pre and post intervention
- Test 7. Cross-sectional analysis with control group(s) for resting/fasting markers
- Test 8. Cross-sectional analysis with control group(s) for responses to stimulation tests

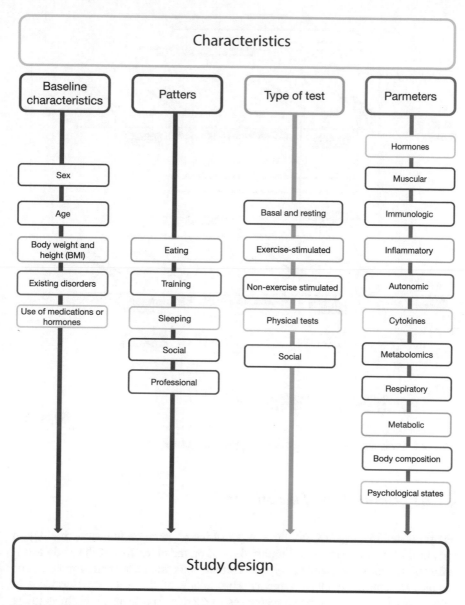

Fig. 14.1 Characteristics that should be present within studies on the endocrinology of physical activity and sport

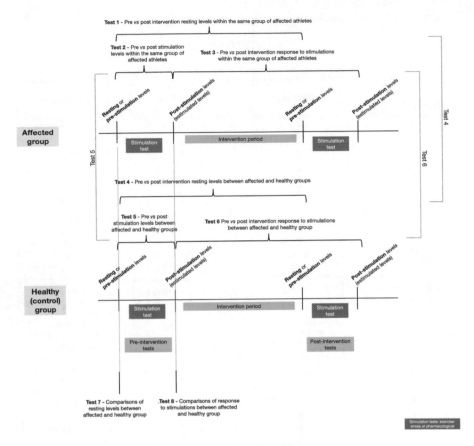

Fig. 14.2 Prevailing types of comparisons employed in athletes

14.2.3   *Chronology of the Studies*

As presented in Chap. 13, the chronology of the research on the endocrinology of physical activity and sport will highly depend on the objectives of the study and is also strictly correlated with the type of comparisons that have been proposed to be employed. The three major types of chronology of the tests are illustrated in Fig. 14.3 and include the acute responses to exercise (protocol 1), changes in the acute responsiveness to exercise in response to an intervention (protocol 2), and chronic hormonal and metabolic changes during a specific period, such as preseason or pre-event (protocol 3).

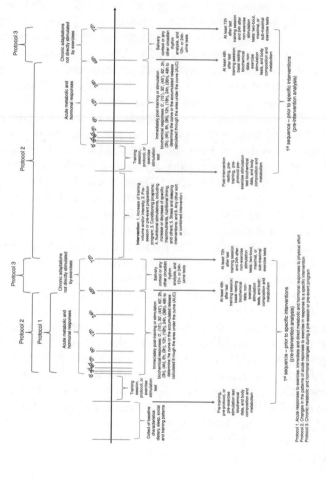

Fig. 14.3 Protocols for the chronology of the research in athletes

14.3   Proposals of Studies on Athletes

14.3.1   From the Sedentary to the Athlete

A particular gap in the understanding is the understanding of the natural course and progression of the endocrine and metabolic adaptations to sports during a conditioning program, from sedentary state until physical conditioning of athletes [1–3, 13]. At which point of the level of conditioning each adaptation occurs, and whether these adaptations are dependent on the type of sport and types of conditioning domains that are improved, within the same population. A proposal to address this question is illustrated in Fig. 14.4 could be termed as "from the sedentary to the athlete." The same proposal could be applied to different populations, of different baseline characteristics, and for different sports, in a manner that these experiments could be further compared post hoc, which will yield precise differences and which variables predict the adaptations disclosed in the studies.

Beautiful studies could be conducted with sedentary controls aiming to become athletes, divided into different types of sports, until they become physically conditioned at the level of amateur or even elite athletes, in a 6- or 12-month program, employing protocols 2 and 3 using the sedentary levels as baseline including exercise-dependent and independent stimulation tests, basal levels, salivary cortisol rhythm, and urinary collect of catecholamines and metanephrines. Tests should be repeated either in 1, 2, 3, 4, 6, 9, and 12 months, or in regular intervals of 2 months, in 1, 2, 4, 6, 8, 10, and 12 months. Regardless of the intervals, tests should be repeated after the first month, since this is the period when the most drastic metabolic and hormonal changes may occur, due to the sudden reshaping of living patterns. Changes in acute responses to exercises and acute responses to stimulations other than exercises and changes in chronic adaptations should be evaluated altogether within the same study, during the conditioning program. Each type of sport to be employed should be conducted with equivalent training characteristics, in similar speed of increase of training level than other sports. Possibly, divide into males and females and between two to three ranges of ages would provide more specific results. Given the increasing population above 50 years old and the growing evidence of multiple benefits from exercising above this age, this population should be considered at least as important. Moreover, the cutoff of 50 years old divides women in before and after menopause, in a relatively precise manner. To date, there are not studies structured at this level to observe the process of conditioning process in a sport-specific manner Figure 14.5 illustrates the proposed protocols for this model of research.

14.3.2   The Endocrine Physiology of the Athlete

While the cardiovascular and musculoskeletal systems have been extensively researched following adequate design and adjusted for confounding variables, which allowed the understanding of how these systems behave according to the conditioning level, type of sport, and sex, the research on the endocrine adaptive

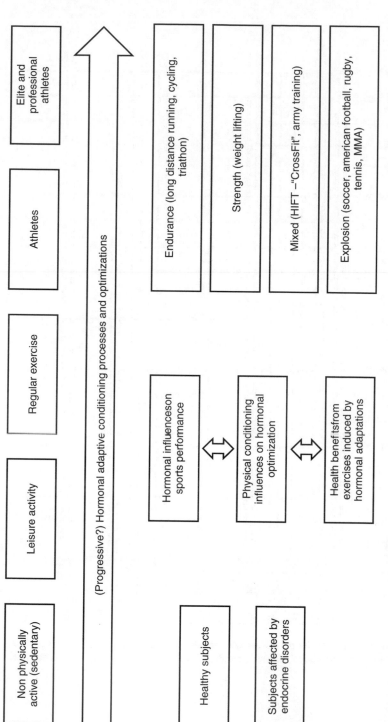

Fig. 14.4 From the sedentary to the athlete: proposal of a model to research the conditioning process in the endocrine system and metabolism

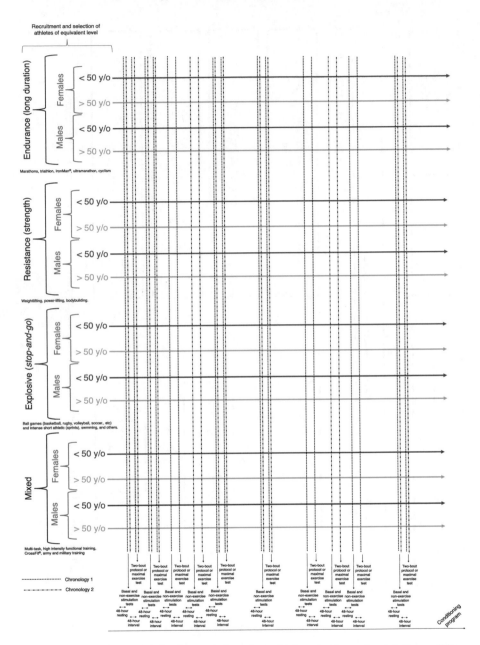

Fig. 14.5 Proposal protocols for the research on the multiple endocrine and metabolic conditioning processes from the sedentary to the athlete state, according to the type of sport, sex, and age

physiology in athletes have been conducted without standardization and uniformization of the design, with heterogeneous choice of parameters, timing of the collects, and population of athletes studied, since most studies have not delineated specific targeted populations and have not specified any further characteristic other than "responses in athletes" aiming to detect sport-, sex-, age-, and conditioning level-dependent hormonal and metabolic responses [1–3].

A well-designed, well-structured, thorough and deep analysis of endocrine and metabolic adaptations according to the type of sport, sex, age, conditioning level, and matched for the training, eating, sleeping and other influencing patterns should be designed in order to understand metabolic and hormonal behaviors, and whether and which of these behaviors are specific to any of the baseline, training, social, eating, or other characteristics [1, 12]. The multiple comparisons according to each characteristic should preferably have uniform methodology to allow further *head-to-head* comparisons and conclusions regarding specific responses according to each aspect and should encompass the major types of tests, including resting levels, exercise-dependent and exercise-independent stimulation tests, physical tests, body metabolism, and composition and assessed for eating, sleeping and social patterns, and psychological states, as detailed in Fig. 14.6.

14.3.3   Endocrine and Metabolic Responses to Interventions in Athletes

In a second moment, after the elucidation of hormonal and metabolic adaptations for each specific population of athletes, the understanding of the effects of each intervention on each of these populations may provide more accurate insights for precise and effective recommendations.

Specific interventions may be employed in: 1. Nutrition, including increase or decrease of carbohydrate, protein, or fat intake with or without caloric compensations, increase or decrease in hydration, or other nutritional interventions; 2. Training, including increase or decrease in training volume, intensity, and other training patterns; 3. Sleeping, including increase or decrease in sleep duration and quality, sleep hygiene, and the use of supplements or specific meals before bed; and 4. Cognitive demands, including reduction or increase of work or study, or mental stress tests.

The effects of each intervention may be measured in basal hormonal, immunologic, muscular, inflammatory, cytokine and other biochemical levels, exercise-dependent and independent responses to stimulations, body composition and metabolism, mood states, libido and cognition (when the intervention is not in cognitive demands), training characteristics and performance (when the intervention is not in training), and sleeping patterns (when the intervention is not in sleeping). In particular, *the* specific hormonal and metabolic consequences according to each sort of nutritional interventions may be of great interest. Figure 14.7 illustrates some of the options to evaluate the influences of specific interventions in athletes.

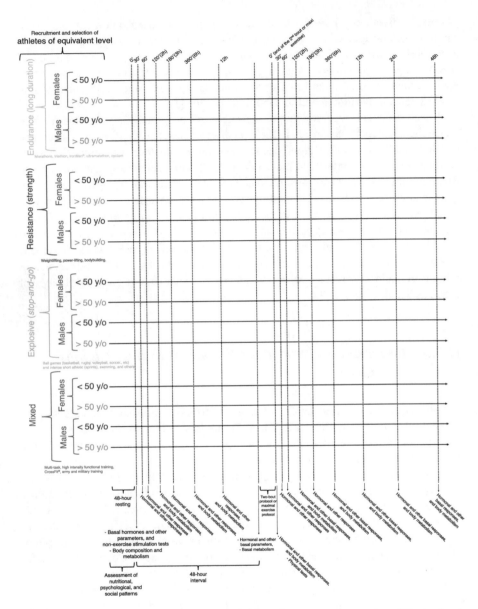

Fig. 14.6 Recommended protocols for the evaluation of sports, sex, age, and conditioning level-specific endocrine and metabolic responses, preferably matched for eating, sleeping and social patterns

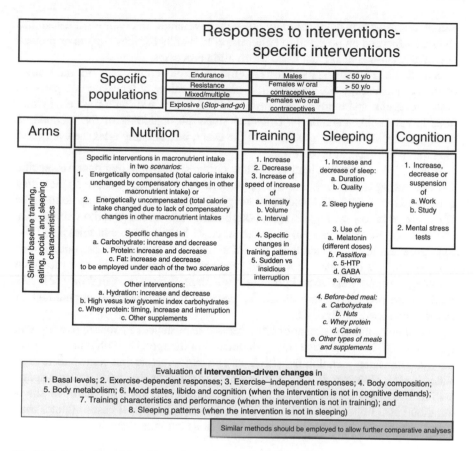

Fig. 14.7 Evaluation of the effects of specific interventions in athletes

14.3.4 Endocrinology and Metabolism on the Extremes

As mentioned in Chap. 7, sports on the extremes are growing and spreading as the sense of competitiveness increases. Despite the potential harms, since these events should not be avoided or prohibited, insights from these extreme circumstances resulted from one or more severe types of deprivation may uncover reactions to specific deprivations.

Studies on these circumstances should be observational, as a real-life study, allowing for variables to occur naturally, once the objective is to detect the collective result of all interventions and consequent deprivations combined.

Several reports on these sports have been published [16–32], although parameters have been majorly measured without a hormonal and metabolic context and not fully adjusted for all variables that could modulate clinical and biochemical behaviors. Hence, these observational studies should be conducted with strict and structured designed in terms of which parameters will be measured, timing of the data collect, and variables that will be followed and adjusted accordingly, without any sort of interference in these naturally associated variables. Interventional researches should be focused on proposal to either improve results for these sports or prevent potential harms resulted from these deprivation processes.

Standardized parameters and tests should be employed in observational studies of the organism exposure to extremes, aiming to allow appropriate comparisons between regular and extreme circumstances and between different extreme situations in sports. Data should be collected since the beginning of the preparation period, alongside with the process until the event, immediately after the event, and then for up to 7 days after the event.

Some of the sports that lead to extreme circumstances and that deserve specific research owing to its important and large number of athletes include fighters that undergo severe dehydration and rapid weight loss processes to participate in fights of categories of lower weight ranges, bodybuilders that interrupt carbohydrate, salt, and water (in this sequence) to be at the stage with the lowest skin thickness [18–21], Ultraman®, ultra-marathons, ultrarunnings, and IronMan® as competitions that overpass all organism energy and repairing capacity [22–26], and mountain climbing, with progressive and prolonged shortage of oxygen availability [27–32]. Some of the proposed protocols for the research in these peculiar populations of athletes are shown in Fig. 14.8.

In common, the abovementioned extreme circumstances in sports, overtraining syndrome (OTS), and relative energy deficiency of the sport (RED-S) share deprivations and low energy availability (LEA) as two major factors that contribute to their states [4–15]. However, it is more reasonable to consider that these extreme sports lead to functional (FOR) or nonfunctional overreaching (NFOR) states or short-duration LEA, rather than OTS or RED-S, although in several cases athletes may face difficulties in the recovery process from these extreme moments.

14.3.5   The Impaired Athlete

The uncovering of the pathophysiological processes that athletes impaired by a range of sports- or deprived-derived diseases requires an appropriate and sufficiently deep understanding of what should be expected for athletes of similar sports, age, sex, and training levels. Besides the typical abnormalities, the exact characteristics expected to be present are inconclusive in each of these dysfunctions, likely due to the blurry and overlapping dysfunctions clinically and biochemically disclosed in athletes of a variety of these dysfunctions. Even differences between the female athlete triad (TRIAD), reframed to RED-S, are still under dispute regarding whether RED-S is an appropriate increase in the understanding of TRIAD, or if the proposed mechanisms in RED-S really fully encompass all TRIAD features [5–11].

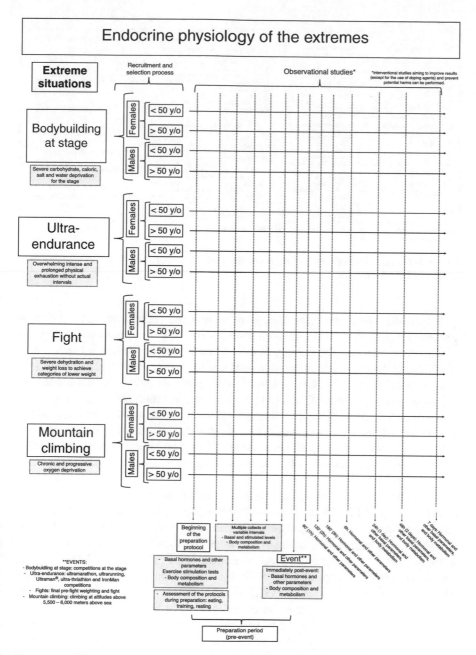

Fig. 14.8 Protocols for the research of the endocrinology and metabolism in sports during extremes

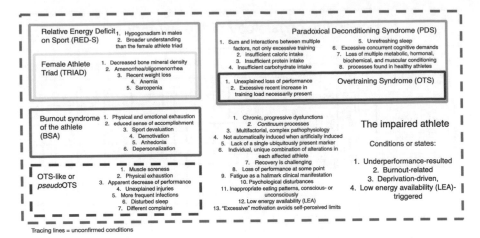

Tracing lines = unconfirmed conditions

Fig. 14.9 Specific characteristics of each impaired athlete-related condition and common hallmarks to all these conditions

Hence, disease-driven researches may narrow the parameters and tests to be employed and preclude from more comprehensive view on its pathophysiology, as well as new insights. Oppositely, assessing the impaired athlete irrespective of labels allows the use of parameters and tests not commonly employed in the research of a specific dysfunction. The assessment of athletes with decreased performance or fatigue for the occurrence of LEA [34], which is the key mechanism that triggers all burnout-related and underperformance-related conditions, without labeling with a specific underlying condition, may unveil insights on the level of similarity between OTS and RED-S [4, 10, 33], and provide substantiation to help understand whether these are same conditions in different sides of a spectrum, or if they are indeed distinct dysfunctions.

The impaired athlete syndrome (IAS) comprises all conditions inherently and primarily triggered by sports direct or indirect effects, which in common result in underperformance, are related to burnout syndrome, driven by one or more deprivation states, and triggered by LEA [34]. These states include OTS [4, 13–15] and its reframed paradoxical deconditioning syndrome (PDS) [12], TRIAD [5, 6, 8, 11] and its expanded concept of RED-S [7, 9, 10], burnout syndrome of the athlete (BSA) [35–39], *pseudo* OTS, and OTS-like conditions, which are characterized by specific features in each of these states, as described in Fig. 14.9.

Researches that are conducted in IAS without labeling specific conditions may naturally allow the evaluation of parameters and tests that are not usually measured when a population of athletes is studied from the perspective of one of these labeled diagnoses.

Researches on all conditions encompassed by IAS should assume that all these states are underperformance-related, burnout-related, deprivation-driven, and LEA-triggered [10, 12], since from this perspective a broader range of aspects has plausibility to be employed in research. The change from a diagnosed-driven analysis to unlabeled analysis is summarized in Fig. 14.10.

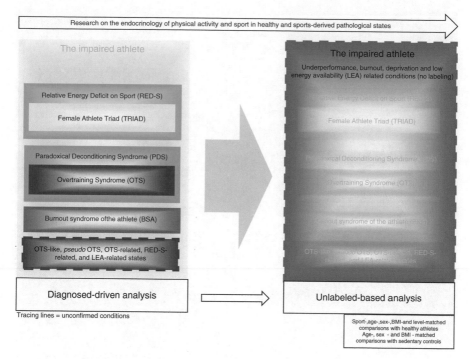

Fig. 14.10   From disease-driven to impaired-driven researches

14.3.6   *Differentiation Between Functional (FOR) and Nonfunctional Overreaching (NFOR), and Overtraining Syndrome (OTS)*

It is noteworthy to reinforce the fact that functional (FOR) and nonfunctional (NFOR) overreaching states, commonly and appropriately alleged to be earlier and less severe stages of OTS, are temporary adaptive states that do not necessarily lead to *maladaptations* and subsequent dysfunctions. Instead, symptoms appear more acutely, are mostly physical, and recover subsequently. The sudden appearance of severe symptoms is a physiological warn that deeper and more important adaptations need to be settled, for which quality resting and eating are necessary and will bring further compensatory improvements resulted from these adaptations, that in these states are functional. The progression from FOR and NFOR to OTS can be considered when symptoms that are in common to all burnout-related states in athletes appear, usually as a consequence of chronically unresolved sequence of FOR and NFOR episodes [4, 33]. Basically, while FOR and NFOR are an acute exhaustion usually triggered by training upload, OTS has multiple similarities to burnout and LEA-derived abnormalities [34], and for this reason, only OTS, but not FOR and NFOR, may be hypothesized to be correlated with RED-S/TRIAD. For instance, in an analogical perspective, equivalent FOR and NFOR states in women would unlikely lead to menstrual dysfunctions and bone loss, which precludes these women from the diagnosis of TRIAD.

The most remarkable differentiations between FOR, NFOR, and OTS are depicted in Fig. 14.11, from a *spectrum* perspective of each aspect, since these states

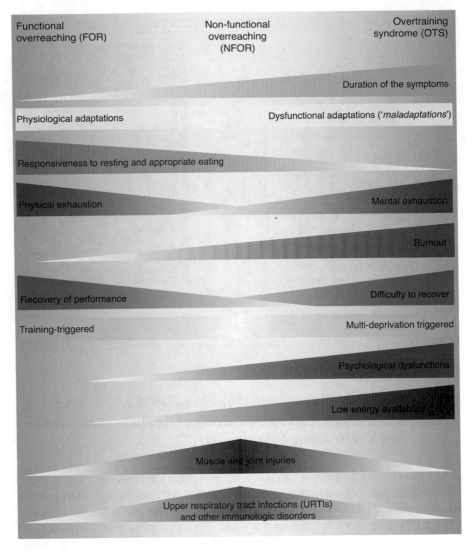

Fig. 14.11 Differences between FOR, NFOR, and OTS

are considered as different stages of a *continuum*. Noteworthy, muscle and joint injuries or soreness and atypical frequency and intensity of infections tend to be absent in both FOR and OTS and are typically identified when athletes progress from unrecovered short-term FOR to NFOR.

14.3.7   Interventions in the Impaired Athlete

Studies on the efficacy of interventions that might help to heal from any of the impaired-related disorders can only yield reliable results when both specific charac-teristics and expected results for both healthy athletes and athletes affected by the dysfunction aimed to be evaluated of the specific sport, sex, age, and conditioning level are unknown, once differences and changes induced by specific interventions on this population of athletes may unveil specific benefits of each type of intervention.

Once this specific type of studies aims to evaluate whether a specific intervention is beneficial to help improve from the impairment that these athletes present, controlling for variables that could influence the results should be strictly controlled. Possibly, proposed interventions may disclose sex-, age-, sport-, or condition-specific responses, whereas lack of responses due to population selection bias may preclude from the identification of potential approaches. Thus, proper characterization of the population of athletes to be addressed is key for accurate conclusion.

A flowchart depicting the essential aspects to be assessed and characterized is presented in Fig. 14.12.

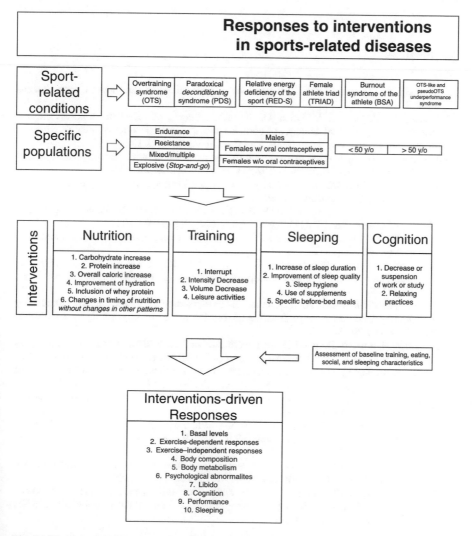

Fig. 14.12 Flowchart for interventional trials on the impaired athlete

14.4 Conclusion

A structured thinking of the design and methodology of the research on the endocrinology of physical activity and sport is key for the improvement of the understanding of the endocrine-adapted physiology in the athlete. Particular proposals for the research on the conditioning processes from the sedentary to the athlete, the endocrine physiology of the athlete, the endocrine and metabolic responses to specific interventions in athletes, the endocrinology and metabolism on the sports of extremes, the assessment of the impaired athlete, the differentiation between functional (FOR) and nonfunctional overreaching (NFOR) and overtraining syndrome (OTS), and the effects of specific interventions in the impaired athlete have been proposed.

References

1. Hackney AC, Constantini NW. Endocrinology of physical activity and sport. Cham: Humana Press; 2020.
2. Kraemer WJ, Ratamess NA, Hymer WC, Nindl BC, Fragala MS. Growth hormone(s), testosterone, insulin-like growth factors, and cortisol: roles and integration for cellular development and growth with exercise. Front Endocrinol (Lausanne). 2020;11:33.
3. Hackney AC. Hypogonadism in exercising males: dysfunction or adaptive-regulatory adjustment? Front Endocrinol (Lausanne). 2020;11:11. Published 2020 Jan 31.
4. Cadegiani FA, Kater CE. Hormonal aspects of overtraining syndrome: a systematic review. BMC Sports Sci Med Rehabil. 2017;9:14.
5. Nattiv A, Loucks AB, Manore MM, et al. American College of Sports Medicine position stand. The female athlete triad. Med Sci Sports Exerc. 2007;39(10):1867–82.
6. Matzkin E, Curry EJ, Whitlock K. Female athlete triad: past, present, and future. J Am Acad Orthop Surg. 2015;23(7):424–32.
7. Mountjoy M, Sundgot-Borgen J, Burke L, et al. The IOC consensus statement: beyond the female athlete triad – relative energy deficiency in sport (RED-S). Br J Sports Med. 2014;48:491–8.
8. De Souza MJ, Williams NI, Nattiv A, et al. Misunderstanding the female athlete triad: refuting the IOC consensus statement on Relative Energy Deficiency in Sport (RED-S). Br J Sports Med. 2014;48:1461–5.
9. Mountjoy M, Sundgot-Borgen J, Burke L, et al. Authors' 2015 additions to the IOC consensus statement: Relative Energy Deficiency in Sport (RED-S). Br J Sports Med. 2015;49:417–20.
10. Mountjoy M, Sundgot-Borgen JK, Burke LM, et al. IOC consensus statement on relative energy deficiency in sport (RED-S): 2018 update. Br J Sports Med. 2018;52:687–97.
11. Tenforde AS, Barrack MT, Nattiv A, et al. Parallels with the female athlete triad in male athletes. Sports Med. 2016;46:171–82.
12. Cadegiani FA, Kater CE. Novel insights of overtraining syndrome discovered from the EROS study. BMJ Open Sport Exerc Med. 2019;5(1):e000542. https://doi.org/10.1136/bmjsem-2019-000542.
13. Cadegiani FA, Kater CE. Novel causes and consequences of overtraining syndrome: the EROS-DISRUPTORS study. BMC Sports Sci Med Rehabil. 2019;11:21.
14. Cadegiani FA, Kater CE. Eating, sleep, and social patterns as independent predictors of clinical, metabolic, and biochemical behaviors among elite male athletes: the EROS-PREDICTORS study. Front Endocrinol. 2020;11:414.

15. Cadegiani FA, Kater CE. Inter-correlations among clinical, metabolic, and biochemical parameters and their predictive value in healthy and overtrained male athletes: the EROS-CORRELATIONS study. Front Endocrinol (Lausanne). 2019;10:858.
16. Gifford RM, O'Leary T, Cobb R, et al. Female reproductive, adrenal, and metabolic changes during an Antarctic traverse. Med Sci Sports Exerc. 2019;51(3):556–67.
17. Anton-Solanas A, O'Neill BV, Morris TE, Dunbar J. Physiological and cognitive responses to an Antarctic expedition: a case report. Int J Sports Physiol Perform. 2016;11(8):1053–9.
18. Steele IH, Pope HG Jr, Kanayama G. Competitive bodybuilding: fitness, pathology, or both? Harv Rev Psychiatry. 2019;27(4):233–40.
19. Spendlove J, Mitchell L, Gifford J, et al. Dietary intake of competitive bodybuilders. Sports Med. 2015;45(7):1041–63.
20. Mitchell L, Hackett D, Gifford J, Estermann F, O'Connor H. Do bodybuilders use evidence-based nutrition strategies to manipulate physique? Sports (Basel). 2017;5(4):76.
21. Chappell AJ, Simper TN. Nutritional peak week and competition day strategies of competitive natural bodybuilders. Sports (Basel). 2018;6(4):126.
22. Knechtle B, Nikolaidis PT. Physiology and pathophysiology in ultra-marathon running. Front Physiol. 2018;9:634.
23. Knechtle B, Chlíbková D, Papadopoulou S, Mantzorou M, Rosemann T, Nikolaidis PT. Exercise-associated hyponatremia in endurance and ultra-endurance performance-aspects of sex, race location, ambient temperature, sports discipline, and length of performance: a narrative review. Medicina (Kaunas). 2019;55(9):537.
24. Knechtle B. Ultramarathon runners: nature or nurture? Int J Sports Physiol Perform. 2012;7(4):310–2.
25. Danielsson T, Carlsson J, Schreyer H, et al. Blood biomarkers in male and female participants after an Ironman-distance triathlon. PLoS One. 2017;12(6):e0179324. Published 2017 Jun 13. https://doi.org/10.1371/journal.pone.0179324.
26. Sousa CV, Aguiar SDS, Olher RDR, et al. Hydration status after an ironman triathlon: a meta-analysis. J Hum Kinet. 2019;70:93–102. Published 2019 Nov 30. https://doi.org/10.2478/hukin-2018-0096.
27. Heggie V. Experimental physiology, Everest and oxygen: from the ghastly kitchens to the gasping lung. Br J Hist Sci. 2013;46(1):123–47. https://doi.org/10.1017/S0007087412000775.
28. Watts PB, Daggett M, Gallagher P, Wilkins B. Metabolic response during sport rock climbing and the effects of active versus passive recovery. Int J Sports Med. 2000;21(3):185–90.
29. Jetton AM, Lawrence MM, Meucci M, et al. Dehydration and acute weight gain in mixed martial arts fighters before competition. J Strength Cond Res. 2013;27(5):1322–6.
30. Matthews JJ, Nicholas C. Extreme rapid weight loss and rapid weight gain observed in UK mixed martial arts athletes preparing for competition. Int J Sport Nutr Exerc Metab. 2017;27(2):122–9.
31. Pettersson S, Berg CM. Hydration status in elite wrestlers, judokas, boxers, and taekwondo athletes on competition day. Int J Sport Nutr Exerc Metab. 2014;24(3):267–75.
32. MacDonald MJ, Green HJ, Naylor HL, Otto C, Hughson RL. Reduced oxygen uptake during steady state exercise after 21-day mountain climbing expedition to 6,194 m. Can J Appl Physiol. 2001;26(2):143–56.
33. Meeusen R, Duclos M, Foster C, European College of Sport Science, American College of Sports Medicine, et al. Prevention, diagnosis, and treatment of the overtraining syndrome: joint consensus statement of the European College of Sport Science and the American College of Sports Medicine. Med Sci Sports Exerc. 2013;45(1):186–205.
34. Logue D, Madigan SM, Delahunt E, et al. Low energy availability in athletes: a review of prevalence, dietary patterns, physiological health, and sports performance. Sports Med. 2018;48:73–96.
35. Smith RE. Toward a cognitive-affective model of athletic burnout. J Sport Psychol. 1986;8:36–50.
36. Raedeke TD. Is athlete burnout more than just stress? A sport commitment perspective. J Sport Exerc Psychol. 1997;19:396–41.

37. Goodger K, Gorely T, Lavallee D, Harwood C. Burnout in sport: a systematic review. Sport Psychol. 2007;21:127–51.
38. Gustafsson H, DeFreese JD, Madigan DJ. Athlete burnout: review and recommendations. Curr Opin Psychol. 2017;16:109–13.
39. Francisco C, Arce C, Vilchez MP, Vales A. Antecedents and consequences of burnout in athletes: perceived stress and depression. Int J Clin Health Psychol. 2016;16(3):239–46.

Chapter 15
Novel Insights in Overtraining Syndrome: Summary and Conclusions

Outline

Novel insights in overtraining syndrome: from the massive amount of new and emerging data on OTS, the author provides an easy-to-understand summarization of the new insights on overtraining syndrome, which will allow readers to consolidate the learnings on the syndrome, as well as why OTS could be termed as *Paradoxical Deconditioning Syndrome* (PDS), and why future researches should go beyond OTS.

Novel insights in overtraining syndrome – Summary and Conclusions: the author shows that OTS can be understood as a complex web of interactions among multiple risk factors, adverse biochemical conditions, and forced dysfunctional adaptations (*maladaptations*), which lead to numerous metabolic, biochemical, and hormonal *deconditionings*, which eventually result in the mitigation of the metabolism and paradoxical decreased performance, also termed as physical "deconditioning," the hallmark of overtraining syndrome. This syndrome should no longer be perceived as caused by excessive training but as a result from the combination of insufficient protein, carbohydrate or caloric intake, poor sleep quality, concurrent other energy demands, particularly the cognitive ones, and lack of recovery. Hence, while the name "overtraining syndrome" only covers one of the risk factors and does not synthetize the hallmark underlying mechanism of this syndrome, the proposed term "paradoxical deconditioning syndrome" (PDS) describes the key feature of the syndrome within one expression.

15.1   New Insights in Overtraining Syndrome in Athletes

Overtraining syndrome (OTS) used to be understood as a dysfunction triggered by excessive training or a relative lack of resting compared to the training load performed, which leads to underperformance not explained by alterations or diseases that could also lead to decreased performance [1–5]. Classical theories on chronic glutamine depletion [1, 6–8], chronic glycogen depletion [1, 6, 9], central fatigue [1, 6, 10, 11], autonomic nervous system [1, 6, 12–14], oxidative stress [1, 6, 15, 16], cytokines [1, 6, 17], and hypothalamus [1, 6, 18] as triggers or underlying mechanisms of OTS pathophysiology have been partially confirmed in clinical studies but have left important gaps in the comprehension of OTS features.

The multiple new findings in OTS [19–27] generated novel insights that allowed the development of more scientifically based concepts and helped better understand the OTS pathophysiology, triggers, and other characteristics [28].

The "new OTS" can be understood as a result from the sum and interactions between multiple excessive efforts and deprivations, including relative caloric and protein malnutrition, unrefreshing sleep, lack of compensatory reduction of training sessions, and concurrent intense physical and cognitive demands. The excessiveness leads to chronic glycogen deprivation, excessive unrepaired damage, impaired immune function, dysfunctional increase of cytokines, and exposure of the brain to chronic mild hypoglycemia, leading to chronic mild neuroglycopenic state. Altogether, these consequences create a hostile tissue environment that is chronically deprived from energy and mechanisms of repair, which forces multiple adaptations to keep surviving and functioning at the same time. Since these corresponding adaptations need to be highly altered to overcome all barriers under these chronic aberrant metabolism, they tend to be overtly dysfunctional (*maladaptations*) and include or result in exacerbated noradrenaline and adrenaline, reduction of their inactivation to metanephrines, paradoxical trend toward fat storage accumulation, dysfunctionally exacerbated proteolysis, loss of the *fight-or-flight* biochemical and consequent clinical response, global blunted metabolic and hormonal responses to stimulations, and prolonged and impaired muscle recovery. These consequences of the *maladaptations* are in common metabolic movements toward energy-saving, anti-anabolic, pro-catabolic, and *hypometabolism* states, multiple losses of the recently unveiled hormonal conditioning processes that occur in athletes [29], and relative dehydration. These abnormalities eventually lead to the main manifestations of OTS, including mental and physical exhaustion, pathological muscle soreness, loss of multiple abilities in physical performance, increased predisposition to overall infections, and burnout-like signs and symptoms. Figure 15.1 summarizes he triggers, pathophysiology, clinical features, and consequences of OTS from the perspective of the new understanding of OTS.

The novel patterns of abnormalities identified in OTS allow to suggest that OTS is neither an overt nor an absent dysfunction but a relative alteration instead. While OTS-affected athletes disclose levels that are in general within the reference ranges, and similar to general populations, these are different when compared to those in sex-, age-, BMI-, and sport-matched healthy athletes. This discovery supports the

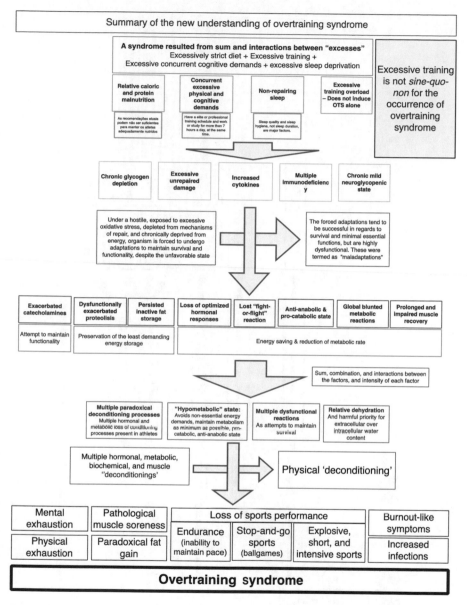

Fig. 15.1 Summary of the triggers, pathophysiology, clinical features, and consequences of OTS, from the perspective of new insights

hypothesis that OTS is a sort of mix of losses of multiple adaptive conditioning processes that athletes undergo and therefore yield different results than those expected for that specific population of athletes. Indeed, clinical, metabolic, and biochemical behaviors in OTS varied from overt dysfunctions to enhanced conditioning processes, as summarized in Fig. 15.2, but the majority of the parameters in OTS disclosed the phenomena of *deconditioning*, either complete or partial.

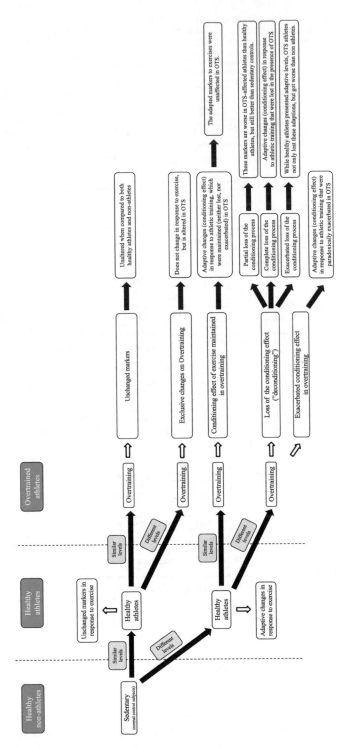

Fig. 15.2 Types of clinical, metabolic, and biochemical behaviors in overtraining syndrome

Collectively, OTS can now be understood as a disease related to burnout syndrome, generated from the combination of chronic deprivations, as part of underperformance-related conditions, and fundamentally triggered by low energy availability (LEA), with important similarities with relative energy deficiency of the sport (RED-S) and its previous term female athlete triad (TRIAD), and burnout syndrome of the athlete (BSA).

OTS is no longer perceived as caused by excessive training but as a result from the combination of insufficient protein, carbohydrate or caloric intake, poor sleep quality, concurrent other energy demands, particularly the cognitive ones, and lack of recovery.

Hence, while the name "overtraining syndrome" only covers one of the risk factors and does not synthetize the hallmark underlying mechanism of this syndrome, the proposed term "paradoxical deconditioning syndrome" (PDS) describes the key feature of the syndrome within one expression.

15.2   Overtraining Syndrome (OTS) or Paradoxical Deconditioning Syndrome (PDS)?

The occurrence of overtraining syndrome arises from chronic energy deprivation, depletion of mechanisms of repair and high levels of oxidative stress, and reactive oxygen species (ROS) without repairing mechanisms, leading to a long-term exposure to a hostile combination. The sum, interactions, severity, and duration of these mechanisms force adaptations to maintain survival and functionality. These adaptations are however highly dysfunctional, termed as *maladaptations. Maladaptations* occur alongside with the chronic energy-depleted environment and high exposure to oxidative stress, across multiple types of tissues and metabolic pathways. Similar to the correlations between the adverse factors that lead to an inhospitable tissue conditions, the sum, intensity, and interactions between these "maladaptive" processes play synergistically and eventually lead to overtraining syndrome.

While the understanding of overtraining syndrome has been redefined as a result from multiple interactions between "maladaptive" processes, its occurrence is now understood to lead to multiple losses of conditioning processes and beneficial adaptations that have been detected to occur in athletes, as a mix of metabolic and hormonal "deconditioning" processes, which in turn leads to physical "deconditioning," with decreased pace, intensity, volume, and explosive capacity in sports performance, associated with an overall state of "hypometabolism," physical and mental exhaustion, loss of libido, paradoxical fat gain and muscle loss, and blunted fight-or-flight reaction response. All these mechanisms aim to prevent further nonessential energy expenditure.

A remarkable learning from the latest studies is that OTS may occur without the presence of excessive training but as a consequence of a sum of chronic shortage of specific macronutrients or overall energy, lack of restoring sleep and resting, and concurrent additional energy demands, particularly cognitive ones.

For these reasons, despite the historical importance, and cultural and spread use, the name "overtraining syndrome" turns out to be imprecise, as excessive training only covers one factor and is not a *sine qua non* aspect of this syndrome. Conversely, the proposed expression "paradoxical deconditioning syndrome (PDS)" synthetizes the key phenomenon and the sine qua non characteristic of OTS.

15.3 Beyond Overtraining Syndrome

The understanding and novel insights on OTS are indistinguishable from the mechanisms, fields, paths, and physiological and dysfunctional aspects that need further researched to be elucidated and clarified and include:

1. The peculiarities of the adaptive hormonal physiology of the athlete, which may vary widely according to the studied population: females and males, age-stratified groups, the absence or presence of comorbidities, and sort of sports, which should be stratified accordingly, to elucidate the mechanisms that underlie the specific benefits of physical activity for each subpopulation.
2. Whether hormonal optimizations acquired with exercises can be correlated with the progressive improvement in sports performance, recovery, and health benefits or if they are separate phenomena.
3. From the recently acquired knowledge and the forthcoming researches, the development of a comprehensive and structured understanding of the adapted physiology of the endocrinology system in physical activity and sport, preferably if specified according to the conditioning level, since the sedentary and leisure activity until the elite athlete, and according to the training patterns.
4. From the understanding of the beneficial hormonal adaptations in athletes, the prediction of novel benefits from exercising for health promotion, disease prevention, and improvement of quality of life.
5. Once benefits and adaptive processes are identified, the proposal of mechanisms that justify how physical activity may attenuate hormonal dysfunctions and their consequences and also which type of sport and at what intensity and volume of training is needed to trigger these benefits.
6. With the elucidation of which hormonal and metabolic ranges are hormonally and metabolically expected in athletes, the identification particularities of the diagnoses of hormonal dysfunctions when these occur in athletes, what distinguishes these from the diagnoses on the general population, and the characterization of the hormonal and metabolic abnormalities in sports-triggered or sports-related diseases.
7. Strategies for the optimization of the management of a variety of endocrine dysfunctions in athletes.
8. Proposal and validation of specific diagnostic tools for overt and sports-specific metabolic and endocrine disorders in athletes.
9. With the learnings of the specific adaptive processes that occur in athletes, as well as the specific benefits from exercising, according to the type of sport,

intensity, volume, duration, regularity of exercise, age, sex, and the recommendations for more precise exercise prescriptions for the enhancement of the metabolic, clinical, cardiovascular, neuromuscular, psychological, physical, and hormonal benefits, as well as for the improvement of sports performance.
10. Strengthen the relationship between the research groups of endocrinology and sports science and medicine and between their correlated societies, for further collaborative researches and partnerships and for continuous improvement of the research on the field of sports endocrinology.
11. From the comprehension of what is physiologically expected in athletes in terms of hormones and metabolism, proposals of more precise detection of androgenic anabolic steroid (AAS) abuse without the need to search for anti-doping agencies and laboratories and to understand how the AAS misuse may affect hormonal, metabolic, and overall health.

Altogether, the physiological functioning of the endocrinology system in athletes according to baseline, training and eating characteristics, and its variances and pathological correspondences in sports-related diseases will help build a systematical understanding of the endocrinology of physical activity and sport, in both health and pathological states, in a similar manner that the field of endocrinology researches and builds the related knowledge in general population.

15.4 The New Overtraining Syndrome: Summary

In summary, overtraining syndrome is a complex web of interactions among multiple risk factors, adverse biochemical conditions, and forced dysfunctional adaptations (*maladaptations*), which lead to numerous metabolic, biochemica,l and hormonal *deconditionings*, which eventually result in the mitigation of the metabolism and paradoxical decreased performance, also termed as physical "deconditioning," the hallmark of overtraining syndrome. Once excessive training currently has a minor role in OTS, and the key characteristic of the syndrome is the paradoxical deconditioning, i.e., oppositely to the expected for athletes, the term paradoxical deconditioning syndrome (PDS) better describes within a short expression the central characteristic of this syndrome and is highly representative of what occurs in these affected athletes.

15.5 Conclusion

The classical understanding and theories on overtraining syndrome (OTS) have been adjusted, amended, and reframed in the novel scientifically based concepts that better explains the major characteristics. OTS can be understood as a result of the sum and interactions between multiple excessiveness and deprivations, which lead to chronic depletion of energy and repairing mechanisms and consequent hostile tissue environment, which in turn force the occurrence of multiple dysfunctional adaptations (*maladaptations*), resulting in multiple clinical and biochemical abnormalities, eventually leading to OTS and its main characteristics of underperformance, exhaustion, and burnout-like manifestations.

OTS was revealed to be neither an overt nor an absent dysfunction, but a relative alteration instead, showing that OTS is a sort of mix of losses of multiple adaptive conditioning processes that athletes undergo (*deconditioning*), and related to burnout, deprivations, and low energy availability (LEA).

Since OTS should no longer be perceived as being exclusively triggered by excessive training, but a combination of insufficient protein, carbohydrate or caloric intake, poor sleep quality, and concurrent other energy demands instead, the popular term "overtraining syndrome" has been proposed to be renamed for "Paradoxical Deconditioning Syndrome (PDS)," since this short expression synthetizes the key pathophysiological characteristic of OTS.

References

1. Meeusen R, Duclos M, Foster C, European College of Sport Science, American College of Sports Medicine, et al. Prevention, diagnosis, and treatment of the overtraining syndrome: joint consensus statement of the European College of Sport Science and the American College of Sports Medicine. Med Sci Sports Exerc. 2013;45(1):186–205.
2. Rietjens GJ, Kuipers H, Adam JJ, et al. Physiological, biochemical and psychological markers of strenuous training-induced fatigue. Int J Sports Med. 2005;26(1):16–26.
3. Stamford B. Avoiding and recovering from overtraining. Phys Sportsmed. 1983;11(10):180.
4. Kreher JB, Schwartz JB. Overtrining syndrome: a practical guide. Sports Health. 2012;4(2):128–38.
5. Budgett R, Newsholme E, Lehmann M, et al. Redefining the overtraining syndrome as the unexplained underperformance syndrome. Br J Sports Med. 2000;34:67–8.
6. Halson SL, Jeukendrup AE. Does overtraining exist? An analysis of overreaching and overtraining research. Sports Med. 2004;34(14):967–81.
7. Walsh NP, Blannin AK, Robson PJ, Gleeson M. Glutamine, exercise and immune function. Links and possible mechanisms. Sports Med. 1998;26(3):177–91.
8. Gastmann UA, Lehmann MJ. Overtraining and the BCAA hypothesis. Med Sci Sports Exerc. 1998;30(7):1173–8.
9. Snyder AC. Overtraining and glycogen depletion hypothesis. Med Sci Sports Exerc. 1998;30(7):1146–50.
10. Davis JM. Carbohydrates, branched-chain amino acids, and endurance: the central fatigue hypothesis. Int J Sport Nutr. 1995;5 Suppl:S29–38.
11. Meeusen R, Watson P, Hasegawa H, Roelands B, Piacentini MF. Brain neurotransmitters in fatigue and overtraining. Appl Physiol Nutr Metab. 2007;32(5):857–64.
12. Kajaia T, Maskhulia L, Chelidze K, Akhalkatsi V, Kakhabrishvili Z. The effects of nonfunctional overreaching and overtraining on autonomic nervous system function in highly trained athletes. Georgian Med News. 2017;264:97–103.
13. Kiviniemi AM, Tulppo MP, Hautala AJ, Vanninen E, Uusitalo AL. Altered relationship between R-R interval and R-R interval variability in endurance athletes with overtraining syndrome. Scand J Med Sci Sports. 2014;24(2):e77–85.
14. Uusitalo AL, Uusitalo AJ, Rusko HK. Heart rate and blood pressure variability during heavy training and overtraining in the female athlete. Int J Sports Med. 2000;21(1):45–53.

15. Margonis K, Fatouros IG, Jamurtas AZ, et al. Oxidative stress biomarkers responses to physical overtraining: implications for diagnosis. Free Radic Biol Med. 2007;43(6):901–10.
16. Tanskanen M, Atalay M, Uusitalo A. Altered oxidative stress in overtrained athletes. J Sports Sci. 2010;28(3):309–17.
17. Smith LL. Cytokine hypothesis of overtraining: a physiological adaptation to excessive stress? Med Sci Sports Exerc. 2000;32:317–31.
18. Urhausen A, Gabriel HH, Kindermann W. Impaired pituitary hormonal response to exhaustive exercise in overtrained endurance athletes. Med Sci Sports Exerc. 1998;30(3):407–14.
19. Cadegiani FA, Kater CE. Hypothalamic-pituitary-adrenal (HPA) axis functioning in overtraining syndrome: findings from Endocrine and Metabolic Responses on Overtraining Syndrome (EROS) - EROS-HPA axis. Sports Med Open. 2017;3(1):45.
20. Cadegiani FA, Kater CE. Growth Hormone (GH) and prolactin responses to a non-exercise stress test in athletes with overtraining syndrome: results from the Endocrine and metabolic Responses on Overtraining Syndrome (EROS) - EROS-STRESS. J Sci Med Sport. 2018;21(7):648–53.
21. Cadegiani FA, Kater CE. Body composition, metabolism, sleep, psychological and eating patterns of overtraining syndrome: results of the EROS study (EROS-PROFILE). J Sports Sci. 2018;36(16):1902–10.
22. Cadegiani FA, Kater CE. Basal hormones and biochemical markers as predictors of overtraining syndrome in male athletes: the EROS-BASAL study. J Athl Train. 2019. https://doi.org/10.4085/1062-6050-148-18.
23. Cadegiani FA, Kater CE, Gazola M. Clinical and biochemical characteristics of high-intensity functional training (HIFT) and overtraining syndrome: findings from the EROS study (the EROS-HIFT). J Sports Sci. 2019;20:1–12. https://doi.org/10.1080/02640414.2018.1555912.
24. Cadegiani FA, Kater CE. Novel causes and consequences of overtraining syndrome: the EROS-DISRUPTORS study. BMC Sports Sci Med Rehabil. 2019;11:21.
25. Cadegiani FA, Kater CE. Eating, sleep, and social patterns as independent predictors of clinical, metabolic, and biochemical behaviors among elite male athletes: the EROS-PREDICTORS study. Front Endocrinol. 2020;11:414.
26. Cadegiani FA, Kater CE. Inter-correlations among clinical, metabolic, and biochemical parameters and their predictive value in healthy and overtrained male athletes: the EROS-CORRELATIONS study. Front Endocrinol (Lausanne). 2019;10:858.
27. Cadegiani FA, Silva PHL, Abrao TCP, Kater CE, et al. J Sports Med (Hindawi Corp). 2020;2020:3937819.
28. Cadegiani FA, Kater CE. Novel insights of overtraining syndrome discovered from the EROS study. BMJ Open Sport Exerc Med. 2019;5(1):e000542. https://doi.org/10.1136/bmjsem-2019-000542.
29. Cadegiani FA, Kater CE. Enhancement of hypothalamic-pituitary activity in male athletes: evidence of a novel hormonal mechanism of physical conditioning. BMC Endocr Disord. 2019;1:117. https://doi.org/10.1186/s12902-019-0443-7.

Index

Printed in the United States
by Baker & Taylor Publisher Services